Proceedings of the

INTERNATIONAL CONFERENCE ON THERMODYNAMICS
CARDIFF, 1970

Comptes rendus de la

CONFERENCE INTERNATIONALE DE THERMODYNAMIQUE
CARDIFF, 1970

L'UNION INTERNATIONALE DE
CHIMIE PURE ET APPLIQUEE

avec

L'UNION INTERNATIONALE DE
PHYSIQUE PURE ET APPLIQUEE

et

L'INSTITUT DE PHYSIQUE ET LA
SOCIETE DE PHYSIQUE DE LONDRES

Comptes rendus de la

CONFERENCE INTERNATIONALE DE THERMODYNAMIQUE

tenue à Cardiff, R.U.
1–4 avril 1970

Editeur pour la Conférence
PETER T. LANDSBERG

LONDON
BUTTERWORTHS

INTERNATIONAL UNION OF
PURE AND APPLIED CHEMISTRY

in conjunction with

THE INTERNATIONAL UNION OF
PURE AND APPLIED PHYSICS

and

THE INSTITUTE OF PHYSICS AND
THE PHYSICAL SOCIETY

Proceedings of the

INTERNATIONAL CONFERENCE ON THERMODYNAMICS

held in Cardiff, U.K.
1–4 April 1970

Conference Editor
PETER T. LANDSBERG

LONDON
BUTTERWORTHS

ENGLAND:	BUTTERWORTH & CO. (PUBLISHERS) LTD. LONDON: 88 Kingsway, W.C.2
AUSTRALIA:	BUTTERWORTH & CO. (AUSTRALIA) LTD. SYDNEY: 20 Loftus Street MELBOURNE: 343 Little Collins Street BRISBANE: 240 Queen Street
CANADA:	BUTTERWORTH & CO. (CANADA) LTD. TORONTO: 14 Curity Avenue, 374
NEW ZEALAND:	BUTTERWORTH & CO. (NEW ZEALAND) LTD. WELLINGTON: 49/51 Ballance Street AUCKLAND: 35 High Street
SOUTH AFRICA:	BUTTERWORTH & CO. (SOUTH AFRICA) LTD. DURBAN: 33/35 Beach Grove

The contents of this book appear in

Pure and Applied Chemistry, Vol. 22, Nos. 3–4 (1970)

Suggested U.D.C. *number:* 536·7(063)

Suggested additional number: 541·11(063)

ISBN 0 408 70101 3

Printed in Great Britain by Page Bros. (Norwich) Ltd., Norwich

PREFACE

This volume contains the papers presented at the International Conference on Thermodynamics which was held at University College, Cardiff, 1 to 4 April 1970. There were about 180 participants, half of them from abroad. For the purposes of the Conference it was desirable to start with the bulk of the papers in section B which, it was felt, would help to allow participants to warm to the subject as a whole. Also, some slight illogicalities were brought about by requirements of the time table, by late papers, etc. In the Proceedings presented here the order of the papers has, therefore, been adjusted.

The Conference contained three lively discussion sessions, which were given focus by signed statements. These were invited by the Organizing Committee who asked for such statements to be 'provocative' in the notices for the meeting. All participants had copies of the statements beforehand. Judging by the attendances, even at the last session (on Saturday morning), this seemed a popular procedure. The contributions at these three discussion sessions were tape recorded, even though they were spontaneous, and a debt is owed to the three gentlemen who kindly agreed to write the reports for the present volume. They were asked to insert in square brackets any comments of their own which they included in the report in their capacity as experts in the field, but which were not raised in the discussion. Apart from this request they were given a free hand to handle the reports as they thought fit, and the result does convey very well the flavour of these sessions.

A second unusual feature of the meeting was that one complete session was devoted to the problems raised by the teaching of the subject. A third unusual feature was that manuscripts were sent in early and proofs were read at the meeting. The willing cooperation of the publishers and authors was a great help in this connection.

The first and decisive encouragement for the idea of this Conference came from the Royal Society, who made a sum of money available. The following organizations also contributed financially:

British Petroleum Company Limited, Imperial Chemical Industries Limited, Monsanto Chemicals Limited, Rolls-Royce Limited, Shell International Petroleum Company Limited, South Wales Electricity Board, Wales Gas Board.

The International Union of Pure and Applied Physics and the International Union of Pure and Applied Chemistry allowed us to use the authority of their names. The Organizing Committee planned the scientific programme, and in addition Dr D. J. Morgan, Dr A. Harvey and a host of other helpers assisted in the actual running of the Conference. The Institute of Physics and the Physical Society helped with the organizational details of the Conference. All these forms of help towards the success of the Conference were given with unstinting generosity for which grateful acknowledgement is made here.

PETER T. LANDSBERG

ORGANIZING COMMITTEE

Chairman: Professor P. T. LANDSBERG (Cardiff)

Professor D. H. EVERETT (Bristol)
Professor F. C. FRANK, F.R.S. (Bristol)
Professor E. J. LE FEVRE (London)
Professor M. L. MCGLASHAN (Exeter)
Dr A. Harvey (Cardiff) (Chairman, Local Committee)
Dr D. J. MORGAN (Cardiff) (Honorary Secretary)*

* Mr W. J. Hornix acted as Honorary Secretary until he went abroad in September 1969.

CONTENTS

(I) = Invited paper

Proceedings of the International Conference on Thermodynamics; Cardiff (UK); 1–4 April 1970

D. Irreversibility and Quantum Mechanics

E. Statistical Thermodynamics in Astrophysics and Relativity

F. Pedagogical Papers

G. Discussion Reports

A. FOUNDATIONS

MAIN IDEAS IN THE AXIOMATICS OF THERMODYNAMICS

P. T. LANDSBERG

Department of Applied Mathematics and Mathematical Physics, University College, Cardiff

ABSTRACT

Recent advances in axiomatic treatments of thermodynamics are surveyed, by considering the new ideas rather than the mathematical technicalities. It is shown that the advance has been considerable, and can be summarized by the remark that the number of primitive concepts needed (for example to arrive at the notion of entropy) has been steadily decreased. The importance and significance of certain mathematical notions, notably those of various forms of order, is emphasized. It is explained in what connections broad continuity assumptions are convenient and indications are given of how these can be replaced by more rigorous procedures. Remarks about extensive properties and about the zeroth law are also included.

1. INTRODUCTION

Among scientists there exists a healthy ambivalence towards axiomatics. On the one hand there is the doubt whether one can arrive at any new science by axiomatizing; on the other, no one likes faulty arguments, and it is in the attempt to eliminate these that one is led in the direction of axiomatization.

Though no professional axiomatizer, I responded favourably to the request to deal with axiomatics here, because I believe that the progress which has recently been made in the understanding of the foundations of thermodynamics has in fact advanced scientific understanding. It is the purpose of this article to remove the thick shell of occasionally very pure mathematics, utilized in this work, in order to lay bare the essential ideas lying behind it. Our concern will be with systems which are free of adiabatic partitions and vacuous spaces, and in which the effects of long range forces, surface tension etc., are all neglected.

Discussions with Dr C. G. Gould (Cardiff), Dr J. B. Boyling (Leeds) and W. J. Hornix, and correspondence with Professor R. D. Luce (Pennsylvania) are gratefully acknowledged.

2. AN EMPIRICAL ENTROPY VIA ORDER RELATIONS

In the axiomatic approach the thermodynamic phase space E of a system dwindles to a set of points $x, y, \ldots \in E$, and each point becomes a state of the system only when our rules of interpretation are applied to the abstract mathematics. If an adiabatic transition is possible from state x to state y, we shall write xRy. Such relations R can pale into abstract objects of set theory, and there be specified merely by their properties. Of these we note six:

(i) Reflexivity: $(\forall x)(x \in E \Rightarrow xRx)$
(ii) Transitivity: $(\forall x, y, z)(x, y, z \in E$ and xRy and $yRz \Rightarrow xRz)$
(iii) Symmetry: $(\forall x, y)(x, y \in E$ and $xRy \Rightarrow yRx)$
(iv) Antisymmetry: $(\forall x, y)(x, y \in E$ and xRy and $yRx \Rightarrow x = y)$
(v) (Strong) connectedness or comparability:

$$(\forall x, y)(x, y \in E \Rightarrow \text{either } xRy \text{ or } yRx \text{ or both})$$

(vi) Conditional connectedness: $(\forall x, y, z)$

$$(x, y, z \in E \text{ and } xRy \text{ and } xRz \Rightarrow \text{either } yRz \text{ or } zRy \text{ or both}).$$

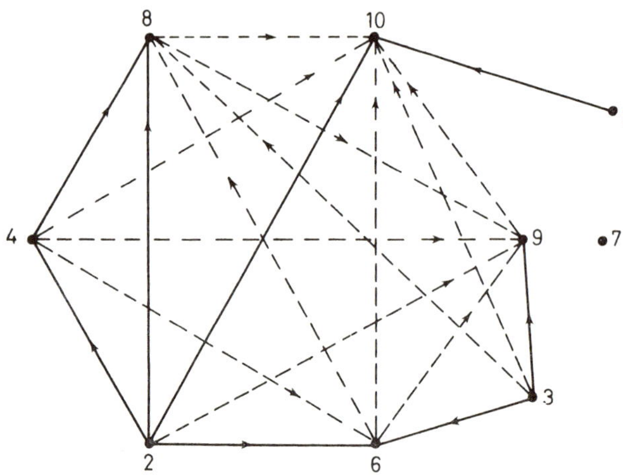

Figure 1. A set of points $E = \{2, 3, 4, 5, 6, 7, 8, 9, 10\}$ with partial preorder, and also exhibiting property (iv). Solid arrows mean that xRy. The set may be interpreted by the rule that xRy if and only if there exists an integer r such that $xr = y$.

The dashed arrows are needed if conditional connectedness is imposed in addition. For example, $2R4$ and $2R6$ then implies $4R6$ even though $4r = 6$ does not hold for any integer r.

The interpretation of R as adiabatic accessibility between states suggests that it is safe to impose conditions (i) and (ii) on R to obtain a *partial preorder* (or *quasiorder*). For states arbitrarily numbered from two to ten this type of order is illustrated in *Figure 1* (solid lines). The relation is seen to lack properties (iii) and (vi) (and therefore also (v)). This is reasonable in the present interpretation. Thus:

Lack of (iii): (xRy) and (not yRx) can both be true for states of different entropy, when there is only a one-way adiabatic link.

Lack of (iv): (xRy) and (yRx) can both be true with $x \neq y$ for distinct states of the same entropy.

Lack of (v): (xRy) and (yRx) can both fail if x is an equilibrium state not connected with others by adiabatic processes.

This possibility of states like No. 7 (*Figure 1*) which are like isolated islands in a sea of adiabatically linked states shows that R is still too general for the simple systems in view here. The effect of conditional connectedness (ref. 1, axiom 2.1.2 (i), pp 31, 126; axiom A.2.2, p. 193) is to add the dotted links in *Figure 1*. Arrows have been affixed to these according to the arbitrary convention xRy if $x > y$ (which rules out contradictions with transitivity).

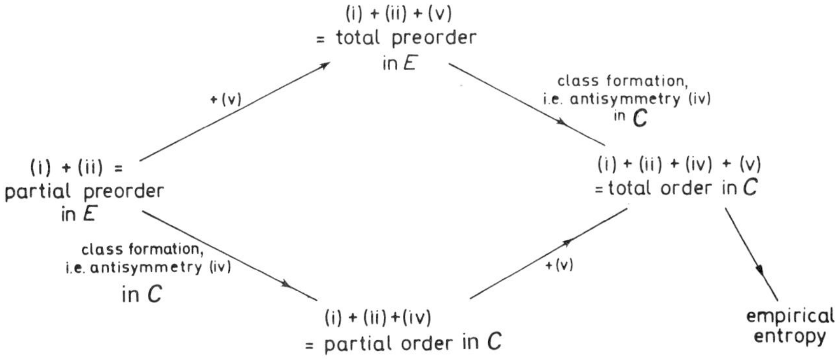

Figure 2. Types of order.

To remove islands like state 7 one needs the stronger *connectedness* and arrives at a *total preorder* (see *Figure 2*). This is exactly in accord with the now usual idea that any pair of states of normal systems are adiabatically linked. This idea originates from the remarks that there are no physical processes which involve such adiabatically isolated states (the processes of class Pl in ref. 2, and in ref. 3, pp 91, 93), and that thermodynamics makes statements about certain sets of points β in phase space[2, 3].

Mutual adiabatic accessibility has the additional property of symmetry (iii). This condition converts partial preorder to *equivalence*, and in the abstract scheme one defines it by

$$x \sim y \Leftrightarrow xRy \text{ and } yRx$$

Such a relation divides the set E into so-called equivalence classes. Each class contains all the states belonging to one and the same entropy. Let the class containing a certain state x be denoted by C_x. Then the properties of these classes are:

(a) $y \in C_x \Rightarrow C_y = C_x$
(b) $y \notin C_x \Rightarrow C_y \cap C_x = 0$
(c) $E = C_x \cup C_y \cup \ldots..$
(d) If (xRy), then $(x' \in C_x, y' \in C_y \Rightarrow x'Ry')$

Since all points of C_x and C_y are therefore ordered in the same sense, one can write

$$xRy \Leftrightarrow C_x \rho C_y$$

where ρ is the ordering relation between the classes. Since R has property (v), so has ρ. But unlike R, ρ has the property (iv). This yields *total order* of the set C of equivalence classes (see *Figure 2*), which are said to form a *chain*.

A simple interpretation of the relation ρ is the relation $\leqslant$ among real numbers, which clearly has the properties (i), (ii), (iv) and (v). This suggests that (subject to additional assumptions) one can associate a real number $\sigma(C_x)$ with any equivalence class such that

$$(C_x \rho C_y) \text{ and } (C_x \neq C_y) \Rightarrow \sigma(C_x) < \sigma(C_y) \tag{2.1}$$

This function σ has the properties of an empirical entropy, and we have arrived at it without mention of work, temperature or phase space. Heat has also not been mentioned, though knowledge of it may (but need not) be assumed to define an adiabatic linkage of states. The realization that the introduction of these concepts can be delayed without errors in logic until after the empirical entropy has been introduced, is one of the results of recent work on the foundations of thermodynamics. Note that the important additive property of the entropy cannot necessarily be attributed to the empirical entropy, which is clearly a weaker concept.

3. PROBLEMS OF CONTINUITY

Current mathematical work on the foundations of thermodynamics tends to exhibit either an emphasis on algebraic and group theoretical properties (Approach A) or on topological and analytical concepts (Approach B). A good combination of them may emerge in due course. A and B will be illustrated by recent attempts to find quick and intuitive (though unrigorous) ways of deriving equations of the form $d'Q = \lambda d\sigma$ for an increment of heat.

(A) *From order.*'One wants to argue[5]

(A1) Given that for distinct C_x, C_y, C_z, if $C_x \rho C_y$, $C_y \rho C_z$, . . . then $\sigma(C_x) < \sigma(C_y) < \sigma(C_z) \ldots$ (monotonic empirical entropy)

(A2) There exists a function λ in E such that $d'Q = \lambda d\sigma$ (integrating factor)

The difficulty is that the inferences (2·1) and (A1) are necessarily valid without additional hypotheses only if the set is denumerable, while (A2) requires a non-denumerable set.

(B) *From Carathéodory's theorem*[6]. In §9 of his paper, Carathéodory used accessibility considerations from a given point in phase space to show that entropy changes in a standard direction in adiabatic processes. This argument

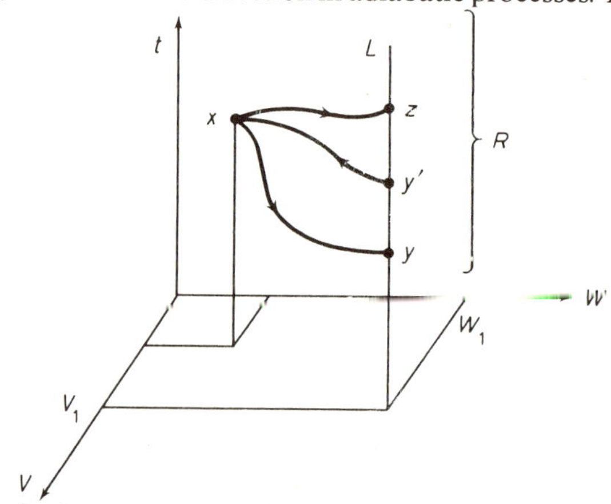

Figure 3. Diagram for the argument showing that L contains one point which is in a special relation to x.

can be extended. Denote deformation coordinates (volume, magnetic field, etc.) by V and W and the thermodynamic coordinate (empirical temperature or internal energy) by t. We seek to get adiabatically from a state x to the line L defined by $V = V_1$, $W = W_1$ (*Figure 3*). Among the range R of accessible states on L will be a state which can be reached by *quasistatic* adiabatic processes. If it is in the position y', one can get from y' to all neighbouring points on L by adiabatic processes $y' \to x \to y$, $y' \to x \to z$. By then changing V and W slightly a whole neighbourhood of y' is seen to be adiabatically accessible from y', contrary to Carathéodory's principle. So the state which is accessible from x by quasistatic adiabatic processes must lie at an endpoint of R, say at y. On changing V and W, y may be expected to generate a surface[7] on which x itself will also lie. Starting with a different state x, other surfaces are found. From these surfaces one can argue to the existence of an empirical entropy σ labelling these surfaces, and eventually to an equation $d'Q = \lambda d\sigma$.

Alternatively, from Kelvin's principle, a quasistatic adiabatic linkage between the state x and the line L is possible only for one point y on L. For if there were two such points y, y' then, choosing t to represent the internal energy, one can construct a cycle $xyy'x$ which violates Kelvin's principle. On the path yy' the energy is changed by a supply of heat Q and in the rest of the cycle it is changed by the performance of mechanical work W, whence $W = Q$. One again arrives at surfaces generated by the single states y as V and W are changed. If two (quasistatic adiabatic) surfaces intersect one would again violate Kelvin's principle, and one would also have a-points [defined in equation (4.1), below] if the surfaces are smooth enough.

Each of the procedures A and B requires continuity assumption, as pointed out in ref. 8 [see also refs. 9, 10]. To axiomatize these, one has to borrow results from pure mathematics.

Method A

Chronologically the first result to be invoked[3,8,9] was the theorem that a chain C is isomorphic to a subchain of the reals, provided[11] C contains a denumerable subset order-dense in C. This enables one to associate with any point $C_x \in C$ a real number $\sigma(x)$ such that for $C_x \neq C_y$

$$C_x \rho C_y \text{ implies } \sigma(C_x) < \sigma(C_y)$$

The natural order of the real numbers thus mirrors the order of the equivalence classes. However, gaps can exist in this representation in the sense that for all C_z such that $C_x \rho C_z$ one might have

$$\sigma(C_z) \geqslant a + b > a \geqslant \sigma(C_x)$$

where $(a, a + b)$ is a non-zero interval.

One can remove these gaps (and ensure continuity) by imposing on C additional conditions which can readily be granted for normal thermodynamic systems (C to be continuous, with a denumerable subset dense in C, and without first or last element). C then becomes similar to the real numbers in their natural order[12]. Thus one can construct an empirical entropy which is continuous in the following sense: If $\sigma(C_a) - \epsilon < \sigma(C_a) < \sigma(C_a) + \epsilon$ there exist elements y and z such that $C_y \rho C_a \rho C_z$ and such that for any x

satisfying $C_y \rho C_x \rho C_z$ we have

$$\sigma(C_a) - \in < \sigma(C_x) < \sigma(C_a) + \in$$

This second procedure does not appear to have been used explicitly, though it is very direct, and yields a continuous empirical entropy without using Carathéodory's axiom.

Method B.

Let N_x denote a *neighbourhood* of *any* element x of E. Then a third way of ensuring continuity derives from this axiom:

$$[(\forall N_x)(\exists x')(x' \in N_x \text{ and not } xRx')], \text{ i.e. } [x \text{ is an 'i-point'}] \qquad (3.1)$$

The notion of neighbourhood in the original set E implies the existence of a topology in E (E becomes a *topological space*), and continuity means that closeness in E according to this topology must be linked to the preorder relation R already defined in E. Thus one requires that any x and y satisfying xRy and not yRx have neighbourhoods N_x, N_y such that

$$x' \in N_x \text{ and } y' \in N_y \Rightarrow x'Ry' \text{ and not } y'Rx'$$

In addition to this continuity condition for R, first used[9] in 1962, a separable topological space E is needed[13, 14]. The separability of E guarantees the existence of a denumerable subset dense in C. To exclude gaps in C, i.e. to make C dense in itself (and in that sense 'continuous') one must assume that E is connected in the topological sense[9, 14].

For different equivalence classes to be represented by a surface and for σ to be also differentiable, E has to be a locally Euclidean space (a 'differentiable manifold') and it is sufficient that quasistatic adiabatic transitions be characterized by a condition $\Sigma X_j(x_1, x_2, \ldots)\mathrm{d}x_j = 0$ where the X_j are differentiable functions. Falk and Jung consistently avoided the mathematical problem of ensuring continuity (ref. 4, pp 120, 125, 131, 142).

4. THE 'OLD' CARATHÉODORY APPROACH

The qualification 'old' in the title of this section is intended to avoid confusion between Carathéodory's own work and its recent developments[14].

The 'old' Carathéodory approach shared with the conventional Clausius treatment a number of demerits[4, 15]: (α) Though strict axiomatics was not intended by these authors, it was always prevented by the occurrence of unstated assumptions which supported stated 'laws' or 'axioms'. (β) The introduction *together* of absolute temperature and entropy was a cause of confusion. Approaches A and B have removed these defects. Only partial removal of the following additional defects appears to have taken place: (γ) The combination of the inaccessibility *axiom* and the *restriction* to simple system led Carathéodory to peculiar results. One of these is that an empirical entropy can be found for an ideal gas without any appeal being made at all to his *axiom* (3.1). (δ) The stipulation that all but one coordinate of the thermodynamic phase space be 'non-thermodynamic' meant that a clear distinction between mechanisms and thermodynamics was not yet part of the formal structure. (ε) Insufficient attention is paid to the fact that thermodynamic functions are defined strictly only for interior points of a phase space.

At the present time, then, these shortcomings of the old Carathéodory approach are becoming clearer, and the few specialists who *are* working on the foundations are leaving this approach in order to develop instead approaches A and B. It is an interesting thought that this is just the time when the nature of the old Carathéodory argument, and how it relates to the Kelvin and Clausius treatments, is becoming clearer[16] so as to endow the old Carathéodory approach with some popularity among a wider group of scientists.

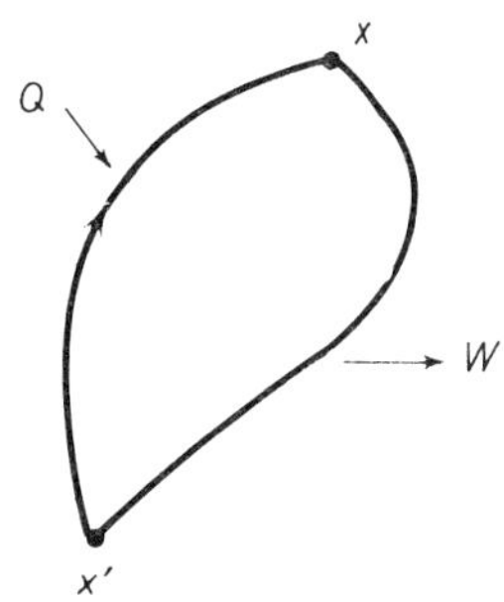

Figure 4. A circuit to relate the principles of Kelvin and Carathéodory.

The key idea here is the straight deduction of Carathéodory's axiom from Kelvin's principle! If Carathéodory's axiom is not satisfied, there exists for at least one state x a neighbourhood N_x of x such that

$$(\forall x')(x' \in N_x \Rightarrow xRx'), \text{ i.e. } x \text{ is an a-point.} \tag{4.1}$$

Keeping the deformation coordinates such as volume and magnetic field fixed, choose a state $x' \in N_x$ such that the transition from x' to x can be performed by adding an amount of heat ($Q > 0$, say) to the system. One can then return the system from x to x' adiabatically so that work W is performed by the system. This means that for the cycle $(x'xx')$ $Q = W$, and heat energy can be completely converted into work. This violates Kelvin's principle. The initial assumptions must therefore have been in error. Carathéodory's axiom (3.1) follows from this contradiction.

Suppose now the violation of Carathéodory's principle, i.e. the existence of the a-point x is granted, while one maintains Kelvin's principle. This feat can be achieved only if the system is such that there are no states $x' \in N_x$ with the required property that the transition from x' to x can occur with $Q > 0$ and fixed deformation coordinates. One must ask: What kind of systems are these? The answer is simple, the existence of a-points signifies that these are purely *mechanical* systems.

These above considerations suggest that Carathéodory's paper should no longer be regarded as an attempt at axiomatics. Instead its contribution is to distinguish between the simpler forms of mechanical and thermal systems in terms of the topology of the phase space:

Simple mechanical systems: All points are a-points.
Simple thermal systems: All points are i-points.

5. METRIC VARIABLES: EXTENSIVITY OR ADDITIVITY

Basic to the notion of length, weight, etc., is the existence of a 'joining' operation. The theory of measurement postulates it so that two equal standard rods joined in a line end to end can be equated (in the sense of equal lengths) to another rod which is then two units long. It is this operation which makes measurement possible. From this notion there then emerges the idea of a metric (i.e. measurable) variable. The quantities measured in this way are additive in this sense: if q_1 is measured as a join of n_1 units of weight, q_2 as a join of n_2 units, then there exists a join (q_1, q_2) of the weights which will be measured as a join of $n_1 + n_2$ units. Weight is said to be additive; alternatively it is said to be extensive.

These notions are deeper than the foundations of thermodynamics, for they lie at the basis of the theory of measurement itself. One must, therefore, expect any axiomatic treatment of thermodynamics to lay bare the need for an operation of joining. While the empirical entropies do not necessarily add on joining two systems, the absolute entropies do, and one must discover what axioms make such extensive quantities possible.

To find an extensive energy one may start with a system in 'energetic isolation'. Possible transitions are then restricted to occur between equilibrium states x, y of equal energy. The corresponding relation will still be denoted by R. In this case $xRy \Rightarrow yRx$, and this, together with transitivity, yields an equivalence relation. It enables the states x to be divided into classes as described in section 2 for adiabatic isolation. However, there is no obvious way of ordering these classes of constant energy in a way which corresponds to the ordering of the entropy classes C_x. The new idea here is to consider two equilibrium systems which are joined, energetically isolated from the surroundings, and allowed to interact. Possible transitions in which the first part of the system loses energy (say) are

$$(x_1, x_2) \to (x_1', x_2'),\ (x_1 x_2') \to (x_1', x_2''),\ (x_1, x_2'') \to (x_1', x_2'''), \ldots$$

The transition $x_1 \to x_1'$ of the first part is here used rather like a measuring rod, and this induces an ordering of the equivalence classes of the second part. A metric energy $U(x)$ results if one associates real numbers $U(x)$ with states x such that

$$U(x_2') - U(x_2) = U(x_2'') - U(x_2') = \ldots\ldots$$

It is found that the derivation of a metric and additive variable is possible by this method whenever the variable is subject to a conservation law[4]. It follows that, by confining attention to quasistatic adiabatic processes, when the entropy is conserved, a metric entropy can also be found.

The additivity of the entropy in Giles's method is more troublesome, no doubt due to the fact that he tries to describe the theory and the rules of interpretation in directly experimental terms. The main ideas are:

(i) One uses states x of systems (not necessarily equilibrium states) which evolve in isolation into other states y: xRy. Adiabatic isolation and equilibrium come much later, so that the partial preorder R has a generalized interpretation.

(ii) The joining operation acts on *states* rather than systems. Being commutative and associative, it is denoted by $+$. It makes the theory of partially preordered semigroups relevant.

(iii) The connection between additivity and conservation laws occurs here through components of content (e.g. energy) $Q(x)$ which are defined by

$$Q(x + y) = Q(x) + Q(y)$$

$$xRy \Rightarrow Q(x) = Q(y)$$

(iv) The connection between additivity and absolute entropy $S(x)$ is defined through the properties

$$S(x + y) = S(x) + S(y)$$

$$xRy \Rightarrow S(x) \leqslant S(y)$$

(v) The axioms are sufficient[17] to establish the existence of an absolute entropy and a family $F(Q)$ of components of content such that

$$xRy \Leftrightarrow S(x) \leqslant S(y) \text{ and } Q(x) = Q(y) \text{ for all } Q \in F(Q).$$

Within this interesting and elegant framework, it has so far proved difficult to add simple axioms[17] which ensure that the set $F(Q)$ is finite and that its members are well behaved. Also the existence of metric variables is established by somewhat complicated procedures[1, 17] involving components of content. Lastly, there are a few difficulties of interpretation and this can be illustrated by the definition of non-equilibrium state (ref. 1, p 83). Using R to denote a 'natural process' x is a *non-equilibrium* state if a state y exists such that

$$xRy, \text{ and not } (yRx) \tag{5.1}$$

But for any *equilibrium* state x, as normally understood, one can construct a state y, e.g. by withdrawing a partition, so as to satisfy (5.1).

An alternative procedure[4, 13, 18] for additivity is to put for the absolute entropy $S(x_0) = 0$, $S(x_1) = 1$ for arbitrary states x_0 and x_1. If one has adequate assumptions concerning joins of systems $A_1, A_2, \ldots A_n$, i.e. about states $(x_1, x_2, \ldots, x_n)$ of such systems, one can require $S(x_1) + S(x_2) = S(x_1') + S(x_2')$ if (x_1, x_2) goes over into (x_1', x_2') by a quasistatic adiabatic process. Such an equation can be used to identify one unknown entropy value. For example, if (x, x) goes over into (x_1, x_1) in this way, then $S(x) = \frac{1}{2}$. By continuing in this way additivity of the entropy can be established.

6. THEORY OF ELEMENTARY INEQUALITIES

Following quasistatic (or reversible) processes and the attendant additivity of thermodynamic functions (§5), we now return to the general (non-static or irreversible) processes of §§2–4. The important relations to be considered are inequalities. We shall denote functions of specified variables by small letters (x), and variables, when they are not considered as functions, by corresponding capital letters (X). Sets of variables $X_1, X_2, \ldots$ will be denoted by vector symbols ($\mathbf{X}$). The following properties are noteworthy:

(α) Monotonic increase of f with X_1:

$$[\mathrm{f}(X_1', X_2, \ldots)]\,\rho\,[\mathrm{f}(X_1, X_2, \ldots)] \Leftrightarrow X_1'\rho X_1 \tag{6.1}$$

where ρ can stand for 'greater than'; it can alternatively stand for 'equal'.

(β) Homogeneity (of degree unity): For all real $a \neq 0$

$$\mathrm{f}(a\mathbf{X}) = a\mathrm{f}(\mathbf{X})$$

(γ_1) Superadditivity and (γ_2) subadditivity:

$$\{\mathrm{f}(\mathbf{X} + \mathbf{Y})\}\,R\{\mathrm{f}(\mathbf{X}) + \mathrm{f}(\mathbf{Y})\}$$

(δ_1) Concavity and (δ_2) convexity:

$$\{\mathrm{f}\,[\tfrac{1}{2}(\mathbf{X} + \mathbf{Y})]\}\,R\{\tfrac{1}{2}\,[\mathrm{f}(\mathbf{X}) + \mathrm{f}(\mathbf{Y})]\}$$

Here R is $\geqslant$ for (γ_1) and (δ_1) and $\leqslant$ for (γ_2) and (δ_2).

Assuming (α), it is possible to invert $\mathrm{F} = \mathrm{f}(X_1, X_2, \ldots)$ to yield

$$X_1 = x_1(F, X_2, \ldots)$$

so that for all $X_2, \ldots$ the function x_1 satisfies

$$\mathrm{F} = \mathrm{f}[x_1(\mathrm{F}, X_2, \ldots), X_2, \ldots].$$

It can now be shown (see Appendix), on putting the function to which a condition applies in brackets behind it, that

$$\text{given (α) as stated in (6.1)} \begin{cases} \beta(\mathrm{f}) \Leftrightarrow \beta(x_1) & (6.2) \\ \gamma_1(\mathrm{f}) \Leftrightarrow \gamma_2(x_1),\ \gamma_2(\mathrm{f}) \Leftrightarrow \gamma_1(x_1) & (6.3) \\ \delta_1(\mathrm{f}) \Leftrightarrow \delta_2(x_1),\ \delta_2(\mathrm{f}) \Leftrightarrow \delta_1(x_1) & (6.4) \end{cases}$$

$$\text{given } \beta(\mathrm{f}) \begin{cases} \gamma_1(\mathrm{f}) \Leftrightarrow \delta_1(\mathrm{f}) & (6.5) \\ \gamma_2(\mathrm{f}) \Leftrightarrow \delta_2(\mathrm{f}) & (6.6) \end{cases}$$

Write $S = s(U, V, N)$ for the entropy, so that X_1 becomes the internal energy. The equivalence of certain thermodynamic statements then emerges as follows:

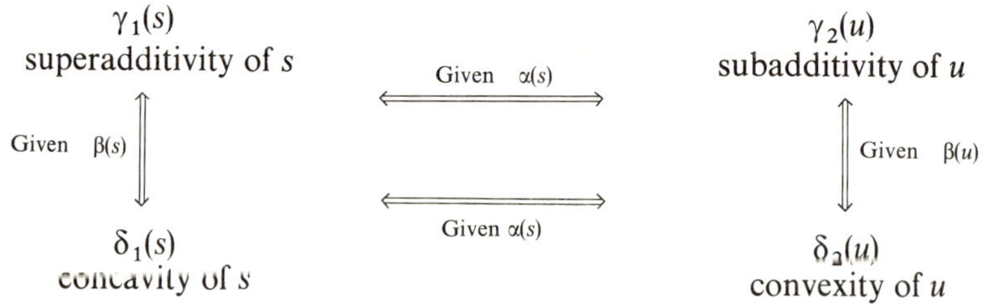

One can with the aid of this scheme again arrive at the continuity of the entropy. Assume from a 'fourth law' (ref. 3, p 142) the homogeneity of the entropy, from a form of the second law the superadditivity of the entropy[19], and from some form of the third law that the entropy is bounded in the interval considered. It then follows:

(i) from the above scheme that the entropy s is concave or $-s$ is convex.

(ii) from the theory of convex functions[20] if V and N are fixed then s is continuous in the interval of U, and has right-hand and left-hand derivatives which are decreasing functions of U.

This set of ideas is of importance in presentations of thermodynamics, to be called approach C, in which the existence of an entropy is assumed. An important part of Gibbsian thermodynamics comes under this heading. The inequalities introduced here are also useful in discussing limiting properties of statistical mechanical ensembles[21].

7. THE ZEROTH LAW

If u_K be the thermodynamic variable for a system K, let $F_{12}(u_1, u_2) = 0$ be a relation specifying thermal equilibrium between systems 1 and 2. Then the zeroth law states that any two of the relations

$$F_{12}(u_1, u_2) = 0,\ F_{23}(u_2, u_3) = 0,\ F_{31}(u_3, u_1) = 0$$

imply the third (transitivity). It is then inferred that functions $t_j(u_j)$ $(j = 1, 2, 3)$ exist such that thermal equilibrium between systems 1 and 2 may be equivalently expressed by

$$t_1(u_1) = t_2(u_2)$$

Such a result assumes certain uniqueness properties of thermal equilibrium (ref. 3, p 11 and ref. 22). Sometimes this caution has not been made explicit, and perhaps scientists can be excused for not always stating that the existence of a thermometer is an assumption (ref. 2, p 373; ref. 3, p 117) in an axiomatic treatment! In fact the following assumption is needed: Suppose that the values of all but one of the variables u_1 are specified. Then for each state of system 2 (for example) there must be a unique value of this remaining variable for which the systems 1 and 2 are in equilibrium.

As an illustration of alternative mathematical situations, let a_{ij}, b_{ij}, be positive numbers, and let

$$F_{ij}(u_i, u_j) \equiv a_{ij}(\theta_i - \theta_j)^2 + b_{ij}(\phi_i - \phi_j)^2$$

It is then true that any two of the equilibrium conditions imply the third but there are now two 'temperatures' since the equilibrium condition for systems 1 and 2 is equivalent to

$$\theta_1 = \theta_2 \text{ and } \phi_1 = \phi_2$$

This example has been suggested by the specification of a line j by the position a_j of a point on it, together with two angles θ_j, ϕ_j in a cartesian coordinate system. The equilibrium condition $F_{ij} = 0$ is then the condition for parallelism of the lines i and j, and transitivity is valid. But a single 'temperature' function is clearly not adequate: there are two such functions θ and ϕ. A system with a built-in adiabatic partition can also have two temperatures.

8. CONCLUSION

Axiomatics does not make new science; this is seen very clearly by the references which were made by Carathéodory[6] and Giles[1] to relativistic thermodynamics (already in the conventional Planck Einstein form by 1907).

Table 1. Concepts needed to arrive at an empirical entropy ($\surd$: concept is needed, x: concept is not needed)

Approach \ Concepts	Heat $d'Q$	Absolute temperature T	Adiabatic processes and equilibrium states
e.g. Clausius (ca. 1850)	$\surd$	$\surd$	$\surd$
Carathéodory (1909)	x	$\surd$	$\surd$
Falk and Jung (1959)	x	x	$\surd$
Giles (1964)	x	x	x

The criticism of this formulation was to come in the middle sixties from the scientists rather than from the axiomatizers. In spite of the limitations of axiomatics, I tried to show that recent foundations research in thermodynamics is nonetheless of importance to scientists, and of intrinsic interest.

Perhaps it will be illuminating to sum up one aspect of this work by saying that there has been an attempt to decrease the number of basic concepts needed. This is illustrated in *Table 1*. The axiomatic schemes indicated there, though not yet fully satisfactory, represent considerable advances. That these advances are surprising shows that entropy retains 'an untarnished lustre of novelty and an aura of unplumbed depth[23]. Perhaps its study may lead to more surprises in the future.

It is regretted that space did not permit adequate discussion of otherwise relevant work in which the existence of entropy is *assumed*[24].

APPENDIX

(See also ref. 19.)

In condition (α') substitute as follows:

$$\mathrm{F}, X_2, \ldots \rightarrow \begin{cases} aF, aX_2, \ldots, & \text{if f satisfies } (\beta) \\ \mathrm{F}' + \mathrm{F}'', X_2' + X_2'', \ldots, & \text{if f satisfies } (\gamma_1) \\ \tfrac{1}{2}(\mathrm{F}' + \mathrm{F}''), \tfrac{1}{2}(X_2' + X_2''), \ldots, & \text{if f satisfies } (\delta_1) \end{cases}$$

In the first case one has from (α') and (β)

$$\mathrm{f}[x_1(a\mathrm{F}, aX_2, \ldots), aX_2, \ldots] = a\mathrm{F} = a\mathrm{f}[x_1(\mathrm{F}, X_2, \ldots), X_2, \ldots]$$
$$= \mathrm{f}[ax_1(F, X_2, \ldots), aX_2, \ldots]$$

whence $x_1(a\mathrm{F}, aX_2, \ldots) = ax_1(\mathrm{F}, X_2, \ldots)$

Since one can also argue backwards from the last equation, this yields (6.2). In the second case one has from (α') and (γ_1):

$$\mathrm{f}[x_1(\mathrm{F}' + \mathrm{F}'', X_2' + X_2'', \ldots), X_2' + X_2'', \ldots]$$
$$= \mathrm{f}[x_1(\mathrm{F}', X_2', \ldots), X_2', \ldots] + \mathrm{f}[x_1(\mathrm{F}'', X_2'', \ldots), X_2'', \ldots]$$
$$\leqslant \mathrm{f}[x_1(\mathrm{F}', X_2', \ldots) + x_1(\mathrm{F}'', X_2'', \ldots), X_2' + X_2'', \ldots]$$

whence

$$x_1(\mathrm{F}' + \mathrm{F}'', X_2' + X_2'', \ldots) \leqslant x_1(\mathrm{F}', X_2', \ldots) + x_1(\mathrm{F}'', X_2'', \ldots).$$

This is part of (6.3). The other relations (6.3), (6.4) are obtained similarly.

Assume next that f satisfies (β) and (γ_1). Then choosing $a = 2$,

$$f[\tfrac{1}{2}(\mathbf{X} + \mathbf{Y})] \geqslant f[\tfrac{1}{2}(\mathbf{X})] + f[\tfrac{1}{2}(\mathbf{Y})] = \tfrac{1}{2}[f(\mathbf{X}) + f(\mathbf{Y})]$$

which is part of (6.5), and the remaining relations are established similarly.

REFERENCES

[1] R. Giles, *Mathamatical Foundations of Thermodynamics*. Pergamon: Oxford (1964).

[2] P. T. Landsberg, *Rev. Mod. Phys.* **28**, 363 (1956).

[3] P. T. Landsberg, *Thermodynamics with Quantum Statistical Illustrations*. Wiley: New York (1961).

[4] G. Falk and H. Jung, in *Handbuch der Physik*, **III** /**2**, p 119. Springer: Berlin (1959).

[5] H. A. Buchdahl, *J. Phys.* **152**, 425 (1958); *Am. J. Phys.* **28**, 196 (1960).

[6] C. Carathéodory, *Math. Ann.* **67**, 355 (1909).

[7] L. A. Turner, *Am. J. Phys.* **28**, 781 (1960); **29**, 40 (1961).

[8] P. T. Landsberg, *Physica Status Solidi*, **1**, 120 (1961).

[9] H. A. Buchdahl and W. Greve, *Z. Phys.* **168**, 386 (1962).

[10] L. A. Turner, *Am. J. Phys.* **30**, 506 (1962).

[11] G. Birkhoff, *Lattice Theory*, p 32. American Mathematical Society: New York (Revised edition, 1948).

[12] A. Abian, *The Theory of Sets and Transfinite Arithmetic*, pp 296, 178, 188. W. B. Saunders: Philadelphia (1965).

[13] J. L. B. Cooper, *J. Math. Anal. Appl.* **17**, 172 (1967).

[14] J. B. Boyling, *Comm. Maths Phys.* **10**, 52 (1968).

[15] P. T. Landsberg, *Bull. Inst. Phys. and Phys. Soc.* p 150 (June 1964).

[16] B. Crawford Jr and I. Oppenheim, *J. Chem. Phys.* **34**, 1621 (1961).
P. T. Landsberg, *Nature, Lond.* **201**, 485 (1964).
J. Dunning-Davies, *Nature, Lond.* **208**, 576 (1965).
U. M. Titulaer and N. G. van Kampen, *Physica*,**31**, 1029 (1965).
M. W. Zemansky, *Am. J. Phys.* **34**, 914 (1966).
M. W. Zemansky, *Heat and Thermodynamics* (5th ed.) McGraw-Hill: New York (1968).

[17] J. J. Duistermaat, *Synthese*, **18**, 327 (1968).
F. S. Roberts and R. D. Luce, *Synthese*, **18**, 311 (1968).

[18] W. J. Hornix, in the Pittsburgh Symposium on 'A critical review of classical and relativistic thermodynamics' (April 1968). To be published.

[19] L. Galgani and A. Scotti, *Physica*, **41**, 150 (1968); **42**, 242 (1969); see also this volume, p.229

[20] G. H. Hardy, J. E. Littlewood and G. Pólya, *Inequalities*, p 91, Theorem 111. Cambridge University Press: London (1934).

[21] e.g. M. E. Fisher, *Arch. f. Rat. Mech. Anal.* **17**, 377 (1964).
R. B. Griffiths, *J. Math. Phys.* **6**, 1447 (1965).
D. Ruelle, in *Fundamental Problems in Statistical Mechanics*, Vol. II (Ed. E. G. D. Cohen). North Holland: Amsterdam (1968).

[22] G. Whaples, *J. Rat. Mech. Anal.* **1**, 301 (1952).

[23] P. T. Landsberg, *Entropy and the Unity of Knowledge*, p 24. University of Wales Press: Cardiff (1961).

[24] L. Tisza, *Generalized Thermodynamics*. M.I.T. Press: Cambridge, Mass. (1966).
H. B. Callen, *Thermodynamics*. Wiley: New York (1961).
E. A. Guggenheim, *Proc. Phys. Soc. Lond.* **79**, 1079 (1962).
M. E. Gurtin and W. O. Williams, *Arch. f. Rat. Mech. Anal.* **26**, 83 (1967); *Z. Angew. Math. Phys.* **17**, 626 (1966).
W. Vinzenz, *Ann. Phys. (Leipzig)*, **21**, 341 (1968).
J. Meixner, *Arch. f. Rat. Mech. Anal.* **33**, 33 (1969).

ON SUBADDITIVITY AND CONVEXITY PROPERTIES OF THERMODYNAMIC FUNCTIONS†

L. GALGANI

Istituto di Fisica dell'Università, Milan, Italy

AND

A. SCOTTI

C.C.R. Euratom, Ispra, Italy and Istituto di Fisica dell'Università, Milan, Italy

ABSTRACT

It is pointed out that the usual basic postulate of increase of entropy for an isolated system, as stated for example by Tisza and Callen, if mathematically formalized, is expressed as a superadditivity property of entropy. This fact has two kinds of implications: (a) it allows one to deduce in a very direct and mathematically clear way stability properties such as $C_V \geqslant 0$ and $K_T \geqslant 0$ and the equivalence of various thermodynamic schemes as expressed for example by the fact that the 'minimum' property of the free energy is a consequence of the 'maximum' property of entropy; (b) it makes it possible to establish a link with foundations research, notably the system developed by Giles, where superadditivity of entropy appears as a consequence of other axioms.

INTRODUCTION

Recent developments in the foundations of thermodynamics are of two kinds. On the one hand[1–4] one can find attempts to make rigorous the connection between the possibility of defining entropy and the classical statements of the second principle. On the other hand[5–6] the concept of entropy is taken for granted and some of its properties are assumed in an axiomatic way, so that attention is turned to deducing rigorous consequences therefrom: these works are on the line of Gibbs.

In statistical thermodynamics both attitudes have their counterpart. The second approach has been intensively investigated particularly since 1963, when Ruelle[7, 8] was able to study in a rigorous way the problem of the so-called 'thermodynamic limit' on the basis of the conditions of stability and strong tempering on the intermolecular potentials. In the technical treatment of the problem the mathematical property of superadditivity was considered and the limit thermodynamic functions turned out to have convexity properties[9–14]. This was just the clue to some improvements in

† Work supported in part by CNR (Consiglio Nazionale delle Ricerche).

thermodynamics itself, by the realization that superadditivity has a deep physical meaning: it expresses formally the basic postulate assumed by Tisza and Callen[15]. This simple property allows one to derive in a very direct way properties such as continuity and differentiability (almost everywhere) of entropy, stability conditions, extremum properties of thermodynamic potentials and Massieu functions, which had been postulated independently or derived in a more complicated or obscure way by those authors.

The way this is done is presented in the first three sections. In section 4 the physical meaning of convexity properties is illustrated and the conclusions are given in section 5.

1. ENTROPY SCHEME

We consider for definiteness and simplicity a system whose equilibrium states are characterized by the values of three extensive quantities: energy U, volume V and number of moles N; its thermodynamic properties are deduced from the entropy S defined through the functional relation

$$S = \mathcal{S}(U, V, N) \tag{1}$$

The assumed properties of $\mathcal{S}$ are those of homogeneity, strict monotonicity in U and superadditivity:

$$\mathcal{S}(\lambda U, \lambda V, \lambda N) = \lambda \mathcal{S}(U, V, N) \qquad (\lambda \text{ real}) \tag{2}$$

$$\mathcal{S}(U_2, V, N) \geqslant \mathcal{S}(U_1, V, N) \Leftrightarrow U_2 \geqslant U_1 \tag{3}$$

$$\mathcal{S}(U_1 + U_2, V, + V_2, N_1 + N_2) \geqslant \mathcal{S}(U_1, V_1, N_1) + \mathcal{S}(U_2, V_2, N_2) \tag{4}$$

Equation 2, commonly employed, is a consequence of the postulate of extensivity of entropy: geometrically it means the $\mathcal{S}$ is a ruled surface. Equation 3 is related to positivity of temperature; it will be used only for the passage to the energy scheme (§2). Equation 4 is the property of superadditivity which expresses physically the increasing property of entropy for isolated systems: for an isolated system of fixed U, V, N the state of unconstrained equilibrium has an entropy greater than all corresponding states of constrained equilibrium. The entropy of the state of constrained equilibrium is represented by the RHS of equation 4, in accordance with the postulate of extensivity. The postulate of increase of entropy, enunciated in words in this form also by Tisza and Callen, was formally exploited only in a way that required consideration of the thermodynamic space of configurations enlarged to include the extensive independent variables of the simple systems constituting the composite constrained one. As it is formalized in equation 4, it just gives a functional relation on the function $\mathcal{S}(U, V, N)$ itself. The power of this relation is shown by the following immediate consequence: by equations 4 and 2, with $\lambda = \frac{1}{2}$, one has

$$2\mathcal{S}\left(\frac{U_1 + U_2}{2}, \frac{V_1 + V_2}{2}, \frac{N_1 + N_2}{2}\right) \geqslant \mathcal{S}(U_1, V_1, N_1) + \mathcal{S}(U_2, V_2, N_2) \tag{5}$$

i.e. $\mathscr{S}$ is a concave function. By known theorems on convex functions it follows that[16]:

(i) $\mathscr{S}(U, V, N)$ is continuous†;
(ii) right and left first partial derivatives of $\mathscr{S}$ always exist; the corresponding differentiated variables are monotonically decreasing (so that they may have at most jump discontinuities in a denumerable set);
(iii) if $\mathscr{S}$ is twice differentiable, then one has

$$Q(x, y, z) \equiv$$

$$\frac{\partial^2 \mathscr{S}}{\partial U^2}x^2+\frac{\partial^2 \mathscr{S}}{\partial V^2}y^2+\frac{\partial^2 \mathscr{S}}{\partial N^2}z^2+2\left\{\frac{\partial^2 \mathscr{S}}{\partial U \partial V}xy+\frac{\partial^2 \mathscr{S}}{\partial U \partial N}xz+\frac{\partial^2 \mathscr{S}}{\partial V \partial N}yz\right\}\leqslant 0$$

In particular, with $\partial\mathscr{S}/\partial U = 1/T$, $\partial\mathscr{S}/\partial V = p/T$, and on the basis of (iii) one can prove the stability conditions:

$$C_V = \left(T\frac{\partial \mathscr{S}}{\partial T}\right)_V \geqslant 0, \qquad K_T = -\frac{1}{V}\left(\frac{\partial p}{\partial V}\right)_T \geqslant 0 \tag{6}$$

which will, however, be derived in a more direct way in §3.

Properties (i) and (ii) are independently postulated by the quoted authors. Tisza derives the stability conditions 6 in a less direct way, while other treatments have been criticized[17].

2. ENERGY SCHEME

By equation 3 it is possible to invert the functional relation 1 and obtain

$$U = \mathscr{U}(S, V, N) \tag{7}$$

where $\mathscr{U}$ is defined by the identity in S, for any V and N,

$$\mathscr{S}[\mathscr{U}(S, V, N), V, N] \equiv S \tag{8}$$

We then have that $\mathscr{U}$ is convex and homogeneous,

$$\mathscr{U}\left(\frac{S_1+S_2}{2}, \frac{V_1+V_2}{2}, \frac{N_1+N_2}{2}\right) \leqslant \tfrac{1}{2}\{\mathscr{U}(S_1, V_1, N_1) + \mathscr{U}(S_2, V_2, N_2)\} \tag{9}$$

$$\mathscr{U}(\lambda S, \lambda V, \lambda N) = \lambda \mathscr{U}(S, V, N) \tag{10}$$

Equation 9 follows from equations 5 and 3, by appropriate repeated use of equation 8

$$\mathscr{S}\left[\mathscr{U}\left(\frac{S_1+S_2}{2}, \frac{V_1+V_2}{2}, \frac{N_1+N_2}{2}\right), \frac{V_1+V_2}{2}, \frac{N_1+N_2}{2}\right] = \frac{S_1+S_2}{2}$$

$$= \tfrac{1}{2}\{\mathscr{S}[\mathscr{U}(S_1, V_1, N_1,), V_1, N_1] + \mathscr{S}[\mathscr{U}(S_2, V_2, N_2), V_2, N_2]\}$$

$$\leqslant \mathscr{S}\left(\frac{\mathscr{U}(S_1, V_1, N_1) + \mathscr{U}(S_2, V_2, N_2)}{2}, \frac{V_1+V_2}{2}, \frac{N_1+N_2}{2}\right)$$

† Strictly speaking this is true only if the function $\mathscr{S}$ is assumed to be measurable.

Equation 10 follows from equations 2 and 3, again using equation 8:

$$\mathscr{S}[\mathscr{U}(\lambda S, \lambda V, \lambda N), \lambda V, \lambda N] = \lambda S = \lambda \mathscr{S}[\mathscr{U}(S, V, N), V, N]$$

$$= \mathscr{S}[\lambda \mathscr{U}(S, V, N), \lambda V, \lambda N]$$

From equations 9 and 10 it now follows that $\mathscr{U}$ is subadditive

$$\mathscr{U}(S_1 + S_2, V_1 + V_2, N_1 + N_2) \leqslant \mathscr{U}(S_1, V_1, N_1) + \mathscr{U}(S_2, V_2, N_2) \quad (11)$$

an inequality that can be physically interpreted in complete analogy to the case of inequality 4, as expressing the decreasing property of energy in isolated systems when internal constraints are released.

We note in passing that it would be possible to derive the subadditivity of $\mathscr{U}$ directly from the superadditivity of $\mathscr{S}$, which is essentially the same procedure used by Gibbs himself, and then get the convexity through homogeneity. Indeed one has

$$\mathscr{S}[\mathscr{U}(S_1 + S_2, V_1 + V_2, N_1 + N_2), V_1 + V_2, N_1 + N_2] = S_1 + S_2$$
$$= \mathscr{S}[\mathscr{U}(S_1, V_1, N_1), V_1, N_1] + \mathscr{S}[\mathscr{U}(S_2, V_2, N_2), V_2, N_2]$$

$$\leqslant \mathscr{S}[\mathscr{U}(S_1, V_1, N_1) + \mathscr{U}(S_2, V_2, N_2), V_1 + V_2, N_1 + N_2]$$

The first way of proceeding can be extended to. Legendre transform representations.

3. LEGENDRE TRANSFORM SCHEMES

The Legendre transform $g(t)$ of a convex function $f(x)$ is given[10] by

$$g(t) = \inf_x \{f(x) - tx\}, \qquad \inf_x \left(\frac{\partial f}{\partial x}\right)^- < t < \sup_x \left(\frac{\partial f}{\partial x}\right)^+ \quad (12)$$

where $(\partial f/\partial x)^{\pm}$ is the right (left) derivative. Indeed if $f(x)$ is differentiable and strictly convex, the infimum is actually a minimum, reached at the unique point $\bar{x}(t)$ such that $\left.\frac{\partial f}{\partial x}\right|_{x=\bar{x}(t)} = t$, so that equation 12 can be reduced to the usual definition of the Legendre transform. Definition 12 is more advantageous, however, because it makes explicit the change of variable and, in addition, because it is meaningful also for general convex functions that may have discontinuities in the derivatives and can be linear in some intervals.

Coming now for definiteness to the case of a convex function of two variables $f(x, y)$,

$$f\left(\frac{x_1 + x_2}{2}, \frac{y_1 + y_2}{2}\right) \leqslant \tfrac{1}{2}\{f(x_1, y_1) + f(x_2, y_2)\}$$

we may consider the partial Legendre transform

$$g(t, y) = \inf_x \{f(x, y) - tx\}$$

for which the following convexity properties are easily established[18, 19]

$$g\left(\frac{t_1 + t_2}{2}, y\right) \geqslant \tfrac{1}{2}\{g(t_1, y) + g(t_2, y)\}$$

$$g\left(t, \frac{y_1 + y_2}{2}\right) \leqslant \tfrac{1}{2}\{g(t, y_1) + g(t, y_2)\}$$

i.e. Legendre transformation inverts the convexity properties in the transformed variables, while leaving unchanged those in the non-transformed variables. Indeed one has:

$$\inf_x \left\{f(x, y) - \frac{t_1 + t_2}{2} x\right\} = \tfrac{1}{2} \inf_x \{f(x, y) - t_1 x + f(x, y) - t_2 x\}$$

$$\geqslant \tfrac{1}{2}\{\inf_x [f(x, y) - t_1 x] + \inf_x [f(x, y) - t_2 x]\}$$

$$\inf_x \left\{f\left(x, \frac{y_1 + y_2}{2}\right) - tx\right\} \leqslant f\left(\frac{\bar{x}(t, y_1) + \bar{x}(t, y_2)}{2}, \frac{y_1 + y_2}{2}\right) - t\frac{\bar{x}(t, y_1) + \bar{x}(t, y_2)}{2}$$

$$\leqslant \tfrac{1}{2}\{f[\bar{x}(t, y_1), y_1] - t\bar{x}(t, y_1) + f[\bar{x}(t, y_2), y_2] - t\bar{x}(t, y_2)\}$$

$$= \tfrac{1}{2}\{\inf_x [f(x, y_1) - tx] + \inf_x [f(x, y_2) - tx]\}$$

where

$$f[\bar{x}(y, t) - t\bar{x}(y, t)] = \inf_x \{f(x, y) - tx\}$$

If the function $f(x, y)$ is concave, the Legendre transform is defined by analogy with equation 12 (using supremum instead of infimum) and the stated theorem on convexity properties continues to hold. So we have the general theorem concerning all possible Legendre transforms of energy (thermodynamic potentials) and entropy (Massieu functions).

Thermodynamic potentials (Massieu functions) are convex (concave) in the extensive variables and concave (convex) in the intensive ones.

From this convexity theorem, if the functions considered are assumed to be twice differentiable, stability properties now follow directly, as anticipated in § 2. Furthermore, observing that Legendre transforms are homogeneous in the extensive variables, one can derive subadditivity or superadditivity in these variables, as was done for energy in § 2. These properties too can be interpreted as expressing extremum properties with respect to states of constrained equilibrium.

4. PHYSICAL MEANING OF CONVEXITY PROPERTIES

In the same way as superadditivity of $\mathcal{S}$ (equation 4) was considered a formalization of the postulate that the state of unconstrained equilibrium has entropy greater than all corresponding states of constrained equilibrium, it seems quite natural to give also an analogous interpretation of the convexity property (equation 5). This interpretation is that the state of constrained equilibrium, in which the constraints are such that all component systems have equal extensive parameters, has entropy greater than all other corresponding states of constrained equilibrium.

Of course, the equivalence of these two statements comes about because of the fact that when all component systems have equal extensive variables the composite system is homogeneous, so that the LHS of equations 4 and 5 are equal. But the physical meaning of the convexity property seems much deeper and more direct than that of superadditivity. Indeed the property of convexity compares states in both of which there are constraints, and asserts that the state of maximum entropy corresponds to equipartition of extensive parameters (homogeneity). In this sense one might think this to be a natural starting point for the consideration of thermodynamics of irreversible processes, which act just in the direction of equalizing extensive parameters in every region.

Finally we remark that convexity properties in the intensive variables for Legendre transforms, which are usually not mentioned, can also be interpreted in this way, and this stands at variance with superadditivity properties, which cannot be proved in this case.

5. CONCLUSION

We have pointed out that the basic postulate of increase of entropy, as expressed by the statement that the state of unconstrained equilibrium of an isolated system has an entropy greater than all corresponding states of constrained equilibrium, if mathematically formalized, is simply expressed as a superadditivity property of entropy.

This remark has two kind of implications. First, it allows one to deduce in a very direct and mathematically clear way:

(i) stability properties such as $C_V \geqslant O$, $K_T \geqslant 0$, where C_V and K_T are the specific heat and the isothermal compressibility respectively;
(ii) the equivalence of various thermodynamic schemes as expressed for example by the fact that the 'minimum' property of free energy is a consequence of the 'maximum' property of entropy.

In particular, property (i) is expressed as a concavity property of entropy and is brought to the general theorem: 'Thermodynamic potentials (Massieu functions) are convex (concave) in the extensive parameters and concave (convex) in the intensive ones'.

Secondly it is possible to establish a link between the basic postulate of the increase of entropy in the form stated above and certain more fundametal discussions of the foundations of thermodynamics, especially that of Giles[3], where superadditivity of entropy for equilibrium states appears as a consequence of other axioms (Theorem 9.1.3).

REFERENCES

[1] H. A. Buchdahl, *Z. Phys.* **152**, 425 (1958).
[2] B. Bernstein, *J. Math. Phys.* **1**, 222 (1960).
[3] R. Giles, *Mathematical Foundations of Thermodynamics.* Pergamon: Oxford (1964).
[4] J. B. Boyling, *Comm. Math. Phys.* **10**, 52 (1968).
[5] L. Tisza, *Ann. Phys.* **13**, 1 (1961).
[6] H. B. Callen, *Thermodynamics.* Wiley: New York (1960).
[7] D. Ruelle, *Helv. Phys. Acta*, **36**, 183 (1963).
[8] D. Ruelle, *Statistical Mechanics, Rigorous Results,* W. A. Benjamin: Amsterdam (1969).
[9] M. E. Fisher, *Arch. Ration. Mech. Anal.* **17**. 377 (1964).
[10] R. B. Griffiths, *J. Math. Phys.* **6**, 1447 (1965).
[11] J. Van der Linden, *Physica*, **32**, 642 (1966).
[12] J. Van der Linden and P. Mazur, *Physica*, **36**, 491 (1967).
[13] J. Van der Linden, *Physica*, **38**, 173 (1968).
[14] L. Galgani, L. Manzoni and A. Scotti, *Physica*, **41**, 622 (1968).
[15] L. Galgani and A. Scotti, *Physica*, **40**, 150 (1968).
[16] G. H. Hardy, J. E. Littlewood and G. Pólya, *Inequalities*, Cambridge University Press: London (1959).
[17] J. T. Lopuszanski, *Acta Physica Polonica*, **33**, 953 (1968).
[18] S. Mandelbrojt, *C.R. Acad. Sci., Paris*, **209**, 977 (No. 11, 1939).
[19] L. Galgani and A. Scotti, *Physica*, **42**, 242 (1969).

THE EXTENSION OF CLASSICAL PHENOMENOLOGICAL THERMODYNAMICS TO OPEN SYSTEMS (THE CHEMICAL POTENTIAL)

W. J. HORNIX

Department of Chemistry, Katholieke Universiteit, Driehuizerweg 200, Nijmegen, The Netherlands

ABSTRACT

The traditional introduction of the chemical potential is through the assumption that the entropy is a differentiable function of U, V and the molar quantities of the chemical components; but entropy and energy functions are defined only for states of closed systems. An alternative introduction is accordingly given here. It meets this difficulty, and is in accordance with recent axiomatizations. An outline of the proposal is given, followed by a critical analysis of the assumptions involved.

Traditional thermodynamics introduces the chemical potential starting from the assumption that the entropy S of an open system can be regarded as a differentiable function of the internal energy U, the volume V and the molar quantities of the different components N_α, N_β, ... However, the definitions of S and U refer to adiabatic linkage, and this presumes closed systems.

Landsberg[1] meets this criticism by the introduction of a "fourth law" which implies that for a certain class of systems (more precisely: for certain sets of states) the entropy is a first order homogeneous function of U, V, N_α, N_β, ... Tisza[2] introduces a phase postulate: a simple system exists potentially in a number of phases, which are spatially homogeneous material extensions, for which a continuous first order homogeneous phase entropy function $S(U, V, N_\alpha, N_\beta, \ldots)$ is defined. Both assume the existence of a function $S(U, V, N_\alpha, N_\beta, \ldots)$ for a precisely defined class of open systems, abandoning an operational definition of entropy and internal energy.

This paper intends to give an alternative, which does not take refuge in an assumption of the above kind, and is in accordance with the operational approach to entropy and internal energy of recent axiomatizations[3–7]: the domains of definition of entropy and energy functions remain restricted to closed systems. The approach is so simple, that it can serve to introduce the chemical potential in undergraduate courses. I will start with an outline, suitable for teaching; afterwards I will justify the assumptions which are involved and in doing so prepare a more formal theory.

Outline of a simple introduction of the chemical potential

Consider a system, which is materially connected with respect to the component α with a homogeneous pure substance α, through a wall permeable exclusively for the component α. The two systems together, denoted by Z_{12}, are closed. The part systems are denoted by $Z_1^\star$ and $Z_2^\star$; the stars indicate that these systems are open (see *Figure 1*).

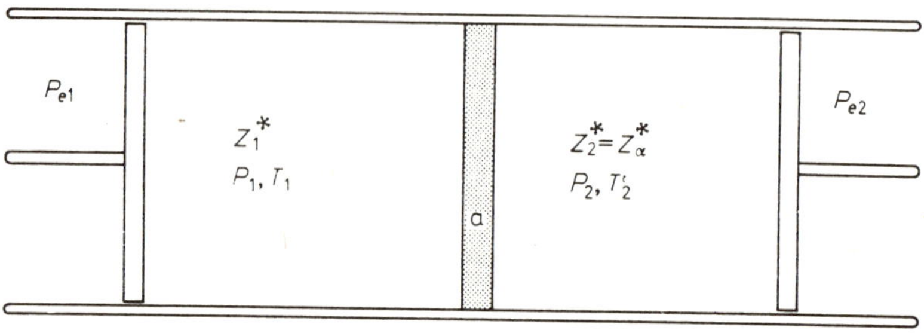

Figure 1. The pressures P_{e1} and P_{e2} of the environment are independently variable; the environment has a unique temperature T_e; a: wall permeable to component α exclusively; all other walls are diathermal, i.e. $T_1 = T_2 = T_e$; the pistons are freely movable, i.e. $P_1 = P_{e1}$, $P_2 = P_{2e}$.

For system Z_{12} an internal energy function U_{12} and an entropy function S_{12} are defined. The volumes V_1 and V_2 of the part systems and U_{12}, or S_{12}, form a complete set of independent variables of the system Z_{12}. (1)
For an infinitesimal quasistatic change of state of system Z_{12}:

$$\mathrm{d}U_{12} = T\,\mathrm{d}S_{12} - P_1\mathrm{d}V_1 - P_2\mathrm{d}V_2 \tag{2}$$

For the homogeneous pure substance $Z_2^\star$ one can write:

$$V_2 = N_\alpha^0 v_\alpha^0 \tag{3}$$
$$v_\alpha^0 = v_\alpha^0(P, T) \tag{4}$$
$$U_2^\star = N_\alpha^0 u_\alpha^0 \tag{5}$$
$$u_\alpha^0 = u_\alpha^0(P, T) \tag{6}$$
$$S_2^\star = N_\alpha^0 s_\alpha^0 \tag{7}$$
$$s_\alpha^0 = s_\alpha^0(u_\alpha^0, v_\alpha^0) = s_\alpha^0(P, T) \tag{8}$$

The functions $U_2^\star$ and $S_2^\star$, with variable N_α^0, are called the "*extended internal energy function*" and the "*extended entropy function*" respectively. The stars indicate that they are not energy functions and entropy functions in the strict sense (i.e. accessibility functions). For constant N_α^0, however, the functions are reduced to energy functions and entropy functions for the closed pure substance α. v_α^0, u_α^0 and s_α^0 are the molar volume, energy and entropy of the pure substance α; the superscript ^{0} indicates that we are concerned with pure substances. Define also:

$$\mathrm{d}U_1^\star = \mathrm{d}U_{12} - \mathrm{d}U_2^\star \tag{9}$$
$$\mathrm{d}S_1^\star = \mathrm{d}S_{12} - \mathrm{d}S_2^\star \tag{10}$$

Suppose that an infinitesimal isothermal quasistatic process of Z_{12} is associated with the transport of a quantity dN_α from Z_2 to Z_1. Then:

$$\begin{aligned} dU_1^\star &= dU_{12} - dU_2^\star = T dS_{12} - P_1 dV_1 - P_2 dV_2 - dU_2^\star \\ &= T dS_1^\star + T dS_2^\star - P_1 dV_1 - P_2 dV_2 - dU_2^\star \\ &= T dS_1^\star - P_1 dV_1 + T d(N_\alpha^0 s_\alpha^0) - P_2 d(N_\alpha^0 v_\alpha^0) - d(N_\alpha^0 u_\alpha^0) \\ &= T dS_1^\star - P_1 dV_1 + N_\alpha^0 (T ds_\alpha^0 - P_2 dv_\alpha^0 - du_\alpha^0) \\ &\quad + (T s_\alpha^0 - P_2 v_\alpha^0 - u_\alpha^0) dN_\alpha^0 \\ &= T dS_1^\star - P_1 dV_1 - g_\alpha^0 dN_\alpha^0 \end{aligned} \tag{11}$$

Now *define* the chemical potential of the component α in the system $Z_1^\star$: $\mu_{\alpha 1}$, as the molar Gibbs free energy g_α^0 of the pure substance $Z_2^\star$ in equilibrium under material connection with the system $Z_1^\star$ (12)

Thus

$$dU_1^\star = T dS_1^\star - P_1 dV_1 + \mu_{\alpha 1} dN_{\alpha 1} \tag{13}$$

This outline will conclude by proving that for arbitrary states z_i and z_j: z_i and z_j are in equilibrium under material connection with respect to component α, or abbreviated "in α-equilibrium", if and only if $\mu_\alpha(z_i) = \mu_\alpha(z_j)$ and $T(z_i) = T(z_j)$.

If $\mu_\alpha(z_i) = \mu_\alpha(z_j)$ and $T(z_i) = T(z_j)$, then for states of a pure substance α in equilibrium with z_i and z_j, say z'_α and z''_α respectively, $g_\alpha^0(z'_\alpha) = g_\alpha^0(z''_\alpha)$ and $T(z'_\alpha) = T(z''_\alpha) = T(z_i) = T(z_j)$. Now $(\partial g_\alpha^0/\partial P)_T = v_\alpha^0 > 0$, therefore $P(z'_\alpha) = P(z''_\alpha)$ and consequently $z'_\alpha = z''_\alpha$. Thus z_i and z_j are in "α-equilibrium" with the same state, and consequently also in mutual α-equilibrium. If z_i and z_j are in mutual "α-equilibrium" then they are in "α-equilibrium" with the same state of a pure substance α, as a consequence of the transitivity of "α-equilibrium". Therefore $\mu_\alpha(z_i) = \mu_\alpha(z_j)$ and $T(z_i) = T(z_j)$†.

A critical analysis of the outline proposed

The above argument contains a number of terms and statements which, in a more rigorous treatment, need explanation and justification.

(a) Every presentation of thermodynamics contains an appeal to the existence of thermal equilibrium, pressure equilibrium, material equilibrium with respect to chemical components α, β (or "α-equilibrium"), etc., with properties which are more or less explicitly defined, e.g. the zeroth law. In a formal approach this implies an assumption concerning the existence of equivalence relations defined on the set of all possible pairs of states of systems to which the relation considered can be applied[3, 7]. Systems are in such an approach sets of (equilibrium) states: $Z_i = \{z_i, z'_i, z''_i, \ldots\}$. The pair of states $z_i z_j$ belongs to the "α-equilibrium equivalence relation" or "α-connection" $C_{\mu\alpha}$, iff the systems Z_i and Z_j, in states z_i and z_j respectively are in equilibrium when materially connected with respect to component α (through a semipermeable wall). This is a rule of interpretation. α-Connection is applicable to all systems which contain the component α. In traditional thermodynamics it is always tacitly assumed that α-equilibrium is reflexive,

† NOTE: Kestin[8], and also Vanderslice *et al.*[9], offer a similar approach. In their derivation the system $Z_2^\star$ is, however, an infinitesimal system, and they assume that $ds_\alpha^0 = 0$, $dv_\alpha^0 = 0$ and $du_\alpha^0 = 0$, which can be criticized.

symmetric and transitive. A further property of "α-equilibrium", "β-equilibrium" etc. is that they imply thermal equilibrium or "θ-equilibrium".

(b) Similarly an appeal is made to the existence of different types of isolation, e.g. adiabatic, energetic or material isolations. In a formal theory[3–7], this will be expressed in existence statements of "accessibility relations", which are equivalence relations defined on the cartesian products $Z_i \times Z_i$. In the case of adiabatic isolation and energetic isolation the systems Z_i are closed systems; the pairs $z_i' z_i''$ contained in the adiabatic isolation relation are reversibly adiabatically accessible; the equivalence classes are the classes of states of equal entropy.

"Material isolation with respect to component α′ can also be expressed as an equivalence relation on $Z_i^\star \times Z_i^\star$, where $Z_i^\star$ are open systems containing this component, and the equivalence classes are the classes of states of equal material content for component α. The term "component" needs careful definition if chemical reactions can occur[2].

(c) In this paper we presuppose that extensive[7] entropy functions $S_i(z_i)$, internal energy functions $U_i(z_i)$, and deformation coordinate functions, e.g. $V_i(z_i)$, defined on closed systems, and absolute temperature and pressure functions $T(z_i)$ and $P(z_i)$ are available, and that the Gibbs fundamental equation for closed systems: $\mathrm{d}U_i = T\,\mathrm{d}S_i - P\,\mathrm{d}V_i$ has been derived. An axiomatization of this fundamental part of thermodynamics on a strictly operational basis is given elsewhere[7]. A further axiomatic development, which leads to the Gibbs fundamental equation for open systems, will now be attempted.

(d) The term "homogeneity of a system" has not so far been defined formally in axiomatizations[2]. A definition requires some preparation: Two states are called "*similar*" if and only if the pair belongs to all applicable connection equivalence relations (i.e. if they are in equilibrium under all possible connections). Thus only states of systems which differ only in extent can be similar. A "*simple system*" is a system $Z^\star$, whose closed parts Z_i, $Z_j \subset Z^\star$, i.e. the equivalence classes of equal material content, are completely specified by $[U_i, V_i]$, $[U_j, V_j]$. A "*homogeneous system*" is a simple system Z^*, such that for closed parts Z_i, $Z_j \subset Z^\star$ the following statement holds: for pairs of similar states z_i', z_j' and z_i'', z_j''

$$\frac{S_i(z_i')\ N\ S_i(z_i'')}{S_j(z_j') - S_j(z_j'')} = \frac{U_i(z_i') - U_j(z_i'')}{U_j(z_j') - U_j(z_j'')} = \frac{V_i(z_i') - V_i(z_i'')}{V_j(z_j') - V_j(z_j'')} = \frac{M_i}{M_j} \tag{14}$$

where M_i and M_j are the masses of the closed systems Z_i and Z_j. If we choose similar states as states of reference for the entropies S_i and S_j and also similar states as states of reference for the internal energies (and possibly for the volumes) then one can write: for all similar states z_i and z_j:

$$\frac{S_i(z_i)}{S_j(z_j)} = \frac{U_i(z_i)}{U_j(z_j)} = \frac{V_i(z_i)}{V_j(z_j)} = \frac{M_i}{M_j} \tag{15}$$

This justifies the introduction of specific entropies, internal energies and volumes for homogeneous systems, and in the case of pure substances, defined below, the introduction of the molar quantities:

$$s_\alpha^0 \equiv \frac{S_{\alpha i}}{N_{\alpha i}^0} \tag{16}$$

$$u_\alpha^0 \equiv \frac{U_{\alpha i}}{N_{\alpha i}^0} \tag{17}$$

$$v_\alpha^0 \equiv \frac{V_{\alpha i}}{N_{\alpha i}^0} \tag{18}$$

An immediate consequence of homogeneity is that the intensities form a complete set of independent variables for closed parts. If not, then similar states z_i' and z_i'' would exist, such that $U_i(z_i') \neq U_i(z_i'')$ or $V_i(z_i') \neq V_i(z_i'')$. But this contradicts

$$\frac{U_i(z_i')}{U_i(z_i'')} = \frac{V_i(z_i')}{V_i(z_i'')} = \frac{M_i}{M_i} = 1 \tag{19}$$

(e) A "*pure substance*" can be defined as a system to which only one material connection is applicable. An alternative definition can perhaps be as follows: A "*pure substance*" is a simple system such that for closed parts:

$$S_i = N_{1i}s_1(P, T) + N_{2i}s_2(P, T) + \ldots \tag{20}$$

$$U_i = N_{1i}u_1(P, T) + N_{2i}u_2(P, T) + \ldots \tag{21}$$

$$V_i = N_{1i}v_1(P, T) + N_{2i}v_2(P, T) + \ldots \tag{22}$$

and

$$N_{1i} + N_{2i} + \ldots = N_i \tag{23}$$

It is essential that there exist two or more functions $s_1, s_2, \ldots,$ and $u_1, u_2, \ldots$ and $v_1, v_2, \ldots$ for certain P, T domains and that S_i, U_i and V_i are linear combinations of these functions, which are completely specified by P and T. The functions are called "*molar entropies, energies, volumes of the phases 1, 2, ...*", and $N_{1i}, N_{2i}, \ldots$ are called the "*molar content of the phases 1, 2, ...*".

(f) Finally the assumption is made that a closed system Z, which is divided into simple parts $Z_1^\star$ and $Z_2^\star$ through a semipermeable wall, has a complete set of independent variables $[U, V_1, V_2]$. The precise status of this assumption is not yet clear. It is possible to consider this assumption as a definition of "semipermeable wall". "Semipermeability" has to be understood as non-permeability with respect to at least one component of the system. The limiting case is non-permeability with respect to all components; the wall is then only diathermal. The attraction of this procedure is that the existence of a semipermeable wall in a given system is decided by means of completely external criteria: there is no need to look inside the system. The ideal of phenomenological thermodynamics to consider the system as a black box, can thus be maintained. The objection that the part volumes V_1 and V_2 imply a look inside the system can be met: what matters are the volume differences ΔV_1 and ΔV_2 and these can be measured without knowledge of the part volume.

The formal introduction of the chemical potential runs as follows: Consider a system Z, which is divided through a semipermeable wall into simple parts $Z_1^\star$ and $Z_2^\star$, the latter being a pure substance α. Let the system undergo an infinitesimal, quasistatic, isothermal change (dU, dV_1, dV_2). Suppose that with this change a change dN_α^0 in the variable N_α^0 of the pure substance $Z_2^\star$ is associated. We shall say, again in "black box language", that $Z_1^\star$ undergoes a change $dN_{\alpha 1} = -dN_\alpha^0$. Then

$$dU_1^\star = dU - dU_2^\star = \ldots = T dS_1^\star - P dV_1 + \mu_\alpha dN_{\alpha 1}.$$

REFERENCES

[1] P. T. Landsberg, *Thermodynamics*, Interscience: New York (1961).

[2] L. Tisza, *Generalized Thermodynamics*, M.I.T. Press: Cambridge, Mass. (1966).

[3] G. Falk and H. Jung, in *Handbuch der Physik*, Band III/2, pp 119–175. Springer: Berlin (1959).

[4] R. Giles, *Mathematical Foundations of Thermodynamics*, Pergamon: Oxford (1964).

[5] J. L. B. Cooper, *J. Math. Anal. Appl.* **17**, 172–193 (1967).

[6] J. J. Duistermaat, *Synthese*, **18**, 311 (1968).

[7] W. J. Hornix, "An axiomatization of classical phenomenological thermodynamics". To be published in the *Proceedings* of the International Symposium "A critical review of the foundations of relativistic and classical thermodynamics" University of Pittsburgh: Pennsylvania (1969).

[8] J. Kestin, *A Course in Thermodynamics*. Blaisdell: Waltham, Mass. (1966).

[9] J. T. Vanderslice, H. W. Schamp and E. Mason, *Thermodynamics*, Prentice-Hall: Englewood Cliffs, N.J. (1966).

GENERAL DEFINITION OF THE PERFECT GAS CONCEPT

G. SÜSSMANN
Sektion Physik der Universität München, Germany

AND

E. HILF
Physikalisches Institut der Universität Würzburg, Germany

ABSTRACT

A differential equation of state is presented for a monatomic gas without virial interactions between its particles. Together with suitable boundary conditions this defines a macroscopic concept of the 'perfect' gas.

INTRODUCTION

Since the time of Boyle, Mariotte and Gay–Lussac many people have considered how an 'ideal' (or 'perfect') gas should be defined. Rather early, the relation

$$p = \nu kT \tag{1}$$

was accepted, p being the pressure, ν the number density (v^{-1}) and T the absolute temperature of the gas. As is well known, this thermal equation of state does not completely determine the thermodynamic properties of the substance. If one also wishes the caloric equation of state, one must known an additional function of one variable, e.g. the molecular heat distribution $c_v(T)$ whence the molecular energy

$$u(T) = \int c_v(T)\mathrm{d}T \tag{2}$$

may be obtained. From both equations of state one derives the molecular entropy

$$s(T, \nu) = \int c_v(T)\, T^{-1}\, \mathrm{d}T - \log p \tag{3}$$

and from these the chemical potential

$$\mu(T, p) = kT + u(T) - Ts(T, p/kT) \tag{4}$$

which contains the whole thermodynamic information about the substance, because it is a thermodynamic potential. An important quantity measuring the 'power of diffusion' is the fugacity $\varphi = \mathrm{e}^{-\mu/kT}$. We call a fluid that obeys equation 1 an *ideal gas*, distinguishing this notion from that of a 'perfect gas', which we are going to describe.

A simple microscopic model that yields equation 1 by classical statistical mechanics (both Galilei or Lorentz invariant), assumes the gas to consist of particles that have no virial interactions which means zero range repulsions, the potential energy of which vanishes in the mean, nevertheless leading to thermal equilibrium. The isochoric heat c_v is determined by the intrinsic dynamics of the particles (molecules). In the simplest case of elementary particles (atoms) one obtains

$$c_v(0) = \tfrac{3}{2}k, \qquad c_v(0) = 3k \tag{5}$$

for the 'non-relativistic' or *low velocity* limit N and the 'extreme-relativistic' or *high velocity* limit E.

The same microdynamic picture yet with quantum instead of classical kinematics no longer reproduces equation 1, although one is still inclined to consider such a gas to be ('ideal' or) 'perfect'. Only in the 'quasiclassical' limit C of *large quantum numbers* is the equation 1 regained asymptotically, whereas in the limit D of *long de Broglie waves* the gas behaviour 'degenerates' completely. These remarks lead to the *microscopic* version[1] of our concept: A *perfect gas* should consist of elementary particles without virial interactions. As temperature and fugacity are unrestricted, all limiting cases (N, E, C, D, and their combinations NC, EC, ND, ED) are contained as limiting cases of the normal situation, which of course is treated by Lorentz invariant quantum statistics. In this respect the perfect gas concept is thermally much broader than that of an ideal gas. On the other hand, it is much narrower with regard to the caloric properties of the gas.

What is the macroscopic equivalent of this fundamental microscopic abstraction? H. Einbinder introduced[2] and P. T. Landsberg proposed[3] the relation

$$p = g\varepsilon \text{ with } \begin{cases} g = \tfrac{2}{3} \text{ for N} \\ g = \tfrac{1}{3} \text{ for E} \end{cases} \tag{6}$$

between pressure and the energy density $\varepsilon \equiv v{:}u$ as a macroscopic characterization of what we call a perfect gas. We shall discuss this definition 6 only in the two indicated limiting cases 6_N or 6_E. So we are left with the problem of interpolating between these two asymptotes.

I. DIFFERENTIAL EQUATION OF STATE

According to our microscopic definition of the perfect gas, we have to consider the well known phase space integrals:

$$v = B\int_0^\infty dw \frac{[w(w+2mc^2)]^{\frac{1}{2}}}{e^{\alpha+\beta w} - \iota}(w + mc^2) \tag{7}$$

$$\varepsilon = B\int_0^\infty dw \frac{[w(w+2mc^2)]^{\frac{1}{2}}}{e^{\alpha+\beta w} - \iota}(w + mc^2)\, w \tag{8}$$

$$p = \tfrac{1}{3}B\int_0^\infty dw \frac{[w(w+2mc^2)]^{\frac{1}{2}}}{e^{\alpha+\beta w} - \iota}(w + 2mc^2)\, w \tag{9}$$

with

$$B \equiv 4\pi c^{-3} h^{-3}(2s + 1) \tag{10}$$

In these equations m, s, w denote the (rest) mass, the spin, and the kinetic energy of the identical particles, respectively (c being the velocity of light *in vacuo* and h Planck's quantum of action). The statistical parameter ι takes the values 0, +1, −1 for Maxwell–Boltzmann, Bose–Einstein, and Fermi–Dirac statistics, respectively. As usual,

$$\alpha \equiv -\frac{\mu}{kT}, \qquad \beta \equiv \frac{1}{kT}, \qquad \gamma \equiv \frac{p}{kT} \tag{11}$$

are the equilibrium variables for particle, energy and volume exchange, respectively.

The partial derivatives of the phase space integrals combine to the following differential relations:

$$\partial_\beta v = \partial_\alpha \varepsilon \tag{12}$$

$$\beta\, \partial_\alpha p + v = 0, \qquad \beta\, \partial_\beta p + \varepsilon + p = 0 \tag{13}$$

and

$$\partial_\beta (3p - \varepsilon) = mc^2\, \partial_\alpha (2\varepsilon - 2p) \tag{14}$$

Of these, equations 12 and 13 are general thermodynamic identities, true for any homogeneous substance. This can be seen from the total differential

$$d\gamma = -v\, d\alpha - \varepsilon\, d\beta \tag{15}$$

of the thermodynamic potential $\gamma(\alpha, \beta)$. Its Maxwell relation 12 is therefore a simple consequence of 13.

On the other hand, the relation 14 is specific for the gas, as indicated by the occurrence of the mass parameter m. We propose to use equation 14 as the defining relation of the *macroscopic* concept of *a generalized perfect gas.* From this derivation it is clear that the microscopic formulation must be contained in the macroscopic one. Furthermore, Landsberg's definition, though it might be more general than the microscopic one, is asymptotically regained from ours, since in the limits of infinite or zero masses, 6_N or 6_E are particular solutions of 14. To what extent our macroscopic definition is more general than the microscopic one, is a question that remains to be examined.

II. GENERAL SOLUTION OF DIFFERENTIAL EQUATION

Our problem now is to find the general solution of equation 14 for the two functions $p(\alpha, \beta)$ and $\varepsilon(\alpha, \beta)$ which are interconnected by the second equation of 13. Because of 15 we may switch to the one potential $\gamma(\alpha, \beta)$ arriving at the second order partial differential equation

$$[(\partial_\beta^2 + 3\beta^{-1}\, \partial_\beta - 3\beta^{-2}) + mc^2(2\,\partial_\beta + 3\beta^{-1})\,\partial_\alpha]\, \gamma = 0 \tag{16}$$

It is homogeneously linear and of hyperbolic type with non-constant coefficients.

Replacing α, β, γ by the dimensionless variables

$$x \equiv \alpha - mc^2\beta, \qquad y \equiv mc^2\beta, \qquad z \equiv B_0^{-1}\beta^3\gamma \tag{17}$$

we achieve a considerable simplification that transforms 16 into

$$(\partial_x^2 - \partial_y^2 - 3y^{-1}\partial_y)\, z = 0 \tag{18}$$

This version immediately exhibits the characteristic curves $x + y = \alpha$ and $x - y = \omega$, where each of the parametrizing constants α and ω may assume any real value.

The next step is to construct a convenient form of the general solution that yields the unknown $z(x, y)$ in terms of arbitrary initial distributions $z(x, \bar{y})$ and $\partial_y z(x, \bar{y})$ at any isotherm $y = \bar{y}$. Using the Green–Riemann method of integration of hyperbolic equations[4], this solution reads

$$\begin{aligned} z(x, y) = {} & \tfrac{1}{2}(y/\bar{y})^{\frac{3}{2}}\,[z(x - y + \bar{y}, \bar{y}) + z(x + y - \bar{y}, \bar{y})] \\ & + \tfrac{1}{2}\textstyle\int_{x-y+\bar{y}}^{x+y-\bar{y}} \mathrm{d}\bar{x}\,[\partial_y z(\bar{x}, \bar{y}) - z(\bar{x}, \bar{y})\,(3\bar{y}^{-1} + \partial_y)]\, R(x - \bar{x}, y, \bar{y}) \end{aligned} \tag{19}$$

where $R(x - \bar{x}, y, \bar{y})$ denotes Riemann's propagator. Its defining properties

$$(\partial_x^2 - \partial_y^2 + 3y^{-1}\partial_y)\, R(x - \bar{x}, y, \bar{y}) = 0$$

and

$$R(y - \bar{y}, y, \bar{y}) = (y/\bar{y})^{\frac{3}{2}} = R(\bar{y} - y, y, \bar{y})$$

are fulfilled by the function

$$R(x - \bar{x}, y, \bar{y}) \doteq (y/\bar{y})^{\frac{3}{2}} P_{\frac{3}{2}}(2q + 1), \quad q = \frac{(y - \bar{y})^2 - (x - \bar{x})^2}{4\bar{y}y} \tag{20}$$

as is shown in the Appendix. Legendre's function $P_{\frac{3}{2}}(1 + 2q) \equiv F(-\frac{3}{2}, \frac{5}{2}; 1; -q)$, primarily defined for $-1 < q \leqslant 1$ by the hypergeometric series, may be continued analytically to all positive values of q.

For infinite values of q one finds[5] the asymptotic expressions

$$P_{\frac{3}{2}}(2q + 1) \approx (8/\pi)\, q^{\frac{3}{2}}, \quad P'_{\frac{3}{2}}(2q + 1) \approx (12/\pi)\, q^{\frac{1}{2}} \quad \text{for } q \to \infty \tag{21}$$

Using this result, we deduce the initial-value representation

$$\begin{aligned} z(x, y) = {} & \tfrac{1}{2} \lim_{\bar{y} \to 0} (y/\bar{y})^{\frac{3}{2}}\,[z(x - y + \bar{y}, \bar{y}) + z(x + y - \bar{y}, \bar{y})] \\ & - \frac{4^{-\frac{1}{2}}}{\pi} \int_{x-y}^{x+y} \mathrm{d}\bar{x} \lim_{y \to 0} \bar{y}^{-3}\,[y^2 - (x - \bar{x})^2]^{\frac{1}{2}}\,\{6yz(\bar{x}, 0) \\ & - [y^2 - (x - \bar{x})^2]\, \partial_y z(\bar{x}, 0)\} \end{aligned} \tag{22}$$

the boundary strip of which has been shifted to the limit of infinite temperature.

III. PHYSICAL SOLUTIONS

This form of the general solution of equation 18 can be converted into the more convenient form

$$z(x, y) = \tfrac{1}{3}\{\textstyle\int_{-\infty}^{x-y} g_-(\bar{x}) + \int_{x+y}^{\infty} g_+(\bar{x})\}[(x - \bar{x})^2 - y^2]^{\frac{3}{2}}\, \mathrm{d}\bar{x} \tag{23}$$

provided the indefinite integrals converge at infinity. Apart from this *zeroth boundary condition*, which is rather weak, the border distributions $g_-(x)$ and $g_+(x)$ may be expressed by a linear function in terms of the initial distributions $z(x, 0)$ and $\partial_y z(x, 0)$.

We are now in a position to formulate our *first boundary condition*

$$g_-(x) = 0 \tag{24}$$

for which no simple physical interpretation is known.

On the other hand, our *second boundary condition* postulates simply that for extreme dilution a generalized perfect gas should behave as an ideal one. Because of 15, 12 and 11 the equation 1, now to read asymptotically only, can be reduced to $\gamma \approx v$ or

$$\partial_\alpha \gamma \approx -\gamma \quad \text{for} \quad \alpha \to \infty \quad \text{with } \beta \text{ fixed} \tag{25}$$

By 17, this can be transformed to $\partial_x z \approx -z$ for $x \to \infty$ with y fixed. Because of the identity $\partial_x^4 z(x, 0) = 2g_+(x)$ we arrive at

$$g_+(x) \approx B\mathrm{e}^{-x} \tag{26}$$

with an integrption constant $B > 0$, the value of which remains undetermined in this context.

We see now that the zeroth boundary condition is already contained in the first and second ones. Inserting 24 and 26 into 23 yields with 17 Jüttner's classical result[6]

$$\gamma(\alpha, \beta) \approx \tfrac{1}{3}B\beta\textstyle\int_0^\infty \mathrm{d}w\, \mathrm{e}^{-\alpha w}\sqrt{[w(w + 2mc^2)]^3} \quad \text{for} \quad \alpha \to \infty \tag{27}$$

is retained. This amounts to the ideal asymptote

$$\gamma(\alpha, \beta) = B\,\mathrm{e}^{-\alpha}\beta^{-1}\,\mathrm{e}^{\beta}K_2(\beta) \quad \text{for } C \tag{28}$$

where K_2 denotes the modified Bessel function. The undetermined integration constant B may be absorbed into the fugacity $\mathrm{e}^{-\alpha}$. Simple consequences of 28 are the 'ideal' gas law 1 and the monotonic property $c_v'(T) > 0$ with the limits 5.

Thus our macroscopic definition of a generalized perfect gas has been fully developed. It is determined by the differential equation of state 18 together with the two implicitly formulated boundary conditions 24 and 26. From 18, 23 and 24 it follows that one function of one variable is arbitrary up to the asymptotic behaviour 26. This situation is similar to that of an ideal gas where one function, $c_v(T)$, remains undertermined.

The microscopic concept of a perfect gas proves to be the choice

$$g_+(x) = B(\mathrm{e}^x - \iota)^{-1} \tag{29}$$

with the value of B given by 10.

ACKNOWLEDGEMENT

We are indebted to Dipl.-Phys. H. Söhn for a valuable hint concerning the solution of our differential equation and Prof. Henneberger for discussions and revising the English text.

APPENDIX

The interpretation of $(\partial_\beta^2 - \partial_y^2 + 2\lambda y^{-1}\partial_y)\, z = 0$ is sketched.

Riemann's initial-value representation of the general solution reads

$$z(x, y) = \tfrac{1}{2}(y/\bar{y})^\lambda \left[z(x - y + \bar{y}, \bar{y}) + z(x + y - \bar{y}, \bar{y})\right] + \tfrac{1}{2}\int_{x-y+\bar{y}}^{x+y-\bar{y}} \mathrm{d}\bar{x} \left[\partial_y z(\bar{x}, \bar{y}) - z(\bar{x}, \bar{y})\,(2\lambda\bar{y}^{-1} + \partial_{\bar{y}})\right] R_\lambda(x - \bar{x}, y, \bar{y})$$

if

$$(\partial_x^2 - \partial_y^2 + 2\lambda y^{-1}\partial_y)\, R(x - \bar{x}, y, \bar{y}) = 0$$

and

$$R_\lambda(y - \bar{y}, y, \bar{y}) = (y/\bar{y})^\lambda = R_\lambda(\bar{y} - y, y, \bar{y})$$

By these conditions the propagator $R_\lambda(x - \bar{x}, y, \bar{y})$ is uniquely determined. It may be calculated from the double power series

$$R_\lambda(x - \bar{x}, y, \bar{y}) = \sum_{\mu=0}^{\infty} \sum_{\nu=0}^{\infty} \frac{(\mu + \nu)!}{\mu!\nu!} \binom{\lambda}{\mu}\binom{\lambda}{\nu} \frac{(x - \bar{x} - y + \bar{y})^\mu (x - \bar{x} + y - \bar{y})^\nu}{(2\bar{y})^{\mu+\nu}}$$

if $(x - \bar{x})^2 + (y - \bar{y})^2 < \bar{y}^2$ or even $|(x - \bar{x}) - (y - \bar{y})| < 2\bar{y}$ and $|(x - \bar{x}) + (y - \bar{y})| < 2\bar{y}$. As is easily verified, $R_\lambda(x - \bar{x}, y, \bar{y}) = (y/\bar{y})^\lambda G_\lambda(q)$ with $q \equiv \tfrac{1}{4}(y\bar{y})^{-1} [(y - \bar{y})^2 - (x - \bar{x})^2]$, if

$$[q(1 + q)\,\partial_q^2 + (1 + 2q)\,\partial_q - \lambda(\lambda + 1)]\, G_\lambda(q) = 0,$$

This is Legendre's differential equation, if $2q + 1 = \tfrac{1}{2}(y\bar{y})^{-1} [y^2 + \bar{y}^2 - (x - \bar{x})^2]$ is taken as argument instead of q. The solution $G_\lambda(q) = P_\lambda(2q + 1)$ is uniquely determined by its value 1 at $2q + 1 = 1$, this being a singular point of the ordinary differential equation.

REFERENCES

1. E. Hilf, *Dissertation*, Frankfurt am Main (1967); *Z. Naturforsch.* (1970) in press.
2. H. Einbinder, *Phys. Rev.* **74**, 803–805, 805–808 (1948); **76** 410 (1949).
3. P. T. Landsberg, *Thermodynamics*, §27. Interscience: New York (1961).
4. R. Courant and D. Hilbert, *Methoden der Mathematischen Physik II*, V, §4. Springer: Berlin (1937).
   E. Kamke, *Differentialgleichungen, II. Partielle Differentialgleichungen*, §9, 43. Akademische Verlagsgesellschaft Geest und Portig K.-G.: Leipzig (1962).
5. W. Magnus, F. Oberhettinger und R. P. Soni, *Formulas and Theorems for the Special Functions of Mathematical Physics*, II and IV. Springer: Berlin (1966).
6. F. Jüttner, *Ann. d. Phys.* **34**, 856–882 (1911).

RESOLUTION OF THE HIERARCHY OF MANY-BODY DISTRIBUTION FUNCTIONS

S. FUJITA

Department of Physics and Astronomy, State University of New York at Buffalo, Buffalo, New York, U.S.A.

ABSTRACT

The rigorous evolution equation for the one-body distribution function f(t) describing an imperfect gas is not closed according to the investigations of Bogoliubov, Born, Green, Kirkwood, Yvon and others (B–B–G–K–Y hierarchy). The analysis of this equation is made in terms of connected diagrams. It is shown that in the bulk limit this equation can rigorously be transformed into an equation which is *closed* with respect to f(t) and which contains a given initial correlation. The latter equation forms a basis of discussing various problems related to irreversibility and transport phenomena.

1. INTRODUCTION

It has been known for many years that the *Boltzmann equation* provides a good description of transport phenomena for a dilute gas of particles interacting with short-range forces. While this equation is a *closed* equation with respect to the one-body distribution function f(**rp**, t), the rigorous evolution equation for f is *not* closed according to the investigations by Bogoliubov, Born, Green, Kirkwood, Yvon, and others[1,2]. Recently much effort has been made to derive and generalize the former equation from the latter by introducing approximations. Unfortunately, most of the approximations previously proposed by various authors seem to be motivated by mathematical tractability rather than physical reasoning. To this category of approximations belongs Bogoliubov's f-functional dependence assumption of many-body distribution functions[1], truncation of the hierarchy through random-phase arguments[3], the factorizability of the initial many-body densities into one-body densities[1,4], and others. Although these assumptions were employed to obtain highly useful results in certain instances, since they were introduced at the beginning of the theories, the validities of the theories and their results are often not clear.

In 1955, Van Hove developed a different approach to the problem[5]. By introducing an infinite-order time-dependent perturbation theory, he attempted to determine the structure of a *collision operator* which describes a general interaction process. Since then, this search for a collision operator has been pursued by many, including Kohn and Luttinger[6], Prigogine, and others[7–9]. In particular, Balescu[8] was successful in determining the collision term for a plasma, which takes account of the dynamic Coulomb screening. This collision term is now known as Balescu–Lenard's term[8,10]. A notable advantage of this approach is that if successful it could clarify the validities of the above-mentioned various assumptions in addition to its own merit of providing a computational method for a transport coefficient.

In 1957 Kubo published a classic work in the theory of transport phenomena[11]. By solving the Liouville equation to first order in the external electric field, he formulated a rigorous closed expression for the electric conductivity in microscopic terms without any guesswork in regard to the collision contribution, which was the crux of the derivation of the Boltzmann equation. This formula is, today, known as the *current correlation function formula*[12]. Concurrently, much effort has been made to formulate other transport coefficients, such as the viscosity coefficient, which are related to thermodynamic perturbations rather than electromagnetic forces[13]. At the present time, it is generally believed that these coefficients, too, can be expressed in the form of time-integrals of current correlation functions. The evaluation of a current correlation function formula is by no means straightforward in spite of its compact expression. Various methods of computation including those already mentioned[1, 2, 4–10] have been proposed[14], each being different from others in spirit and degree of sophistication[15].

In this paper it will be demonstrated that in the *bulk limit* where N (particle number) $\rightarrow \infty$, Ω (volume) $\rightarrow \infty$ while N/Ω remains finite, there exists a rigorous closed equation for the one-body distribution function f. It is done in the following steps. The hierarchy equation of lowest order 2.10 contains an integral of the product of the pair potential v and the two-body distribution function f_2. This integral is analysed in terms of connected diagrams and is shown to be expressible in terms of the one-body distribution function f and initial correlation functions χ. Since the latter, χ, are to be given as an initial condition, the equation obtained is closed with respect to f. It is, however, highly non-linear and non-Markoffian. In obtaining this closed equation, no approximations other than those which can be justified in the bulk limit, are introduced.

From the closed equation one can derive a generalized Boltzmann equation, which is closed in f, which can rigorously describe linear and non-linear transport coefficients, and which no longer depends on the initial condition. This means in particular that Bogoliubov's conjecture of the closure in f is in fact correct although his two explicit assumptions mentioned earlier for achieving this closure are not. The analysis in terms of connected diagrams is also useful in the practical calculation of transport coefficients. In fact it has been shown earlier[16] that it allows one to develop the formal density expansion of a transport coefficient in an unambiguous manner. Unfortunately this density expansion is in general divergent, see below. A serious restrictive feature of the connected diagram analysis is that it applies only to a system obeying classical statistics. Although a gas of monatomic molecules in a certain temperature range should fall in this category, systems of great interest such as an electron gas at high density which obeys Fermi–Dirac statistics, cannot be treated by the present method. This shortcoming can be overcome by working with double-time Green's functions $(g^{>}, g^{<})$ in place of the single-time one-body distribution function f. The existence of a closed set of equations for $(g^{>}, g^{<})$ which rigorously describe transport coefficients can be established[17], but will not be discussed in the present paper. The divergence difficulty of the density expansion of a transport coefficient mentioned earlier can be also overcome by working with Green's functions, which will be discussed elsewhere.

Obviously, the essential point in the present theory is the demonstration of the existence of a closed evolution equation. Previously this was done for a system of interacting quantum molecules obeying classical statistics[18]. The same technique can be extended to a classical imperfect gas. Since the technique is rather involved and since the classical system can be handled with less conceptual complexities, we shall review the essential steps of the demonstration for a classical gas.

2. B–B–G–K–Y HIERARCHY

Let us consider a system of particles interacting with pair forces, characterized by the Hamiltonian

$$\begin{aligned} H &\equiv \sum_j \frac{1}{2m} p_j^2 + \lambda \sum_{j>k} v(\mathbf{r}_j - \mathbf{r}_k) \\ &\equiv \sum_j h_0^{(j)} + \lambda \sum_{j>k} v^{(jk)} \\ &\equiv H_0 + \lambda V \end{aligned} \tag{2.1}$$

where h_0 and v are respectively the kinetic energy and pair potential energy. The one-body and two-body distribution functions will be defined by:

$$\begin{aligned} \mathrm{f}(\mathbf{rp}, t) &\equiv (N!)^{-1} \iint (\Pi\, \mathrm{d}^3 r_j)(\Pi\, \mathrm{d}^3 p_j)\, \hat{\mathrm{f}}(\mathbf{rp})\, \rho(\mathbf{r}_1\mathbf{p}_1, \mathbf{r}_2\mathbf{p}_2, \ldots, t) \\ &\quad \mathrm{Tr}\,\{\hat{\mathrm{f}}(\mathbf{rp})\,\rho\} \\ \mathrm{f}_2(\mathbf{rp}, \mathbf{r}'\mathbf{p}', t) &\equiv \mathrm{Tr}\{\hat{\mathrm{f}}_2(\mathbf{rp}, \mathbf{r}'\mathbf{p}')\}\, \rho(t) \end{aligned} \tag{2.2}$$

$$\begin{aligned} \hat{\mathrm{f}}(\mathbf{rp}) &\equiv \frac{1}{\Omega} \sum_j \delta^{(3)}(\mathbf{r} - \mathbf{r}_j)\, \delta^{(3)}(\mathbf{p} - \mathbf{p}_j) \equiv \sum_j \hat{\mathrm{f}}^{(j)}(\mathbf{rp}) \\ \hat{\mathrm{f}}_2(\mathbf{rp}, \mathbf{r}'\mathbf{p}') &\equiv \frac{1}{\Omega^2} \sum_{j \neq k} \delta^{(3)}(\mathbf{r} - \mathbf{r}_j)\, \delta^{(3)}(\mathbf{p} - \mathbf{p}_j)\, \delta^{(3)}(\mathbf{r}' - \mathbf{r}_k)\, \delta^{(3)}(\mathbf{p}' - \mathbf{p}_k) \end{aligned} \tag{2.3}$$

$\rho(t)$ is the A-body distribution function which obeys the Liouville equation

$$\begin{aligned} i\frac{\partial \rho(t)}{\partial t} &= i \sum_k \left\{ \frac{\partial H}{\partial \mathbf{r}_k} \cdot \frac{\partial}{\partial \mathbf{p}_k} - \frac{\partial H}{\partial \mathbf{p}_k} \cdot \frac{\partial}{\partial \mathbf{r}_k} \right\} \rho(t) \\ &\equiv \mathscr{H} \rho(t) \end{aligned} \tag{2.4}$$

Here the operator denoted by $\mathscr{H}$ will be called a *Liouville operator*; it is a differential operator which is generated from a given Hamiltonian H and which is convenient in the description of the time development of the system.

Differentiating $f(\mathbf{rp}, t)$ with respect to t and using (2.4), one obtains

$$\begin{aligned} \frac{\partial f(\mathbf{rp}, t)}{\partial t} &= -i Tr\{\hat{\mathrm{f}}(\mathbf{rp})\, \mathscr{H}\, \rho(t)\} \\ &= -i \sum_j \mathrm{Tr}\{\hat{\mathrm{f}}^{(j)}(\mathbf{rp}) h_0^{(j)} \rho(t)\} - i\lambda \sum_{j \neq k} \mathrm{Tr}\{\hat{\mathrm{f}}^{(j)}(\mathbf{rp}) v^{(jk)} \rho(t)\} \end{aligned} \tag{2.5}$$

where the script $(\mathscr{h}_0, \mathscr{v})$ are the Liouville operators corresponding to (h_0, v), e.g.

$$h_0^{(j)} \equiv i\sum_k \left\{\frac{\partial h_0^{(j)}}{\partial \mathbf{r}_k}\cdot\frac{\partial}{\partial \mathbf{p}_k} - \frac{\partial h_0^{(j)}}{\partial \mathbf{p}_k}\cdot\frac{\partial}{\partial \mathbf{r}_k}\right\}$$

$$\equiv -i\frac{\mathbf{p}_j}{M}\cdot\frac{\partial}{\partial \mathbf{r}_j}. \tag{2.6}$$

The first term in the third member of 2.5 can be written as

$$-i\sum_j \mathrm{Tr}\{\hat{\mathrm{f}}^{(j)}(\mathbf{rp})\,h_0^{(j)}\rho(t)\} = -\frac{\mathbf{p}}{M}\cdot\frac{\partial}{\partial \mathbf{r}}\mathrm{f}(\mathbf{rp}, t)$$

$$\equiv -ih_0(\mathbf{rp})\,\mathrm{f}(\mathbf{rp}, t) \tag{2.7}$$

In a similar manner the second term can be written as

$$-i\lambda\sum_{j\neq k}\mathrm{Tr}\,\{\hat{\mathrm{f}}^{(j)}(\mathbf{rp})\ \mathscr{v}^{(jk)}\rho(t)\}$$

$$= -\lambda\int\int \mathrm{d}^3r_2\,\mathrm{d}^3p_2\frac{\partial v(\mathbf{r}-\mathbf{r}_2)}{\partial \mathbf{r}}\cdot\frac{\partial}{\partial \mathbf{p}}\mathrm{f}_2(\mathbf{rp}, \mathbf{r}_2\mathbf{p}_2, t)$$

$$\equiv -i\lambda\mathrm{tr}^{(2)}\{\mathscr{v}^{(12)}\mathrm{f}_2^{(12)}(t)\} \tag{2.8}$$

where the symbol $\mathrm{tr}^{(2)}$ means the integration with respect to the phase-space variables of the second particle. Therefore one can rewrite 2.5 as

$$\left[\frac{\partial}{\partial t} + ih_0(\mathbf{rp})\right]\mathrm{f}(\mathbf{rp}, t) = -i\lambda\,\mathrm{tr}^{(2)}\{\mathscr{v}^{(12)}\,\mathrm{f}_2^{(12)}\} \tag{2.9}$$

This equation can be written in a more familiar fashion

$$\left[\frac{\partial}{\partial t} + \frac{\mathbf{p}}{M}\cdot\frac{\partial}{\partial \mathbf{r}}\right]\mathrm{f}(\mathbf{rp}, t) = \int\int \mathrm{d}^3r'\mathrm{d}^3p'\frac{\partial v(\mathbf{r}-\mathbf{r}')}{\partial \mathbf{r}}\cdot\frac{\partial \mathrm{f}_2(\mathbf{rp}, \mathbf{r}'\mathbf{p}', t)}{\partial \mathbf{p}}. \tag{2.10}$$

This is the lowest-order equation of the B–B–G–K–Y hierarchy. The notations, more abstract and more concise than usual, which were used in the derivation of 2.10 will be found convenient in the following development of the theory.

3. CLOSED EVOLUTION EQUATION FOR DISTRIBUTION FUNCTION f

The Hamiltonian H and therefore the corresponding Liouville operator $\mathscr{H}$ are independent of time. The formal solution to (2.4) is

$$\rho(t) = \exp(-it\mathscr{H})\,\rho(0) \equiv \exp(-it\mathscr{H})\,\rho \tag{3.1}$$

In general a function of an operator is defined as a polynomial or a power series, e.g.

$$\exp(-it\mathscr{H}) \equiv 1 - it\mathscr{H} + \tfrac{1}{2}(-it)^2\mathscr{H}^2 - \ldots \tag{3.2}$$

This operator $\exp(-it\mathscr{H})$ is a function of the coupling parameter λ, and can be expanded in a power series of λ (perturbation series):

$$\exp(-it\mathscr{H}) = \exp(-it\mathscr{H}_0)\{1 + \sum_1^\infty (-i\lambda)^k \int_0^t d\tau_1 \int_0^{\tau_1} d\tau_2 \ldots \int_0^{\tau_{k-1}} d\tau_k$$

$$\times \mathscr{V}(\tau_1)\mathscr{V}(\tau_2)\ldots\mathscr{V}(\tau_k)\} \tag{3.3}$$

$$\mathscr{V}(t) \equiv \exp(it\mathscr{H}_0)\mathscr{V}\exp(-it\mathscr{H}_0) \tag{3.4}$$

A many-body distribution function ρ to be specified at the initial time $t = 0$ contains all the information about the inhomogeneity and particle correlation of the system at $t = 0$. The latter are, however, more conveniently described by the reduced distribution functions f, $f_2, \ldots$, which may be defined through 2.2 with the use of time-independent ρ. The particle correlation can be more appropriately described by the *correlation functions* defined by:

$$\begin{aligned} f^{(1)} &\equiv f(\mathbf{r}_1\mathbf{p}_1) \equiv \chi(\mathbf{r}_1\mathbf{p}_1) \equiv \chi^{(1)} \\ f_2^{(1,2)} &\equiv f_2(\mathbf{r}_1\mathbf{p}_1, \mathbf{r}_2\mathbf{p}_2) \equiv \chi^{(1)}\chi^{(2)} + \chi_2(\mathbf{r}_1\mathbf{p}_1, \mathbf{r}_2\mathbf{p}_2) \\ &\equiv \chi^{(1)}\chi^{(2)} + \chi_2^{(12)} \\ f_3^{(123)} &\equiv \chi^{(1)}\chi^{(2)}\chi^{(3)} + \chi^{(1)}\chi_2^{(23)} + \chi^{(2)}\chi_2^{(31)} + \chi^{(3)}\chi_2^{(12)} + \chi_3^{(123)} \end{aligned} \tag{3.5}$$

It is clear that one can specify the initial condition by giving $\chi \equiv f, \chi_2, \ldots,$ $\ldots$ rather than ρ. In fact this specification is obviously more realistic.

Let us now consider

$$\begin{aligned} &-i\lambda \sum_{j \neq k} \mathrm{Tr}\,\{\hat{f}^{(j)}(\mathbf{rp})\ \ v^{(jk)}\rho(t)\} \\ &= -i\lambda \sum_{j \neq k} \mathrm{Tr}\,\{\hat{f}^{(j)} v^{(jk)} \exp(-it\mathscr{H})\rho\} \end{aligned} \tag{3.6}$$

which appeared in 2.5 and which was transformed into an integral involving f_2 in 2.10. We expand $\exp(-it\mathscr{H})$ in a perturbation series by means of 3.3, and regard ρ as the N-body reduced distribution function f_N, and expand the latter as

$$f_N^{(1,\ldots,N)} = \prod_1^N \chi^{(j)} + [\chi_2^{(12)} \prod_3^N \chi^{(j)} + (\text{similar terms obtained by permutations})]$$

$$+ \chi_3^{(123)} \prod_4^N \chi^{(j)} + \ldots \tag{3.7}$$

the Nth equation of 3.5. We represent terms in the expansion of 3.6 by diagrams as follows.

We draw N horizontal solid lines for the N particles. The operator $\hat{f}^{(j)}$ is represented by the open circle at the left end of the particle line j, and $v^{(jk)}$ by a vertical dotted line, called a *potential bond*, connecting the pair of particle lines (j, k) at their left ends. Corresponding to

$$\begin{aligned} v^{(jk)}(\tau) &\equiv \exp(i\mathscr{H}_0\tau)\, v^{(jk)} \exp(-i\mathscr{H}_0\tau) \\ &= \exp[i\tau(h_0^{(j)} + h_0^{(k)})]\, v^{(jk)} \exp[-i\tau(h_0^{(j)} + h_0^{(k)})] \end{aligned} \tag{3.8}$$

we draw a potential bond (j, k) at t (time) $= \tau$, where the time is measured from the right to the left. The χ_l, $l \geqslant 2$, are indicated by broken lines, called *correlation bonds*, connecting the right ends of the particle lines. In this way we can represent all the expansion terms in one-to-one correspondence. Some typical diagrams are drawn in *Figure 1*.

A diagram is said to be *connected* if the set of particles describing the

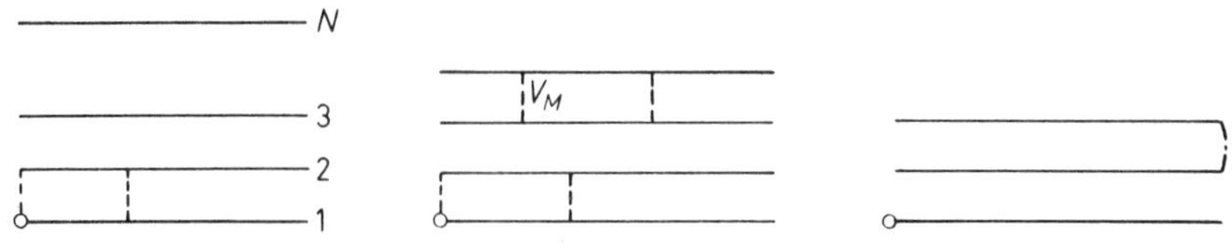

Figure 1. Diagrams representing components of statement 3.6. Diagram a is connected and b and c are disconnected.

potential and/or correlation bonds cannot be separated out into two or more subsets. Otherwise the diagram will be called a *disconnected* one. For example, the diagram b is disconnected.

None of the disconnected diagrams contribute to 3.6. This can be proved with the aid of the following two theorems:

Theorem U Any diagram containing an M-type potential bond contributes nothing.

A *potential bond of M-type* is any bond, like the V_M in diagram b, which sees nothing but the two free particle lines on its left. A particle line segment is said to be *free* if the diagram is broken into two by cutting it. This theorem is proved as follows. Let us suppose that an M-type potential bond connects the pair of lines (j, k) at $t = \tau$. The contribution of the diagram will then contain a factor (see 3.8)

$$\mathrm{tr}^{(j)}\mathrm{tr}^{(k)}\{\exp\,[i\tau h_0^{(j)} + i\tau h_0^{(k)}]\; v^{(jk)}\, \mathrm{g}^{(jk\ldots)}\} \tag{3.9}$$

where g is a certain function of the variables corresponding to the particles (j, k) and possibly others. This quantity 3.9 can be decomposed into vanishing integrals of the following three types:

$$\begin{aligned} &\mathrm{tr}^{(j)}\{h_0^{(j)}\mathrm{g}_1^{(jk\ldots)}\} \equiv \iint \mathrm{d}^3 r_j \mathrm{d}^3 p_j h_0^{(j)} \mathrm{g}_1(jk\ldots) = 0 \\ &\mathrm{tr}^{(k)}\{h_0^{(k)}\mathrm{g}_2^{(jk\ldots)}\} = 0 \\ &\mathrm{tr}^{(j)}\mathrm{tr}^{(k)}\{v^{(jk)}\mathrm{g}_3{}^{(jk\ldots)}\} = 0 \end{aligned} \tag{3.10}$$

which can be simply shown by integration by parts.

Theorem II Any diagram containing a correlation bond with one or more free lines on its left contributes nothing.

Diagram c contains a correlation bond with a free line and yields a vanishing contribution. This is because such a correlation bond contributes a vanishing factor of the form

$$\mathrm{tr}^{(1)}\{\chi_l^{(1,\ldots l)}\} = 0, \qquad l \geqslant 2 \tag{3.11}$$

which can be in turn proved from the definition 3.5. In fact, from the second equation 3.5, $\mathrm{tr}^{(1)}\{\chi_2^{(12)}\} = \mathrm{tr}^{(1)}\{\mathrm{f}_2^{(12)}\} - \mathrm{tr}^{(1)}\{\chi^{(1)}\}\,\chi^{(2)} = n\chi^{(1)} - n\chi^{(1)} = 0$; such proof can be extended to the case of higher $l > 2$. This theorem is valid rigorously in the bulk limit.

A disconnected diagram has, by construction, a potential bond and/or a correlation bond, of the types referred to in Theorems I and II, and yields a vanishing contribution. Disregarding all disconnected diagrams, we have only to deal with connected diagrams containing the open circle at $t = t$. We may simplify the drawings by omitting particle lines without potential bonds. For example, we may represent diagram a in *Figure 1* as one component of diagram a in *Figure 2* where in addition we leave out indices for particle lines. By such an *unindexed diagram* we shall imply a collection of particle-indexed diagrams of the same structure.

The power of these two theorems is not limited to the elimination of the disconnected diagrams. In fact it allows elimination of a large number of

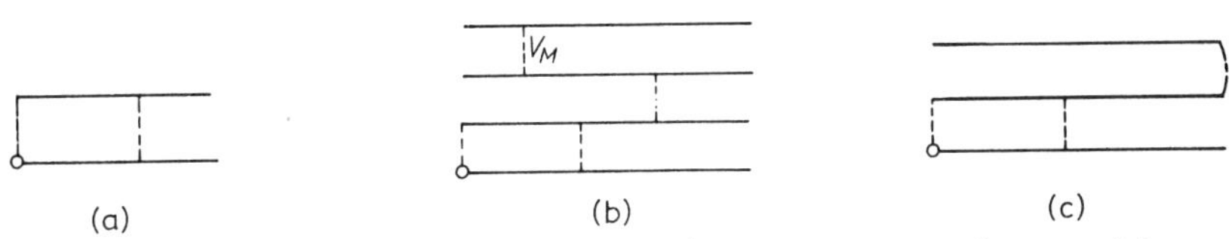

Figure 2. The unindexed diagram a represents the collection of indexed diagrams of the same structure as that of diagram a in *Figure 1*. Connected diagrams b and c do not contribute because they contain an M-type potential and a correlation bond with a free line, respectively.

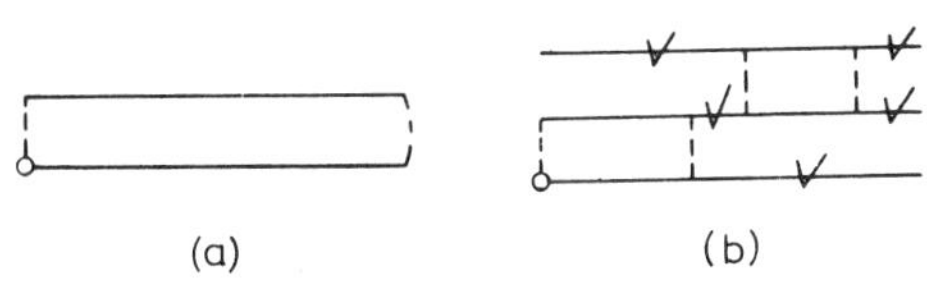

Figure 3. Diagram a contains a d-part and b, two g-parts.

connected diagrams, too. For example, diagrams b and c in *Figure 2* are connected diagrams but they contain respectively an M-type potential bond and a correlation bond with a free line, and thus contribute nothing.

A connected diagram will in general contain several free line segments. Some free segments are indicated by check marks $\surd$ in *Figure 3*. A diagram will contain a certain number of those parts which consist of non-free line segments, potential bonds and correlation bonds, and which are connected by free segments. Such a part will be called a *d-part* or *g-part* according to whether or not it contains a correlation bond. Diagram a in *Figure 3* has a d-part and diagram b two g-parts.

If a diagram should contain a g-part suspended by two free segments corresponding to the same particle or a d-part standing to the right of a free line segment, it could be *reduced* by suppressing the g- or d- part. Otherwise the diagram is called *irreducible*. In the process of reduction, only the particle line which is marked by the open circle should not be suppressed. With this rule, the reduction becomes unique. Conversely, reducible diagrams can be obtained from an irreducible diagram by *dressing* its free particle lines with g- and/or d- parts.

We have so far considered those diagrams representing 3.6. We may represent by similar diagrams the expansion of the one-body distribution function

$$\mathrm{f}(\mathbf{rp}, t \equiv \sum_j \mathrm{Tr}\{\hat{\mathrm{f}}^{(j)} \exp(-it\mathscr{H})\,\rho\}. \tag{3.12}$$

The only difference will be that we should omit the potential bond at $t = t$ which appeared in the representation of 3.6. Analysing in a similar manner we can easily see that any diagram giving a non-trivial contribution is a connected one containing a number of g- and/or d- parts, one of which contains the open circle representing the $\hat{\mathrm{f}}^{(j)}$. Typical diagrams are drawn in *Figure 4.* It is immediately seen that all such diagrams (except one) are reducible to the unique diagram b in *Figure 4*, which is free from any potential or correlation bond.

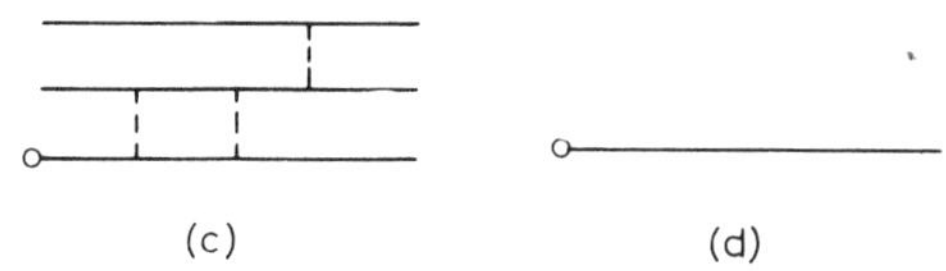

Figure 4. Diagram a, representing a component of $\mathrm{f}(\mathbf{rp}, t)$, is uniquely reducible to diagram b. Conversely, diagram a may be obtained from b by dressing the free line with g-parts.

The diagrams drawn here appear to represent the past history of those particles which contribute to the change in $\mathrm{f}(t)$ at $t = t$, and the g- and d- parts describe the effects of interaction processes.

Let us consider an irreducible diagram containing a g-part. This contribution can always be expressed in the form of a certain operator g acting on the product of the one-body distribution function corresponding to the system without the interparticle potential ($\lambda = 0$).

$$\mathrm{f}_0(t) \equiv \mathrm{Tr}\{\hat{\mathrm{f}} \exp(-it\mathscr{H}_0)\,\rho\}$$

$$= \sum_j \mathrm{tr}^{(j)}\{\hat{\mathrm{f}}^{(j)} \exp(-ith_0^{(j)})\,\chi^{(j)}\} \tag{3.13}$$

$$= \sum_\gamma \Big(\prod_2^{m(\gamma)} \mathrm{tr}^{(j)}\Big)\, \mathrm{g}(\gamma) \sum_1^{m(\gamma)} \mathrm{f}_0^{(k)} \equiv \Pi \mathrm{g}\mathrm{f}_0 \tag{3.14}$$

where $m(\gamma)$ is the number of free lines at the right of a chosen irreducible structure γ.

For example, the contribution of the irreducible diagram a in *Figure 2* can be written down as

$$(-i\lambda)^2 \sum_{jk} \mathrm{tr}^{(k)}\{v^{(jk)} \exp[-it(h_0^{(j)} + h_0^{(k)})] \int_0^t \mathrm{d}\tau\, v^{(jk)}(\tau_1) \exp(i\tau_1 h_0^{(j)})$$

$$\times \exp(i\tau_1 h_0^{(k)})\, \mathrm{f}_0^{(j)}(\tau_1)\, \mathrm{f}_0^{(k)}(\tau_1)\} \tag{3.15}$$

Consider now a reducible diagram which upon reduction gives rise to an irreducible diagram. The former can be constructed from the latter by dressing the free lines on the RHS.

By construction the two sets of particles involved in the evolution of any two of the originally free lines are separated from each other. Furthermore, the structures of all those subdiagrams which upon reduction give rise to a free line can be seen to be identical with the structures of all the connected diagrams for f(t) in the bulk limit. In this identification it is important to notice that the dressing of the free-particle line should be made always to the right, i.e. in the direction of decreasing time since dressing made otherwise would necessarily introduce an M-type potential as seen in the diagram b in *Figure 2*, and therefore would give no contribution. These analyses lead us to write for the contribution of all the irreducible diagrams containing g-parts and the reducible diagrams generated from them

$$\sum_{\gamma} \Big(\prod_{2}^{m(\gamma)} \mathrm{tr}^{(j)} \Big) \mathrm{g}(\gamma) \prod_{1}^{m(\gamma)} \mathrm{f}^{(k)} \equiv \sum \mathrm{g}\Pi\mathrm{f} \tag{3.16}$$

which is obtained by simply replacing every f_0 in 3.14 with f and by summing over all irreducible diagrams.

The irreducible diagrams containing d- parts and the reducible diagrams generated from them can be analysed in a similar manner. Their contribution may be symbolically written as

$$\sum \mathrm{d}\, \Pi\, \mathrm{f}\, \Pi\, \chi \tag{3.17}$$

which may or may not contain the factors in f but must include one or more of the initial correlation functions χ_l, $l \geqslant 2$.

Thus, we obtain for 3.6

$$-i\lambda \sum_{j \neq k} \mathrm{tr}^{(j)} \mathrm{tr}^{(k)} \{\hat{\mathrm{f}}^{(j)} v^{(jk)} \rho(t)\} = \sum \mathrm{g}\Pi\mathrm{f} + \sum \mathrm{d}\Pi\mathrm{f}\Pi\chi \tag{3.18}$$

Using this and (2.7), we can rewrite (2.5) as

$$\left(\frac{\partial}{\partial t} + ih_0\right) \mathrm{f}(t) = \sum \mathrm{g}\Pi\mathrm{f} + \sum \mathrm{d}\Pi\mathrm{f}\Pi\chi \tag{3.19}$$

This is an evolution equation which holds rigorously in the bulk limit. Since the correlation functions χ are to be given as the initial condition, this equation is a closed equation with respect to the distribution function f in contrast to the hierarchy equation 2.5 from which it is derived. The equation has, however, infinitely many terms, most conveniently defined in connected diagrams; and it is non-linear and non-Markoffian.

ACKNOWLEDGEMENTS

The author wishes to thank Dr A. Lodder and Mr C. L. Ko for reading the manuscript and for their constructive comments. The author gratefully acknowledges support in part by the National Science Foundation, NSF GP-9040.

REFERENCES

[1] N. Bogoliubov, *Problemi Dynamitchesky Theorie Statisticheskey Physike.* OGIS: Moscow (1946); English translation by E. K. Gora, *Studies in Statistical Mechanics*, Vol. I, p. 1. Edited by J. H. de Boer and G. E. Uhlenbeck. North Holland Publ. Co.: Amsterdam (1962).

[2] J. Yvon, *La Théorie Statistique des Fluides et l'Equation d'Etat.* Hermann: Paris (1935).
J. G. Kirkwood, *J. Chem. Phys.* **14**, 180 (1946) and **15**, 72 (1947).
M. Born and H. S. Green, *General Kinetic Theory of Liquids.* Cambridge University Press: London (1949).
[3] D. Bohm and D. Pines, *Phys. Rev.* **92**, 609 (1953).
[4] R. Brout, *Physica*, **22**, 509 (1956).
R. Brout and I. Prigogine, *Physica*, **22**, 621 (1956).
[5] L. Van Hove, *Physica*, **21**, 517 (1955) and **23**, 441 (1957).
[6] W. Kohn and J. M. Luttinger, *Phys. Rev.* **108**, 590 (1957) and **109**, 1892 (1958).
[7] I. Prigogine, *Non-Equilibrium Statistical Mechanics.* Interscience: New York (1962).
[8] R. Balescu, *Statistical Mechanics of Charged Particles.* Interscience: New York (1963).
[9] S. Fujita, *Introduction to Non-Equilibrium Quantum Statistical Mechanics.* W. Saunders: Philadelphia (1966).
[10] A. Lenard, *Ann. Phys. (N.Y.)*, **10**, 390 (1960).
[11] R. Kubo, *J. Phys. Soc. Japan*, **12**, 570 (1957).
[12] See also H. Nakano, *Progr. Theor. Phys.* **15**, 77 (1956).
[13] M. S. Green, *J. Chem. Phys.* **22**, 898 (1954).
H. Mori, *J. Phys. Soc. Japan*, **11**, 1029 (1956).
R. Kubo, M. Yokota and S. Nakajima, *J. Phys. Soc. Japan*, **12**, 1203 (1957).
J. A. McLennan Jr, *Phys. Rev.* **115**, 1405 (1959).
J. A. McLennan, *Physics of Fluids* **3**, 493 (1966).
J. M. Luttinger, *Phys. Rev.* **135**, A1505 (1964).
N. Hashitsume and S. Fujita, *J. Math. Phys.* **5**, 1572 (1964).
[14] D. Montgomery and D. A. Tidman, *Plasma Kinetic Theory.* McGraw-Hill: New York (1964).
L. P. Kadanoff and G. Baym, *Quantum Statistical Mechanics.* W. Benjamin: New York (1962).
W. E. Brittin (Editor), *Lectures in Theoretical Physics*, 9C: *Kinetic Theory.* Gordon and Breach: New York (1967).
Yu. L. Klimontovich, *Statistical Theory of Non-Equilibrium Processes in a Plasma.* Massachusetts Institute of Technology: Cambridge, Mass. (1967).
[15] For a critical and comparative review, see T. Y. Wu; *Kinetic Equations of Gases and Plasmas.* Addison-Wesley: Reading, Mass. (1966).
[16] S. Fujita, *Proc. Nat. Acad. Sci. Wash.* **56**, 16 (1966).
[17] S. Fujita, *J. Phys. Soc. Japan*, **27**, 1096 (1969).
[18] S. Fujita, *J. Phys. Soc. Japan*, **26**, 505 (1969).

B. PHASES, SURFACES AND THERMODYNAMIC LIMIT

THE THERMODYNAMICS OF PHASE EQUILIBRIUM: FROM THE PHASE RULE TO THE SCALING LAWS

L. TISZA

Department of Physics, Massachusetts Institute of Technology

ABSTRACT

In recent years the author has advanced a conceptual structure based on the generalization of Gibbsian thermodynamics and statistical mechanics. The purpose of this paper is to bring this theory up-to-date by harmonizing it with the recent developments in the theory of critical phenomena.

---

1. INTRODUCTION

At the turn of the century thermodynamics and statistical mechanics had their well defined logical structure. Thermodynamics was supposed to be macroscopic and non-statistical; its microscopic foundations was expected to be given by means of 'reduction' to statistical mechanics.

The rich developments of this century wrought havoc with these neat categories, and present research pays, at best, lip-service to the traditional logical structure.

One would like to hope that science can create order in the chaos of experience, but unfortunately, successful scientific activity tends to create a chaos of its own. A logical structure, to be of real use must be flexible enough to cope with this complex situation. If such a structure can be developed at all, this is likely to happen in successive approximations, in terms of manageable steps.

It seems to me that a modernized form of Gibbsian thermodynamics can provide the point of departure for such a programme. A few years ago I published a volume entitled *Generalized Thermodynamics*[1] to report on progress achieved along these lines†.

I am grateful for the opportunity to summarize my thoughts at this meeting, particularly because significant developments occurred since the completion of my report which render its updating desirable.

I am alluding to the remarkable expansion in the exploration of the phenomena in the neighbourhood of the critical point[2–4]. The critical point was discovered a little over a hundred years ago by Andrews, and was presumably so named because of its critical role with respect to the possibility of condensing a gas which appears 'permanent' at higher temperatures. Such points are 'critical', however, for a deeper reason as well. In their

† I shall use the term GTD to designate generically any of the theories within this overall structure.

critical states substances reach the limits of their thermodynamic stability, and many of their measured properties exhibit a singular behaviour, the quantitative description of which poses extreme demands on experimentalist and theorist alike. The intensive activity of the last few years came about thanks to breakthroughs in experimental and theoretical techniques. Within theory, these involve the treatment of singular functions by means of the so-called critical point exponents[2, 3].

Since critical points play an important role in GTD, the emergence of these new techniques enables me to sharpen my argument and eliminate unsatisfactory approximations formerly used out of expediency.

I have to confine myself to a somewhat impressionistic sketch of a number of ideas without any pretence of proving specific statements. However, I made an effort to line up the ideas so as to bring out their interdependence, and would like to hope that the parsimony in detail will further this end.

One of the problems arising in this connection is taking a new look at the relation of statistical mechanics to thermodynamics. My thesis is that we are dealing with two complementary aspects of the structure of matter which have to be used jointly in a carefully dovetailing pattern. At the outset statistical mechanics starts with structureless permanent point particles and thermodynamics with cells localized in space time. Both pictures are capable of refinement and their ultimate relation should be inferred from the careful analysis of experience, rather than from some preconceived opinion about what is more 'fundamental'.

The structure of GTD has a hierarchic character built up in a step-by-step procedure. Section 2 is devoted to the outline of a theory denoted by MTE, an abbreviation suggested by 'macroscopic thermodynamics of equilibrium'. Actually, MTE denotes a precisely defined deductive system developed in ref. 1, whereas, following current practice, the term thermodynamics is used in a somewhat vague generic sense. The situation is similar for STE (statistical thermodynamics of equilibrium) discussed in Section 3, along with some quantum mechanical considerations. Up to this point the rigour of the developments compares favourably with that expected in classical thermodynamics. The pace is changed somewhat in Section 4 in which approximate methods are admitted in order to handle some detailed properties of critical points which are beyond the reach of the rigorous methods.

The discussion is limited to time independent phenomena. Although GTD can account for a variety to time dependent processes, the situation has not matured to the point of admitting a concise presentation.

In view of the fact that I am attempting to survey a vast range of subjects, I beg to be excused for not supplying a thorough bibliography. The material of sections 2 and 3 is developed in detail in ref. 1. My minor excursion into the recent theory of critical point exponents is adequately documented in refs 2–4.

2. MACROSCOPIC THERMODYNAMICS OF EQUILIBRIUM (MTE)

We start with the fundamental equation

$$U = U(X_1, X_2, \dots X_{r+1}) \tag{1}$$

expressing the energy in terms of the extensive variables which for the time being we choose to be the entropy S, the mole numbers of the c independent components and the volume V. To be more precise, we assume that we are dealing with what Gibbs called the primitive fundamental equation corresponding to a single homogeneous phase.

We see that

$$r = c + 1 \tag{2}$$

We single out the volume $V = X_{r+1}$ as a 'scale factor' and define the densities:

$$x_k = X_k/V, u = U/V \tag{3}$$

The intensities conjugate to the X_k are defined:

$$P_k = \partial u/\partial X_K \tag{4}$$

Although superficially it does not seem to offer anything new if we solve equation 1 for the entropy and write the fundamental equation in the entropy scheme

$$S = S(X_1, X_2, \ldots X_{r+1}) \tag{5}$$

with the appropriate definition of intensities:

$$\pi_k = \partial s/\partial X_K \tag{6}$$

the parallel use of the two schemes is important because they serve to describe reversible and irreversible processes respectively. While the energy scheme is more convenient in the theory of phase equilibrium, the entropy scheme is indispensable for a smooth transition to STE.

The choice of densities and intensities for the description of thermodynamic systems seems simple enough from the point of view of an elementary theory. It is a common experience that thermodynamic systems can be 'scaled'. Although this property is, strictly speaking, inconsistent with the discrete structure of matter it is nevertheless assumed in the standard thermodynamic formalism. I like to express this by saying that one assumes the validity of the *principle of scale invariance.* This principle along with the conservation laws and the extremal principles, form the backbone of the formalism of MTE. The traditional tests of thermodynamics pay little if any attention to this principle, maybe because of reluctance to consider principles of limited validity. Yet, in spite of its fundamental importance, scale invariance is limited for more than one reason. Such limitations arise because of surface effects, and as a consequence of long range forces; a serious breakdown sets in at atomic scales. Many interesting problems arise from the requirement of handling situations in which one or the other of these restrictions becomes effective. We shall consider some of these questions below.

Scale invariance, or the physical homogeneity of a phase, finds expression in the mathematical homogeneity of the fundamental equation. This leads to

$$U = \sum_{1}^{r+1} X_k P_k \tag{7}$$

and

$$P_{r+1} = u - \sum_1^r x_k P_k. \tag{8}$$

Assuming that equation 2 can be solved for x_k we obtain

$$P_{r+1} = \omega(P_1, P_2, \ldots P_r) \tag{9}$$

Note that

$$x_k = -\,\partial\omega/\partial P_k \tag{10}$$

We examine the response of the system around equilibrium in terms of the relations:

$$\delta P_i = \sum_1^r u_{ik}\delta x_k \tag{11}$$

where

$$u_{ik} = \partial^2 u/\partial x_i \partial x_k = \partial P_i/\partial x_k \tag{12}$$

and

$$\delta x_i = \sum_1^r \omega_{ik}\delta P_k \tag{13}$$

where

$$\omega_{ik} = \partial\omega/\partial P_i \partial P_k = \partial x_i/\partial P_k \tag{14}$$

The matrices $\|u_{ik}\|$ and the reciprocal $\|\omega_{ik}\|$ are called the stiffness and the compliance matrix respectively.

A thermodynamic system is in a state of normal stability if the stiffness matrix is positive definite. The matrix can be brought to diagonal form with the diagonal elements:

$$\lambda_k = \frac{D_k}{D_{k-1}} = \left(\frac{\partial P_k}{\partial x_k}\right)_{P_1, P_2, \ldots P_{k-1}} \qquad k = 1, 2, \ldots r \tag{15}$$

Here the D_ks are the principal minors of the stiffness matrix. Normal stability requires that all the λ_k be positive.

We note that the use of densities and intensities for the specification of the system are entirely equivalent, provided

$$D_r = \frac{\partial(P_1, P_2, \ldots P_r)}{\partial(x_1, x_2, \ldots x_r)} \neq \begin{cases} 0 \\ \infty \end{cases} \tag{16}$$

and equations 11 and 13 respectively are soluble. The case $D_r = 0$ is precluded in states of normal stability, but this does indeed occur at critical points, and $D_r = \infty$ is realized near absolute zero.

In order to explain the meaning of this 'breakdown' of the theory we resort to the artifice of associating intensities with infinite reservoirs, and describe systems in terms of their densities. Under these conditions the breakdown of conditions 16 makes excellent sense. The case $D_r = 0$ is understood by recognizing that at critical points the densities of the system are subject to abnormally large fluctuations.

The nature of the anomaly near absolute zero is also easily understood. At sufficiently low temperatures the entropy of a system becomes constant, and below such a characteristic temperature the extensive parameters are no longer uniquely associated with the temperature of the environment. This is the basis for the notorious difficulties of low temperature thermometry.

In order to account for the more general situations in which densities and intensities do not uniquely determine each other, we have the plausible option of assuming this relation to be statistical. This idea is followed up in the next section.

Meanwhile we conclude this section by pointing out that the Gibbs phase rule follows at once from equation 9. Consider, indeed, a heterogeneous system of f phases in equilibrium. The intensities have to satisfy an equation of the type 9 for each phase. The space of intensities has the dimension

$$\delta = r + 1 - f = c + 2 - f \geqslant 0 \tag{17}$$

where δ is called also the number of thermodynamic degrees of freedom.

It can be shown (see ref. 1, p 155) that critical states cannot arise unless two distinct phases become identical at the point in question. It follows from here that a one-component system can have only an isolated critical point. Thus the classical theory cannot account for lambda-points which form a line in the P/T diagram. The necessary extensions of the theory will be discussed in the next section.

3. STATISTICAL AND QUANTUM MECHANICAL CONSIDERATIONS

The considerations of the last section suggest the development of a statistical theory in which the extensive variables of a system are considered as random variables. Randomness enters the picture because of the coupling between system and reservoir involving the exchange of the quantities X. The random variables are assumed to be statistically independent from each other in the following sense: the values of the same quantity measured at discrete instances of time are independent from each other, and so are values associated with different systems coupled to the same reservoir. These requirements are the statistical expressions of the state of equilibrium. For a detailed discussion of this approach I refer to Tisza and Quay[5] who have shown that the elaboration of this picture yields under very general and at the same time realistic assumptions the grand canonical distribution function (d.f.)

$$\mathrm{d}F(X\,|\,\pi) = \mathrm{d}G(X)\exp[-\Phi(\pi) - \Sigma\pi X] \tag{18}$$

The vertical bar indicates that the d.f. is conditioned by the intensities of the reservoir. $G(X)$, a function depending only on the properties of the system, is the so-called structure function. Its differential dG is the number of linearly independent eigenfunctions of the Schrödinger equation. The function Φ depends on the properties of the system and on the intensities $\pi_1, \ldots, \pi_r$ of the reservoir; $\Sigma\pi X$ is short for $\sum_{T} \pi_K X_K$. All the functions involved contain the volume of the system as a constant parameter. The normalization of $\mathrm{d}F$ yields

$$e^{\Phi} = \int e^{-\Sigma X\pi} \mathrm{d}G(X) \tag{19}$$

This equation has a pivotal role, exhibiting the connection of all the relevant major theories. Thus suppose we start with the Hamiltonian of the system involving the phase space coordinates of structureless particles:

$$H = H(q_1, \dots q_f, p_1, \dots p_f) \tag{20}$$

Then proceeding from here to the Schrödinger equation we arrive at the structure function $G(X)$. Equation 19 then yields the entire formalism of MTE. We have indeed for the entropy

$$\begin{aligned} S &= k(\Phi + \Sigma X\pi) \\ &= k\int \mathrm{d}F \ln \mathrm{d}F/\mathrm{d}G \end{aligned}$$

and the potential of the grand canonical d.f.

$$\Omega = V\omega = -kT\Phi$$

Note that whereas the structure function may have discontinuities in the case of a discrete spectrum, the Laplace–Stieltjes transform 19 yields an absolutely continuous differentiable Φ. (Only the critical points call for special consideration.)

Contrary to the traditional foundations of thermodynamics on the basis of statistical mechanics, there is no need to assume that the system has a large number of degrees of freedom in order to arrive at the continuous differential–geometrical representation of thermodynamics. Of course, the large number of degrees of freedom does appear in the reservoir.

The formalism generated along the lines indicated has been called STE, short for statistical thermodynamics of equilibrium. Its main point can be simply stated: The formalism of MTE remains valid even in the statistical case, provided we replace the macroscopic extensive parameters with the canonical averages of the corresponding random variables.

I briefly note that the formalism admits also another interpretation. We consider the system as a measuring device, as a sensor that explores the intensities of the environment. Since in STE, the connection between the measured extensive quantities and the intensities is statistical, the latter can only be *estimated* from the former. The elaboration of this idea leads to interesting thermodynamic results[5].

I wish to point out here a curious analogy. As we go from MTE to STE we can no longer attribute simultaneous sharp values to the extensive and intensive variables. This uncertainty is governed by Boltzmann's constant just as the somewhat similar uncertainty of quantum mechanics is governed by Planck's constant.

The connection between the mechanical and thermodynamic formalisms provided by equation 19 still leaves many questions unanswered. It is a major challenge for statistical mechanics to prove that, assuming reasonable intermolecular forces, one can actually justify the scale invariance of MTE in the so-called thermodynamic limit. A great deal has been achieved in this direction, but the discussion of these results is outside the scope of this paper.

Although the procedure outlined above brings about the transition from the mechanical to the thermodynamic variables introduced above, the choice of the latter is not wide enough to account for the situations of greatest current interest. Thus we have no parameters as yet to describe, say, crystal symmetry. To achieve this further enrichment we have to use the well known Born–Oppenheimer (B–O) approximation. I like to call this procedure rather the B–O *transformation*, because it achieves the transition from the particle Hamiltonian to other Hamiltonians specified in terms of the spatial configuration of the nuclei. Please note the plural in this statement. It is indeed of capital importance that the B–O transformation leads often to different 'branches', say corresponding to white tin and grey tin, and each of these leads through equation 19 to one of the primitive fundamental equations that is to be used in the determination of heterogeneous equilibrium.

This approach opens up a new avenue for introducing additional parameters into the fundamental equation in a systematic fashion. In statistical mechanics the criterion for choosing such parameters is: take additive invariants (or also permutational invariants in the case of identical particles). This means in practice, the number of particles, momentum and angular momentum. In the present case we are authorized to take also translational invariants. (See p 186 of ref. 1). We obtain thus the important parameter of long range order, called quasi-thermodynamic, because it has no conjugate intensity.

Another important extension of the theory is to magnetic (and electric) systems. It is easy to join the couple M, H (magnetic moment and field) in analogy to density and chemical potential, to the formalism. It is a surprising aspect of recent studies[2–4] that this analogy is valid in a quantitative sense which goes considerably beyond the requirements of the thermodynamic analogy. It is therefore worth while to point out that this precise analogy is obtained only under carefully selected experimental conditions. Whereas pressure, temperature and chemical potential are constant over a heterogeneous system in equilibrium, this is true for the magnetic field only for special geometrical conditions. This is connected with the fact that the long range dipolar forces interfere with scale invariance. We note that the Ising model, although expressed originally in the language of magnetic systems, does not contain the disturbing dipolar interaction. Among its most important applications are cases in which the magnetization is interpreted as a quasi-thermodynamic order parameter.

4. CRITICAL POINTS

From the experimental point of view critical points in the phase diagram correspond to states in which two distinct modifications, such as liquid and vapour, or domains of opposite magnetization merge into a homogeneous phase.

According to thermodynamic theory critical points are *stable* states in which the determinant of the stiffness matrix vanishes. In terms of the symbols introduced in Section 2 we have two equivalent statements:

$$D_r = 0, \lambda_r = (\partial P_r/\partial X_r)_{P_1, \ldots P_{r-1}} = 0 \tag{21}$$

States satisfying these relations are called *spinodal*. In general, they are on the boundary of the metastable and unstable regions, and qualify as critical equilibrium points only when they are on the boundary of the normal region of stability. Here the system escapes instability by splitting into distinct modifications, and we see that the empirical definition is equivalent to the one based on thermodynamic stability.

This equivalence is, in fact, a theorem and enables us to predict that the compliance coefficients are singular. (This conclusion cannot be reached from the empirical definition alone.) In the case of a one component fluid we have

$$\lambda_1 = \left(\frac{\partial T}{\partial s}\right)_\rho = \frac{T}{c_v}, \qquad \lambda_2 = \left(\frac{\partial \mu}{\partial \rho}\right)_T = \frac{1}{\rho^2 \kappa_T} \tag{22}$$

where c_v and κ_T are the specific heat and the compressibility respectively. At the critical point $\lambda_2 = 0$, $D_2 = 0$ implies

$$\alpha, \kappa_T, c_p \rightarrow \infty \tag{23}$$

where α is the expansion coefficient. These conclusions are borne out by experiment. It is noteworthy that they follow from the vanishing of the determinant of the stiffness matrix, and there is no need to assume that the elements themselves vanish. The latter situation arises in the case of an accidental degeneracy with $\lambda_1 = \lambda_2 = 0$. Such a degenerate situation actually prevails in the Ising model, but in this case the degeneracy comes about because of the symmetry between the states of opposite magnetization.

Accordingly, when I discussed these matters some ten years ago, I duly stressed the difference between the Ising model and the real fluid[6]. At that time it was believed that the c_v of fluids is finite at the critical point. The situation was entirely reversed as, shortly thereafter, more precise measurements showed that the specific heat c_v is logarithmically divergent, in close analogy with the Ising model. We can hardly avoid the conclusion that the two cases exhibit similar symmetries. While the symmetry between the coexisting liquid and vapour modifications is by no means evident, such a symmetry has actually been predicted by the well known lattice gas model of Lee and Yang[9]. It is a surprise, however, that this model should prove not only manageable, but also more realistic than the van der Waals model of pairwise interacting particles.

In order to give justice to the hidden symmetry of the fluid it is important to use the density ρ, with the chemical potential μ as the conjugate variables instead of the more customary V, P[7]. Note that in MTE and STE the choice of variables is not entirely conventional, because this choice determines the type of exchange process that underlies the coupling of systems.

At this point we have arrived at the conclusion of the rigorous theory. Details have to be put in from experiment, from statistical mechanical calculations and also from approximate methods within GTD. The following is hardly more than a list of well known procedures with a few evaluating remarks.

(i) Van der Waals theory

This is a wonderful *tour de force* arriving from a simple particle picture at the gross features of a fluid system. However, the method of Maxwell construction has been incorrectly identified with the rigorous Gibbs theory. Differences of principle are pointed out on p 161 of ref. 1.

(ii) Internal field theory

By introducing an internal field depending on the magnetization, P. Weiss generalized the concept of intensity in order to cope with the limitations of scale invariance. This method was greatly developed by Landau and a further improved modern version is found in ref. 2.

The proper microscopic model to interpret the molecular field theories is the cellular model. Assuming the cells to be statistically independent corresponds to the trivial scale invariant case. The internal field theory is one step better, and takes an average interaction effect of neighbouring cells into account. What is neglected is the correlation of actual states. The inclusion of the correlations is achieved in the next approximation.

(iii) Ornstein–Zernike theory

This theory takes intercellular correlations into account and provides an excellent description of critical light scattering. Corrections required by the best experiments seem to be no more than marginal.

(iv) The Ising model

This is the cellular model simplified to make rigorous calculations possible. It has been developing into the prototype for most critical points. It was Onsager's rigorous calculations that led to the discovery of the critical point exponents as the proper analytical tools to account for critical phenomena.

(v) Critical point exponents

The introduction of this technique represents a turning point in the thermodynamic theory of critical points. Under normal conditions the thermodynamic fundamental equation is an extremely smooth function and the method of power series expansion is entirely justified. When using this method in his theory of continuous transitions, Landau did not suspect that these transitions might be singular. In contrast, my own approach was always centred around the importance of singularities, but lacking the proper technique I chose to confine myself in GTD to results of a topological nature, leaving details to statistical mechanics. However, in the last few years the new technique developed to a point where it can no longer be ignored by students of thermodynamics. In view of the wealth of results opened up for study I can do no more than whet your appetite for further study.

The underlying mathematical idea is simple enough. In order to make power series expansion applicable we modify as follows:

$$f(x) \approx |x|^{\lambda}(1 + ax + \ldots)$$

where x tends to zero at the critical point and λ is the critical point exponent. Logarithmically divergent functions are associated with an exponent $\lambda = 0$. The relevant variable is often the temperature expressed in terms of the reduced variable: $t = (T_c - T)/T_c$. The same symbols are used for the thermodynamically analogous exponents for fluids, magnetic systems and more general order–disorder transformations. Here are some examples:

$$
\begin{array}{lll}
 & & c_v \sim |t|^{-\alpha} \\
(\rho_G - \rho_L) & \text{or} & M \sim |t|^{\beta} \\
\rho^2 \kappa_T = \left(\dfrac{\partial \rho}{\partial \mu}\right)_T & \text{or} = & \left(\dfrac{\partial M}{\partial H}\right)_T \sim |t|^{-\gamma}
\end{array}
\tag{24}
$$

In the case of α and γ it is conventional to use primed indices below and unprimed above the critical point.

The first question that comes to mind is: how good are such representations? I confine myself to referring to *Figure 1* taken from a recent paper by M. Giglio and G. B. Benedek[8]†.

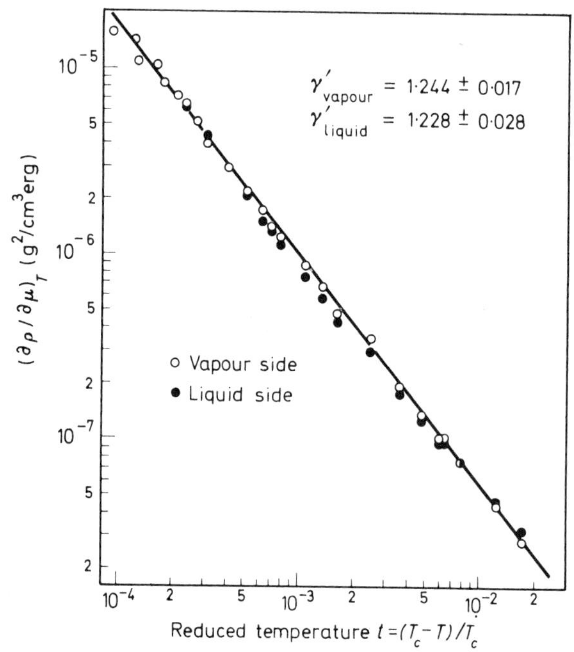

Figure 1. Plot of $(\partial\rho/\partial\mu)_T$ along the vapour and liquid sides of the coexistence curve of xenon as a function of the reduced temperature $t = (T_c - T)/T_c$. Open circles show vapour side; closed circles show liquid side.

The second point is that inserting the representations 24 into the standard thermodynamic formalism yields inequalities such as the Rushbrooke inequality:

$$\alpha' + 2\beta + \gamma' \geqq 2 \tag{25}$$

† I am indebted to the authors for letting me use their remarkably accurate results. They pointed out to me in a personal communication that the quality of their plot depends very markedly on the use of the variables ρ, μ. The constancy of the two exponents and their coincidence is lost in the plot of the conventional compressibility. This confirms the suggestion of Chase and Tisza[7].

Such inequalities are borne out by experiment, and indeed they have proved helpful in spotting errors in the determination of exponents.

More recently a number of so-called scaling laws have been proposed which tend to bring about a greatly increased coherence in the field. I confine myself to referring to the scaling law proposed by Kadanoff[2] which can be explained most simply in the present context.

Kadanoff considers an Ising model of lattice constant a_0 and defines a cell of size La_0 satisfying the relation

$$1 \ll L \ll \xi/a_0 \tag{26}$$

where ξ is the coherence length. Kadanoff argues that the properties of the site and cell will be identical provided we scale the magnetic field H and the reduced temperature as follows:

$$\begin{aligned} \tilde{H}/H &= L^x \\ \tilde{t}/t &= L^y \end{aligned} \tag{27}$$

$\tilde{H}$, $\tilde{t}$ refer to the intensities effective in the cell description, whereas H, t refer to the site. The main result is that all nine exponents mentioned above can be expressed in terms of x and y. The situation in this respect is remarkably good. Small discrepancies only arise in connection with the exponent of the correlation length describing, say, the extent of the short range order of spins on the paramagnetic side of the critical point. However, this pehnomenon is already a long way from the traditional domain of thermodynamics. Assuming that the scaling law will continue to hold up under the close scrutiny it is being subjected to, we can rationalize its meaning within GTD as follows: We known that the scale invariance of the macroscopic system must be limited, because any subdivision must stop at the single site; the scaling law is a generalization of the principle of scale invariance, to which it can be reduced if x and y are set equal to zero.

CONCLUSIONS

Generalized thermodynamics is a flexible framework accommodating many theories beyond those traditionally considered. Thus open systems and their fluctuations can be rigorously handled without invoking specific microscopic models. The extension of Gibbsian phase theory is consistent with the recent results concerning the phenomena in the neighbourhood of the critical points. The technique of critical point exponents made many of the classical procedures of approximation obsolete. However, this affects only various approximations to the main theory. Thus the shortcomings of the van der Waals theory are often ascribed incorrectly to the Gibbs theory. Although the latter is also in need of corrections and additions, when all this is done, the basic structure of the theory is seen to be considerably strengthened.

REFERENCES

[1] L. Tisza, *Generalized Thermodynamics*. M.I.T. Press: Cambridge, Mass. (1966).
[2] L. P. Kadanoff *et al. Rev. Mod. Phys.* **39**, 395 (1967).
[3] M. E. Fisher, *Rep. Progr. Phys.* **30**, 615 (1967).
[4] P. Heller, *Rep. Progr. Phys.* **30**, 731 (1967).
[5] L. Tisza and P. M. Quay, *Ann. Phys.* (*N.Y.*), **25**, 48 (1963).
[6] L. Tisza, *Ann. Phys.* (*N.Y.*), **13**, 1 (1961).
[7] L. Tisza and C. E. Chase, *Phys. Rev. Letters*, **15**, 1 (1965).
[8] M. Giglio and G. B. Benedek, *Phys. Rev. Letters*, **23**, 1145 (1969).
[9] T. D. Lee and C. N. Yang, *Phys. Rev.* **87**, 410 (1952).

THERMODYNAMICS OF SURFACE PHENOMENA

J. C. MELROSE

Mobil Research and Development Corporation,
P.O. Box 900, Dallas, Texas 75221, U.S.A.

ABSTRACT

The thermodynamic treatment of the interfacial region corresponding to two fluid phases in contact is discussed. The features of this analysis which are reviewed include the classical treatment of Gibbs and the extensions to this treatment which are due to Buff. It is shown that these extensions are essential if the logical structure of the analysis is to be regarded as complete.

1. INTRODUCTION

The thin, nonhomogeneous region separating two homogeneous bulk phases in contact constitutes an interface. It is generally recognized that an adequate thermodynamic treatment of such a region must be based on the work of Gibbs[1]. Nevertheless, the literature contains a number of proposals for modifying various features of this treatment. It is, in fact, remarkable that no other important contribution of Gibbs to the understanding of the equilibrium states of heterogeneous substances has given rise to so many reservations and attempts to develop alternative treatments.

The proposed modifications are usually concerned with one or more of several concepts which are characteristic of the Gibbsian treatment. The first of these concepts involves the notion of a mathematical dividing or reference surface, located within or very near the nonhomogeneous interfacial region. This surface serves to define the geometrical configuration of the interfacial region and also partitions the volume of the system between the two bulk phases. A second feature of the Gibbs treatment which is occasionally challenged is the definition of the chemical potentials which are appropriate to the components present in an interfacial region. Finally, the nature of the interdependence between the parameter referred to as the interfacial or surface tension and the geometrical configuration itself has been the source of much conflicting opinion.

It is one of the purposes of this review to discuss in detail the principal area of conceptual difficulty in the Gibbsian treatment. It is concluded that in this instance the classical approach of Gibbs, when properly interpreted, plays an essential role in providing generality as well as precision to the resulting formalism. A further purpose will be to discuss the recent contributions of Buff[2] to the general phenomenological theory of fluid–fluid interfacial regions. On the basis of these contributions, which involve concepts not traditionally regarded as thermodynamic in nature, it is possible to provide a substantial extension of the Gibbsian treatment. In fact, it is seen that the various difficulties which have given rise to proposed modifications are in part due to the incomplete nature of the Gibbsian approach.

A detailed exposition of the Gibbsian or strictly thermodynamic treatment of interfacial regions is to be found in the treatise by Defay *et al.*[3]. The review articles by Buff[4] and by Ono and Kondo[5] include, in addition to the thermodynamic analysis, accounts of the manner in which the condition for hydrostatic equilibrium provides an essential feature of the phenomenological theory. The treatment of Ono and Kondo, like that of Tolman[6], Koenig[7] and Hill[8], is restricted to interfaces of spherical shape, whereas Buff's treatment[2,4] applies to surfaces of nonspherical configuration. The following discussion will rely extensively on a recent review[9] of Buff's work.

2. THE DIVIDING SURFACE AND GIBBS EXCESS QUANTITIES

The thermodynamic analysis which will be developed is intended to describe a system such as is represented in *Figure 1*. Two homogeneous fluid phases α and β are taken to be in contact, and external macroscopic fields are

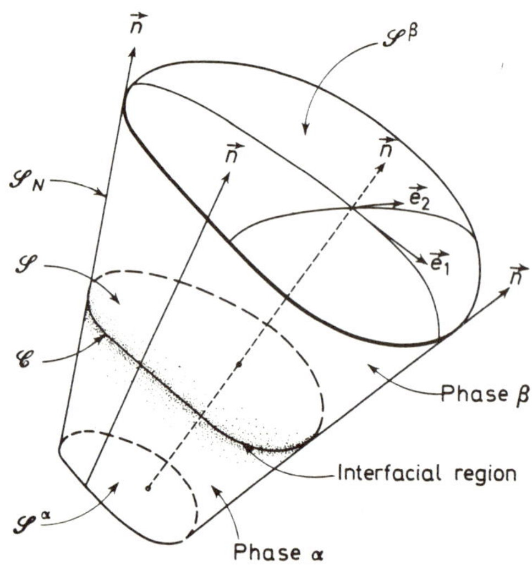

Figure 1. System with curved fluid–fluid interfacial region

assumed to be absent. Within the thin, nonhomogeneous region which separates α and β a mathematical surface, denoted as $\mathscr{S}$, is placed. This surface will in general be curved, since its geometrical shape or configuration is determined by the configuration of the interfacial region. The nature of the physical conditions which determine the latter will emerge at a further stage of the analysis.

Whatever the configuration of the interfacial region may be, the configuration of $\mathscr{S}$ is taken to be such that the variation of the point density of matter along a normal to $\mathscr{S}$ is sensibly invariant over the entire set of normal directions. This is possible only if the interfacial region is assumed to be homogeneous in a two-dimensional sense. The position of $\mathscr{S}$ with respect to the variation in the density in any given normal direction through the interfacial region is arbitrary. Consequently, various conventions may be used in order to fix its position with respect to an origin selected on a given normal. It is assumed, following Buff[2,4], that the various positions so chosen then establish a set of surfaces which are parallel in the mathematical sense[10,11].

Without loss of generality, the external boundaries of the system may be considered to be determined, first, by specifying a closed curve $\mathscr{C}$ lying within $\mathscr{S}$. The set of normals to $\mathscr{S}$ passing through the curve $\mathscr{C}$ then forms a surface $\mathscr{S}_N$. Secondly, two surfaces, $\mathscr{S}^\alpha$ and $\mathscr{S}^\beta$, parallel to and lying on either side of $\mathscr{S}$, are specified to be sufficiently far from $\mathscr{S}$ that in their vicinity the point densities of matter are uniform. Thus, $\mathscr{S}^\alpha$ and $\mathscr{S}^\beta$ lie entirely within the homogeneous phases α and β, respectively. The matter enclosed within the three surfaces, $\mathscr{S}^\alpha$, $\mathscr{S}^\beta$ and $\mathscr{S}_N$, then constitutes the system $\mathscr{B}$. This system is considered to be an open system in the usual thermodynamic sense.

A basic assumption underlying the Gibbsian approach just outlined is that the densities of the extensive thermodynamic quantities vary in a continuous manner along a coordinate which is normal to the interfacial region. The parameters in question are the internal energy, U, the entropy, S, and the number of moles of each component, $N_i (i = 1 \ldots \kappa)$. If M is taken as denoting each of these extensive quantities and $\hat{M}$ as the density, i.e. the limit of (M/V) as the volume, V, vanishes, the corresponding surface excess quantity can be written as

$$m = \int_{\lambda^\alpha}^{\lambda^\beta} \{\hat{M}(\lambda) - \hat{M}^{\alpha\beta}(\lambda^0)\} B(\lambda, \lambda^0)\, \mathrm{d}\lambda \tag{2.1}$$

Here λ measures the distance along the normal coordinate from some arbitrary origin, and λ^0 then specifies the position of the Gibbs dividing or reference surface $\mathscr{S}$ (see *Figure 2*). The area of the reference surface $\mathscr{S}$ is denoted by Ω. The quantity $\hat{M}^{\alpha\beta}(\lambda^0)$ is given by

$$\hat{M}^{\alpha\beta}(\lambda^0) = \{1 - A(\lambda)\} \hat{M}^\alpha + A(\lambda)\hat{M}^\beta \tag{2.2a}$$

where $A(\lambda)$ is a unit step function,

$$A(\lambda) = 0 \quad \text{for} \quad \lambda < \lambda^0; \quad A(\lambda) = 1 \quad \text{for} \quad \lambda \geqslant \lambda^0 \tag{2.2b}$$

and $\hat{M}^\alpha$ and $\hat{M}^\beta$ are the densities characteristic of the regions of the bulk phases in which homogeneity can be assumed. Clearly, the limits of the integration in equation 2.1 need only be extended to such regions, i.e. to the surfaces $\mathscr{S}^\alpha$ and $\mathscr{S}^\beta$.

The quantity $B(\lambda, \lambda^0)$ in equation 2.1 is a factor accounting for the change in area of the surface on which the function $\hat{M}(\lambda)$ is defined and is given by

$$B(\lambda, \lambda^0) = 1 + (\lambda - \lambda^0)J + (\lambda - \lambda^0)^2 K \tag{2.3}$$

Here, J and K are the mean and Gaussian curvatures[10,11] characterizing an infinitesimal area, $\delta\Omega$, on the reference surface $\mathscr{S}$. A restriction on the variation of the mean curvature J from point to point on such a surface will be intro-

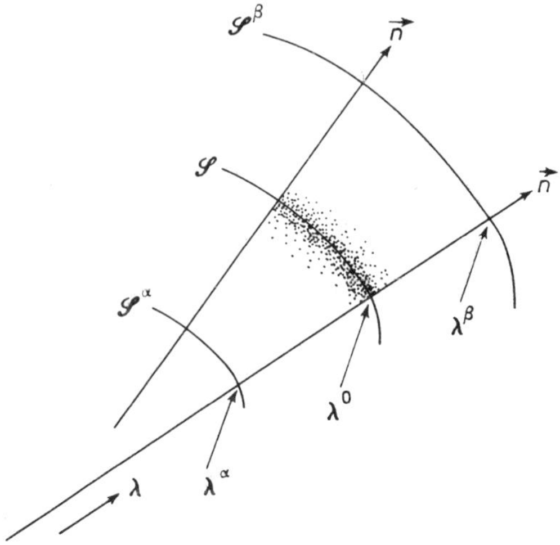

Figure 2. Normal coordinate for parallel surfaces

duced later. This restriction and the small magnitude, as compared with J^{-1} and $K^{-\frac{1}{2}}$, of the distance over which $\hat{M}$ and $\hat{M}^{\alpha\beta}$ differ appreciably are sufficient to ensure that the surface density, m, will be very nearly uniform over the entire interfacial region. Thus, the two-dimensional homogeneity assumed for the density profile $\hat{M}(\lambda)$, where $M = N$, the total number of moles, holds to a high degree of approximation for each of the densities, m.

In order to specify the volumes V^α and V^β into which the reference surface $\mathcal{S}$ partitions the total volume of the system, integrals similar to that of equation 2.1 are required:

$$V^\alpha = \int_0^\Omega \int_{\lambda^\alpha}^{\lambda^0} B(\lambda, \lambda^0)\, \mathrm{d}\lambda\, \mathrm{d}\Omega \tag{2.4a}$$

$$V^\beta = \int_0^\Omega \int_{\lambda^0}^{\lambda^\beta} B(\lambda, \lambda^0)\, \mathrm{d}\lambda\, \mathrm{d}\Omega \tag{2.4b}$$

If now two sets of extensive parameters denoted as M^α and M^β are defined by the following:

$$M^\alpha = V^\alpha \hat{M}^\alpha; \quad M^\beta = V^\beta \hat{M}^\beta \tag{2.5}$$

it is clear that, if the area Ω is sufficiently small that J and K are constant, the total quantity M must be given by

$$M = M^\alpha + M^\beta + m\Omega \tag{2.6}$$

Corresponding to the various extensive quantities denoted by M, equation 2.1 provides a definition of the surface densities of interfacial energy, u, entropy, s, and number of moles of each component, Γ_i $(i = 1 \ldots \kappa)$. The location of the reference surface may now be specified by setting any one of these quantities equal to zero. Each of these possibilities then constitutes a convention. For a one-component system, it is convenient to choose the convention, $\Gamma = 0$. In *Figure 3* the application of this convention to a hypothetical variation of the mass density through the interfacial region is shown. For multicomponent systems, various conventions involving either one of the Γ_i, or a judicious combination of Γ_i, may be used[12].

It is seen that the use of the reference surface concept permits the total quantity M to be partitioned among three quantities. The sets of parameters represented by M^{α} and M^{β}, supplemented by the volumes V^{α} and V^{β}, respectively, constitute sets of homogeneous functions of order one. Similarly, the quantities $m\Omega$, supplemented by the area Ω, form a set of homogeneous functions of order one. Such mathematical properties are essential in developing a rigorous thermodynamic analysis of the system which has been defined. At this stage of the analysis, however, the treatment is rigorous only if the area of $\mathscr{S}$ is taken to be very small.

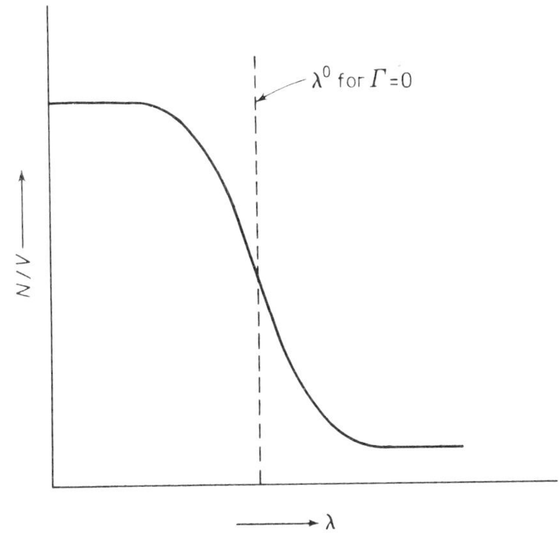

Figure 3. Schematic density profile for interfacial region

The motivation behind the procedures just described has often been interpreted in such a way that alternative formulations have been sought. In particular, the use of two dividing surfaces has been proposed. By this means a volume and a thickness of the interfacial region are defined. If the Gibbs approach is interpreted to imply the assignment of mass, energy and entropy to a fictitious surface, the use of two dividing surfaces would, of course, provide a more satisfactory approach. On the other hand, such an interpretation is not required. The dividing surface is not intended to represent the physical interface. It is instead merely a mathematical device and serves as a reference or fiducial surface. By means of such a surface the continuous variations of the densities of the extensive thermodynamic parameters can be compared with a discontinuous variation, yielding surface densities having required mathematical properties.

In fact, the apparent advantage of using two dividing surfaces is obviated by the corresponding need to introduce two conventions for locating the surfaces. One such convention would be to choose the second dividing surface to coincide with the first. The Gibbs formulation is then recovered. Also, it is not possible to assign a magnitude to the interfacial thickness by any means other than by adopting two conventions. Thus, this distance cannot have the physical significance which is intended.

3. THE DIVIDING SURFACE AND THERMODYNAMIC WORK

In order to interrelate the various thermodynamic quantities denoted by M by a fundamental Gibbsian equation, it is necessary to develop a suitable formulation of the differential work which can be performed on the system $\mathscr{B}$. Clearly, this formulation will involve the displacement of the three surfaces, $\mathscr{S}^\alpha$, $\mathscr{S}^\beta$ and $\mathscr{S}_N$, which form the external boundaries of $\mathscr{B}$. In addition, the reference surface $\mathscr{S}$ is again found to play an essential role in the analysis.

Before proceeding to express the work received by the system in terms of geometric variables, however, it is useful to consider the nature of the thermodynamic potential directly related to such variables. Thus, according to the first law, for adiabatic changes in which the system $\mathscr{B}$ is closed, the differential work received by the system, $\mathrm{d}W$, is given by

$$\mathrm{d}W = \mathrm{d}U; \mathrm{d}S = \mathrm{d}N_i = 0 \tag{3.1}$$

More generally, if T denotes the temperature of the system and μ_i $(i = 1 \ldots \kappa)$ the set of chemical potentials associated with the components of the system,

$$\mathrm{d}U = T\,\mathrm{d}S + \sum_{i=1}^{\kappa} \mu_i\,\mathrm{d}N_i + \mathrm{d}W \tag{3.2}$$

Although W is not itself a function of the thermodynamic state, there exists a thermodynamic potential, denoted as Φ, which is obtained if equation 3.2 is simply integrated, using Euler's theorem on homogeneous first-order functions. Thus, integration of equation 3.2 yields

$$\Phi = U - TS - \sum_{i=1}^{\kappa} \mu_i N_i \tag{3.3}$$

The potential Φ is the free energy available for mechanical work at constant T, μ_i. If P denotes the pressure in a homogeneous fluid phase, the potential Φ for either of the homogeneous regions in the system $\mathscr{B}$ is given by

$$\Phi^j = -P^j V^j \qquad (j = \alpha, \beta) \tag{3.4a}$$

The corresponding expressions for the thermodynamic work done on such regions are:

$$\mathrm{d}W^j = -P^j\,\mathrm{d}V^j \qquad (j = \alpha, \beta) \tag{3.4b}$$

The generalization of equations 3.4 to the entire system, as indicated above, involves the displacement of the surfaces forming the boundaries of $\mathscr{B}$. Appropriate geometrical transformations can then be introduced by a method[9] analogous to that used previously for spherical interfaces[8, 13]. For an infinitesimal area, $\delta\Omega$, of the reference surface $\mathscr{S}$, this procedure yields the following expression for the work received,

$$\mathrm{d}(\delta W) = -P^\alpha\,\mathrm{d}(\delta V^\alpha) - P^\beta\,\mathrm{d}(\delta V^\beta) + \gamma\,\mathrm{d}(\delta\Omega) + c(\delta\Omega)\,\mathrm{d}J \tag{3.5}$$

Here, the intensive quantities γ and c are *defined* by the relationships:

$$\delta\Phi = -P^{\alpha}\,\delta V^{\alpha} - P^{\beta}\delta V^{\beta} + \gamma\,\delta\Omega \tag{3.6}$$

$$P^{\alpha} - P^{\beta} = \gamma J + c(2K - J^2) \tag{3.7}$$

It should be emphasized that the argument leading to equations 3.5 to 3.7 involves no more than two essential steps[5,14]. First, it is recognized that the variation $d(\delta W)$ can be expressed solely in terms of the variations in a set of three independent geometrical variables. These variables are sufficient to specify the three types of volume change capable of doing work, i.e. changes resulting from the displacement of the boundary surfaces, $\mathscr{S}^{\alpha}$, $\mathscr{S}^{\beta}$ and $\mathscr{S}_N$. Secondly, suitable geometrical transformations are employed in order to express these variations in terms of the variables appearing in equation 3.5. In particular, at no stage is any principle of mechanics explicitly invoked.

On the other hand, the Gibbs reference surface $\mathscr{S}$ is directly involved in the geometrical transformations which are used. Thus, each of the five geometrical variables in equations 3.6 and 3.7 is defined only in terms of a particular reference surface. Hence, the magnitude of each quantity depends on the convention used to specify the location of $\mathscr{S}$. The form of equation 3.7 suggests that one such convention is given by the condition, $c = 0$. The surface which corresponds to this convention is known as the 'surface tension'.

Since it is assumed that Φ is a function of state, i.e. Φ is determined for a given equilibrium state of the system, it follows from equation 3.6 that the interfacial tension, γ, is not a pure function of state but depends on the location of the reference surface $\mathscr{S}$. In fact, it can be shown that the parameter c is related to this dependence by the following expression,

$$c = (2K - J^2)^{-1}(\partial\gamma/\partial\lambda^0)_{S,\,N_i,\,\mathscr{S}^{\alpha},\,\mathscr{S}^{\beta},\,\mathscr{S}_N} \tag{3.8}$$

Thus, it follows from equation 3.7 that, unless $J = K = 0$, the function γ has a minimum when λ^0 corresponds to the surface of tension. For the case of a planar interfacial region, γ is invariant with respect to the location of the reference surface $\mathscr{S}$.

Before proceeding further, two major difficulties may be noted. First, unless it is assumed at the outset that the reference surface is characterized by uniform curvatures, J and K, over a finite area, Ω, equations 3.5 to 3.7 provide no basis for regarding γ and c as uniform over such an area. That is, the analysis as developed to this point is restricted to interfacial regions for which the set of parallel reference surfaces is either planar, spherical or cylindrical in configuration. A second difficulty arises in connection with the location of the surface of tension. Within the context of the thermodynamic definition of work provided by equations 3.5 to 3.7, the location of the surface of tension is only specified by the condition $c = 0$. This convention, however, does not provide any information as to the proximity of this surface to the physical interface. In contrast, for a one-component system the convention $\Gamma = 0$, as required by equation 2.1, definitely locates the surface $\mathscr{S}$ within the nonhomogeneous interfacial region.

4. THE STRESS TENSOR AND CONDITION FOR HYDROSTATIC EQUILIBRIUM

Both of the difficulties noted above arise because the Gibbsian thermodynamic approach is actually incomplete. The additional physical principle which is required in order to extend the traditional treatment is provided by an explicit consideration of the basic theorem of hydrostatics. The earlier literature devoted to the hydrostatic treatment of interfacial regions has been reviewed by Bakker[15]. However, the application of the condition for hydrostatic equilibrium to the critical problems noted, leading to a general phenomenological treatment of fluid–fluid interfacial regions, was first carried out by Buff[2,4,5,9].

The basic concept utilized in developing the hydrostatic analysis is that of the stress tensor. This quantity provides a convenient mathematical form for representing the forces or stress vectors acting on a small volume element of matter. In the case of a fluid phase at rest, the off-diagonal components vanish, and if the fluid is also isotropic, as in a bulk homogeneous fluid, the diagonal components are identical and, except for sign, are equal to the fluid pressure. In the nonhomogeneous interfacial region, on the other hand, the diagonal components of the stress tensor are not identical. Clearly, these components are best defined in terms of tangential and normal directions corresponding to a Gibbs reference surface $\mathscr{S}$. The fluid character of the interfacial region ensures isotropy in a two-dimensional sense, and consequently the two tangential components will be equal. Thus, with respect to any reference surface $\mathscr{S}$, the stress tensor, $\boldsymbol{\sigma}$, can be written as

$$\boldsymbol{\sigma} = \sigma_T(\mathbf{e}_1\mathbf{e}_1 + \mathbf{e}_2\mathbf{e}_2) + \sigma_N\mathbf{nn} \tag{4.1}$$

Here, σ_T and σ_N represent the tangential and normal components, respectively, $\mathbf{e}_1$ and $\mathbf{e}_2$ are orthogonal unit tangent vectors, and $\mathbf{n}$ is the unit normal vector.

The condition for mechanical or hydrostatic equilibrium can now be easily expressed. This condition follows from the equation of motion of a fluid which is at rest and not subject to body forces, and therefore represents the principle of momentum balance for such a fluid. It is of interest to note that the explicit use of this condition is not ordinarily required in thermodynamics. For the system under consideration, the condition requires that the divergence of the stress tensor vanish,

$$\nabla \cdot \boldsymbol{\sigma} = 0 \tag{4.2}$$

If now the continuous variations of σ_T and σ_N through the interfacial region are defined in terms of the λ coordinate and associated with the corresponding variation of the mean curvature J of the reference surface, equations 4.1 and 4.2 yield a differential equation. The integration of this equation can be carried out without reference to a discontinuous variation in the components σ_T and σ_N. However, in order to arrive at a relationship corresponding to that obtained from the thermodynamic analysis, i.e. equation 3.7, such a variation must be introduced. Using the unit step function given by equation 2.2b, this variation can be defined as

$$\sigma^{\alpha\beta}(\lambda^0) = -\{1 - A(\lambda)\}P^\alpha - A(\lambda)P^\beta \tag{4.3}$$

Also required are definitions of the quantities γ and c in terms of integrals similar in form to those given by equation 2.1 but involving the variation of the tangential component of the stress tensor. These definitions are given by[2, 4, 9]:

$$\gamma = \int_{\lambda^\alpha}^{\lambda^\beta} \{\sigma_T(\lambda) - \sigma^{\alpha\beta}(\lambda^0)\} B(\lambda, \lambda^0)\, d\lambda \tag{4.4}$$

$$c = \int_{\lambda^\alpha}^{\lambda^\beta} \{\sigma_T(\lambda) - \sigma^{\alpha\beta}(\lambda^0)\}(\lambda - \lambda^0)\, L(\lambda, \lambda^0)\, d\lambda \tag{4.5a}$$

where

$$L(\lambda, \lambda^0) = 1 + JK(J^2 - 2K)^{-1}(\lambda - \lambda^0) \tag{4.5b}$$

Thus, using equations 4.4 and 4.5, it is found that the integration of equation 4.2 yields a form which is identical with equation 3.7. Furthermore, when equation 4.4 is differentiated with respect to λ^0 and equations 4.5 introduced, equation 3.8 is recovered. We may conclude, therefore, that the relationship expressed by equation 3.7 constitutes the condition for mechanical or hydrostatic equilibrium which is applicable to the system $\mathscr{B}$.

These results have a number of important applications. The first is the linking of the condition for mechanical equilibrium to the parameters appearing in the expression for the differential of thermodynamic work, equation 3.5. Also, equations 4.4 and 4.5 provide the appropriate interpretation of γ and c as Gibbs surface excess properties, defined in terms of a Gibbs reference surface.

In this connection, it is seen from equation 4.4 that the interfacial tension, γ, is not restricted to its traditional role as a scalar thermodynamic variable, but has also a vectorial interpretation. Thus, γ represents the direct macroscopic force which an interfacial region exerts on its surroundings[16]. The confusion which has existed in the past with respect to this interpretation of γ has led to a number of incorrect views relating to macroscopic surface phenomena. This confusion may also be in part responsible for the widely-held notion that the parameter γ should be incorporated in a 'true' chemical potential term characterizing the components within the interfacial region.

A further application of equations 4.5 is concerned with establishing the location of the surface of tension with respect to the physical interface. Since this particular reference surface is defined by the convention $c = 0$, equation 4.5a yields

$$\lambda^0 = \frac{\int_{\lambda^\alpha}^{\lambda^\beta} \{\sigma_T(\lambda) - \sigma^{\alpha\beta}(\lambda^0)\} L(\lambda, \lambda^0)\lambda\, d\lambda}{\int_{\lambda^\alpha}^{\lambda^\beta} \{\sigma_T(\lambda) - \sigma^{\alpha\beta}(\lambda^0)\} L(\lambda, \lambda^0)\, d\lambda}, \quad \text{for } c = 0 \tag{4.6}$$

This result shows that the surface of tension is indeed located within the interfacial region.

Another important consequence follows from extending to the function $\sigma_T(\lambda)$ the same assumption of two-dimensional homogeneity which was applied previously to the densities represented by $\hat{M}(\lambda)$. On the basis of equation 4.4, then, the interfacial tension γ may be taken as approximately uniform over the entire area Ω of any reference surface, as in the case of the excess quantities, m, defined by equation 2.1.

These conclusions may be used to arrive at some further important results. First, since the pressures P^α and P^β are invariant with respect to the choice of a normal direction through the reference surface $\mathscr{S}$, it follows from equation 3.7 that the surface of tension is characterized by a uniform value of the product of γ and the mean curvature J. Consequently, J is very nearly uniform for this particular reference surface. As has been seen, the surface of tension lies within the extremely thin nonhomogeneous region, and therefore other reference surfaces lying within the physical interface will vary only slightly from this condition. This constitutes the restriction on the variation of J over the area Ω of any reference surface, to which reference was made previously (Section 2). Also, the equilibrium configuration of the physical interface is now seen to be characterized by a condition of essentially uniform mean curvature.

A further consequence of the two-dimensional homogeneity assumed for the function $\sigma(\lambda)$ is that, to nearly the same degree of approximation, the parameter c may be regarded as uniform over a finite area Ω. To within this degree of approximation, then, it becomes possible to extend the validity of equations 3.5 to 3.7 to a finite area, Ω, of any reference surface.

5. FUNDAMENTAL EQUATIONS FOR EXCESS QUANTITIES

The fundamental Gibbsian differential equation which is applicable to the system $\mathscr{B}$ may now be written. Extending equation 3.5 to finite area and combining this result with equation 3.2, we obtain

$$dU = T\,dS + \sum_{i=1}^{\kappa} \mu_i\,dN_i - P^{\alpha}\,dV^{\alpha} - P^{\beta}\,dV^{\beta} + \gamma\,d\Omega + c\Omega\,dJ \tag{5.1}$$

Similarly, equation 3.6 may be combined with equation 3.3. Differentiating the resulting expression and applying equation 5.1 then gives a Gibbs–Duhem equation in the form.

$$V^{\alpha}\,dP^{\alpha} + V^{\beta}\,dP^{\beta} = S\,dT + \sum_{i=1}^{\kappa} N_i\,d\mu_i + \Omega\,d\gamma - c\Omega\,dJ \tag{5.2}$$

These results, together with equation 3.7, the condition for hydrostatic equilibrium, constitute the thermodynamic formalism suitable for analysing a variety of physical effects involving the system $\mathscr{B}$. In many instances, however, the effect of temperature, pressure and chemical composition on the interfacial tension is of primary concern. It is then useful to develop a formalism more directly applicable to this situation. Following Defay *et al.*[3], the Gibbs excess properties, as defined by equation 2.1 and to which this formalism applies, may be regarded as characterizing a 'non-autonomous system'.

In developing this formalism, it is necessary to introduce explicitly the conditions for thermal and material equilibrium to which the various states of the system $\mathscr{B}$ are subject. Thus, the temperature and chemical potentials which were introduced in writing equation 3.2 also apply[1,3] to each of the homogeneous fluid regions α and β,

$$T = T^{\alpha} = T^{\beta} \tag{5.3a}$$

$$\mu_i = \mu_i^{\alpha} = \mu_i^{\beta} \qquad (i = 1 \ldots \kappa) \tag{5.3b}$$

We then write equations 3.2 and 3.3 for each of the homogeneous regions. Introducing equations 3.4, these expressions yield the Gibbs–Duhem forms

$$V^j \, dP^j = S^j \, dT + \sum_{i=1}^{\kappa} N_i^j \, d\mu_i \qquad (j = \alpha, \beta) \tag{5.4}$$

Subtracting equations 5.4 from equation 5.2, taking into account the appropriate relationships given by equation 2.6, and dividing by the area of the reference surface then gives

$$d\gamma = - s \, dT - \sum_{i=1}^{\kappa} \Gamma_i \, d\mu_i + c \, dJ \tag{5.5}$$

Similarly, subtracting the two forms of equation 3.3 applicable to α and β from the expression obtained by combining equations 3.3 and 3.6 yields

$$u = Ts + \sum_{i=1}^{\kappa} \mu_i \Gamma_i + \gamma \tag{5.6}$$

These results define the formalism which interrelates the various Gibbs excess properties. For a planar interface, the final term in equation 5.5 is eliminated. Since in this case the intensive state of the system is determined by κ variables, equation 5.5 still includes on the RHS one more term than is required. This is, of course, remedied by introducing the convention by which the Gibbs reference surface is located. For example, if component 1 is taken to be the solvent and if the $\Gamma_1 = 0$ convention is then adopted, equation 5.5 yields

$$d\gamma = - s \, dT - \sum_{i=2}^{\kappa} \Gamma_i \, d\mu_i \tag{5.7}$$

This is the Gibbs adsorption equation. Its derivation has been the subject of much discussion in the literature[17–19].

6. CURVATURE DEPENDENCE OF INTERFACIAL TENSION

The degree of approximation which is involved in extending the treatment developed in Sections 2–4 to finite areas is clearly related to the dependence of the interfacial tension on the mean curvature J. However, before discussing the nature of this dependence, it will be instructive to consider briefly the assumptions which are inherent in the experimental procedures used in measuring γ.

Nearly all conventional methods of measuring interfacial tension are based on the condition for hydrostatic equilibrium, equation 3.7. Implicitly, it is assumed that the appropriate dividing surface is the surface of tension. Due to the effect of the gravitational field, the pressure difference, $P^\alpha - P^\beta$, varies with the vertical coordinate. As a result, the mean curvature J also varies with this coordinate. However, the curvatures involved are very small, and hence it is assumed that γ is independent of the magnitude of J.

In systems of this type the boundary curve $\mathscr{C}$ is determined by the configuration of the surface of a solid phase, e.g. a cylindrical capillary, within which the system is enclosed, and by the condition for hydrostatic equilibrium applicable to the three-phase confluent zone[4, 16]. The solid surface must be sufficiently smooth, as well as homogeneous with respect to composition and state of stress. Then, if its configuration is symmetrical with respect to the vertical axis, the interfacial region will be such that any reference surface $\mathscr{S}$ will also be symmetrical in this sense, i.e. a surface of revolution. Under these circumstances, numerical methods can be employed to obtain solutions to equation 3.7. It then becomes possible to relate measurements of the interface configuration to the fluid densities and to the interfacial tension.

In contrast to the situation involved in measuring γ, there are many cases of physical interest in which the curvatures are sufficiently high that gravitational distortion can be neglected. If the curvatures are extremely high, the question as to whether γ is curvature dependent becomes a matter of some importance. As indicated above, this dependence is also of interest in assessing the magnitude of the approximation which is involved in extending the thermodynamic treatment to finite areas.

Since for a system which includes a curved interfacial region there are $\kappa + 1$ independent intensive variables, one of the terms in equation 5.5 may again be eliminated by introducing a convention. For a one-component system, the curvature dependence is then represented by either of two relationships which follow from this expression,

$$\Gamma = 0 \text{ convention: } \rightarrow (\partial\gamma/\partial J)_T = c \tag{6.1a}$$

$$c = 0 \text{ convention: } \rightarrow (\partial\gamma/\partial J)_T = -\Gamma\ (\partial\mu/\partial J)_T \tag{6.1b}$$

The further analysis of the problem, based on equations 6.1, is due primarily to Tolman[20], Koenig[7] and Buff[2]. This analysis utilizes a thermodynamic parameter which characterizes the thickness of the interface and is defined as the distance between the two dividing surfaces in question,

$$\Delta\lambda = \lambda^0_{\Gamma=0} - \lambda^0_{c=0} \tag{6.2}$$

It is then found that for the case of spherical interfaces ($J = 2K^{\frac{1}{2}}$), the derivatives in equations 6.1 can be expressed by means of a polynomial in the variable $\Delta\lambda$. Retaining only the first-order term, it can be shown that

$$\left(\frac{\partial \log \gamma}{\partial \log J}\right)_{T,\Gamma=0} = \left(\frac{\partial \log \gamma}{\partial \log J}\right)_{T,c=0} \simeq -(\Delta\lambda)\, J_{c=0} \tag{6.3}$$

As Buff[2,4] has pointed out, the variable $\Delta\lambda$ is in principle also curvature dependent. Consequently, the integration of equation 6.3 yields only the first-order correction to the interfacial tension appropriate to a planar interface. Denoting the latter as γ_∞ and the corresponding interfacial thickness parameter as $\Delta\lambda_\infty$, this integration gives

$$\gamma \simeq \gamma_\infty(1 - \Delta\lambda_\infty J) \tag{6.4}$$

Since various lines of indirect evidence suggest that the thickness of an interfacial region corresponds to no more than five or six molecular diameters[3–5], a reasonable estimate of the absolute magnitude of $\Delta\lambda_\infty$ is 2 to 4 Å. If this estimate is accepted, equation 6.4 indicates that for spherical interfaces, the effect of curvature becomes appreciable only if the radius of curvature is less than about 500 Å.

The sign of $\Delta\lambda_\infty$ can also be predicted. For a one-component system the surface of tension lies on the higher density or liquid side of the $\Gamma = 0$ surface[2,4]. If phase α is taken to be the liquid phase, $\Delta\lambda_\infty$ will then be positive in sign. The surface tension of a very small liquid drop is then, according to equation 6.4, less than the tension of a planar interface. Conversely, if phase α is the vapour phase, $\Delta\lambda_\infty$ is negative. The surface tension for a small bubble of vapour is correspondingly enhanced.

The experimental techniques used in measuring γ involve curvatures such that the correction term in equation 6.4 is of the order of 0·001 per cent or less. Thus, the assumptions ordinarily used in interpreting measurements of γ introduce errors of negligible magnitude. On the other hand, for applications involving interfaces of high curvature, these corrections can become important[9]. It should be noted that the estimate provided by equation 6.4 is applicable only to interfaces which are spherical in form. That is, the first-order treatment of curvature dependence so far developed does not necessarily extend to interfaces of more complicated shape.

7. CONCLUSIONS

The thermodynamic treatment of fluid–fluid interfacial regions which has been outlined above utilizes several essential concepts. The first is the notion of a Gibbsian dividing or reference surface. In addition, the mathematical description of curved surfaces provided by differential geometry can be used to extend the Gibbsian treatment of excess properties to interfacial regions which are neither flat, cylindrical, nor spherical. The thermodynamic work associated with changes in interfacial configuration is precisely formulated only through the use of these techniques. If the condition for hydrostatic equilibrium is then explicitly taken into account, the required interpretation of the parameters defined by this formulation can be established. Thus, by introducing this principle, the logical structure of the thermodynamic treatment applicable to fluid interfacial systems is completed.

ACKNOWLEDGEMENT

Appreciation is expressed to Mobil Research and Development Corporation for permission to publish this paper.

REFERENCES

1 J. W. Gibbs, *Scientific Papers*, Vol. I, pp. 219–331. Longmans: London (1906); Dover reprint: New York (1961).
2 F. P. Buff, *J. Chem. Phys.* **23**, 419 (1955); **25**, 146 (1956).
3 R. Defay, I. Prigogine, A. Bellemans and D. H. Everett, *Surface Tension and Adsorption*. Wiley: New York (1966).
4 F. P. Buff, *The Theory of Capillarity* in *Handbuch der Physik*, Vol. X, pp 281–304. Springer: Berlin (1960).
5 S. Ono and S. Kondo, *Molecular Theory of Surface Tension in Liquids* in *Handbuch der Physik*, Vol. X, pp 134–280. Springer: Berlin (1960).
6 R. C. Tolman, *J. Chem. Phys.* **16**, 758 (1948).
7 F. O. Koenig, *J. Chem. Phys.* **18**, 449 (1950).
8 T. L. Hill, *J. Chem. Phys.* **19**, 1203 (1951).
T. L. Hill, *J. Phys. Chem.* **56**, 526 (1952).
9 J. C. Melrose, *Ind. Eng. Chem.* **60** (No. 3), 53 (1968).

[10] L. P. Eisenhart, *Treatise on the Differential Geometry of Curves and Surfaces*. Ginn: New York (1909); Dover reprint: New York (1960).
[11] E. Kreysig, *Differential Geometry*. University of Toronto Press: Toronto (1959).
[12] E. A. Guggenheim and N. K. Adam, *Proc. Roy. Soc. A*, **139**, 218 (1933).
[13] S. Kondo, *J. Chem. Phys.* **25**, 662 (1956).
[14] E. A. Guggenheim, *Research, Lond.* **10**, 478 (1959).
[15] G. Bakker, *Kapillarität und Oberflächenspannung* in *Handbuch der Experimentalphysik*, Vol. VI. Akademische Verlagsgesellschaft: Leipzig (1928).
[16] F. P. Buff and H. Saltsburg, *J. Chem. Phys.* **26**, 23, 1526 (1957).
[17] F. O. Koenig and R. C. Swain, *J. Chem. Phys.* **1**, 723 (1933).
[18] E. A. Guggenheim, *J. Chem. Phys.* **4**, 689 (1936).
[19] G. Scatchard, *J. Phys. Chem.* **66**, 618 (1962).
[20] R. C. Tolman, *J. Chem. Phys.* **17**, 333 (1949).

NON-HYDROSTATIC THERMODYNAMICS OF CHEMICAL AND PHASE CHANGES

A. G. McLELLAN

Department of Physics, University of Canterbury, Christchurch, New Zealand

ABSTRACT

The problem of chemical or phase equilibrium under non-hydrostatic stresses is discussed. It is shown how to formulate a formal thermodynamic theory, the essential step being the correct definition of the mechanical coordinates so that they describe not only the change of shape but the work done in such processes as deformation, chemical action, crystallization or solution at a surface, or a phase change. A useful Gibbs function is defined to give the conditions of equilibrium for the above processes. The theory agrees with experiments on the α–β quartz transition and on the de-twinning of quartz.

1. INTRODUCTION

The problem of chemical or phase equilibrium under non-hydrostatic stresses is of considerable interest in the earth sciences as well as in solid state physics generally, and it has been the source of some considerable controversy[1–7]. The discussion centres on the possibility of the treatment of such equilibrium by formal thermodynamic procedures. For instance Kamb[2] stated 'that it is not possible usefully to associate a chemical potential or Gibbs free energy with a nonhydrostatically stressed solid'. Verhoogen[6], however, defined a Gibbs free energy by analogy with, and by extension of, the hydrostatic case and thus not rigorously deduced.

It will be shown how it is possible to formulate a formal thermodynamic theory and that the essential step is the correct definition of mechanical coordinates which must satisfy the requirement that they determine the work done on a system in a change which may be due to deformation, chemical action, crystallization or solution, or a phase change. Using the correct coordinates, a useful Gibbs function may be defined and used to give easily the conditions of equilibrium for such systems as a stressed solid in contact with a solution of the solid, quartz at the transition surface for the α–β transition and similar phase changes, coexsisting crystal twins, stressed solids into which homogeneous diffusion of a chemical component may occur. The application of this theory justifies the empirical theory of Thomas and Wooster[8] inferred from experimental work on de-twinning of quartz, and agrees with experimental work on the α–β quartz transition[9, 10].

2. DEVELOPMENT OF THE THEORY

When a phase changes its shape and bulk, this may be due to the material in it being deformed, or its structure being altered as in a phase change, or to the material being added to either at the external surfaces by crystallization from solution, or by homogeneous diffusion.

Although it is not necessary to the argument at this stage it is desirable to distinguish these processes as to whether they are coherent or not. Coherent here is used in the sense that a process is coherent if the atoms or ions of the basic solid material which were neighbours remain neighbours during the change. Pure deformations, homogeneous diffusion, phase changes such as α–β quartz, twinning of the Dauphiné type in quartz are all coherent, whereas crystallization from solution is incoherent.

In this paper the treatment will be confined to the case of infinitesimal deformations but it may easily be extended to that of finite deformations[6].

In the infinitesimal case, the virtual work done on the phase in a virtual change is given by

$$\delta W = T_{\alpha\beta} \int_{A_0} \tfrac{1}{2}\{\delta u_\alpha n_\beta + \delta u_\beta n_\alpha\}\, \mathrm{d}A_0 \tag{1}$$

where $T_{\alpha\beta}$, the stress tensor, is always taken as uniform throughout the phase. $\mathbf{u}$ is the vector displacement of the surface element $\mathrm{d}A_0$ occurring in the change, $\mathbf{n}$ is the unit outward drawn normal at $\mathrm{d}A_0$, and A_0 is the surface area of the reference state of the phase. Here and throughout the paper the summation convention is used for Greek suffices.

We thus define

$$V_{\alpha\beta} = \tfrac{1}{3}V_0\delta_{\alpha,\beta} + \int_{A_0} \tfrac{1}{2}\{u_\alpha n_\beta + u_\beta n_\alpha\}\, \mathrm{d}A_0 \tag{2}$$

and thus

$$\delta W = T_{\alpha\beta}\, \delta V_{\alpha\beta} \tag{3}$$

since the integration in 1 and 2 is over a fixed surface A_0. The first term in 2 is added as a constant of integration so that

$$V_{\alpha\alpha} = V \tag{4}$$

and this leads to the correct formulae in the hydrostatic case.

It is important to realize that $\mathbf{u}$ must be determined at every stage by the criterion that 1 and hence 3 must be satisfied. In the case of crystallization or solution at an external surface in contact with a solution, the pressure of the fluid is normal and if a mass δN is crystallized on a plane face it is easy to see that

$$\begin{aligned} \delta W &= -\delta N v P \\ &= \delta N v n_\alpha n_\beta T_{\alpha\beta} \end{aligned} \tag{5}$$

where $\mathbf{n}$ is the unit normal to the face and v is the specific volume of the solid, and hence

$$\delta V_{\alpha\beta} = \delta N v n_\alpha n_\beta \tag{6}$$

In an example where a solid phase in the shape of a cube is in equilibrium with two different solutions under different pressures, one of which is in contact with a face normal to the y axis and the other in contact with a face normal to the x axis, a virtual change may be imagined where the shape of the cube becomes that of a rectangular prism by the uniform solution of mass δN of the solid from the face whose outward normal is in the y direction, and the uniform crystallization of the same mass on the face whose outward normal is in the x direction. In this case it is easy to see that although the total mass is constant, and the temperature and stress remain the same, the $V_{\alpha\beta}$ change.

That is

$$\delta V_{yy} = -v\,\delta N, \qquad \delta V_{xx} = v\,\delta N \tag{7}$$

all other coordinates being unchanged.

Thus in general

$$V_{\alpha\beta}(T, T_{\alpha\beta}, \lambda N_1, \lambda N_2 \ldots) \neq \lambda V_{\alpha\beta}(T, T_{\alpha\beta}, N_1, N_2 \ldots) \tag{8}$$

although for a definite process such as described by equation 6 $V_{\alpha\beta}$ has an extensive property. This property leads to no difficulty in practice.

For a pure infinitesimal deformation, it is easy to see that

$$\delta V_{\alpha\beta} = V_0\,\delta e_{\alpha\beta} \tag{9}$$

where $e_{\alpha\beta}$ is the infinitesimal strain tensor.

For a coherent phase change, the structure alters in a definite fashion, for instance in the α–β quartz transition both forms have a unit cell which is a right-angled prism with a rhombus base of angles 60° and 120°, and when the transition occurs the prism axes retain the same direction, however the a and c lattice constants change in length. If we assume that these changes are infinitesimal in this case, it is easy to see that if mass δN of α-quartz changes to β-quartz at constant temperature and stress, then the change of dimensions and the work done are given by

$$\begin{aligned} \delta V_{zz} &= \delta N v(\Delta c/c) \\ \delta V_{xx} &= \delta V_{yy} = \delta N v(\Delta a/a) \end{aligned} \tag{10}$$

all other mechanical coordinates remaining constant. If these changes cannot be described by the infinitesimal deformation theory, the treatment is easily modified[6].

For homogeneous diffusion of a chemical component d into a solid at constant temperature and stress, the change of shape and bulk will be given by

$$\delta V_{\alpha\beta} = (\partial V_{\alpha\beta}/\partial N_{\mathrm{d}})_{T, T_{\alpha\beta}}\,\delta N_{\mathrm{d}} \tag{11}$$

As a final example, Dauphiné twinning of quartz may be considered. Here a twin may be produced by small rotations of the SiO_4 tetrahedra to give a structure which may be obtained by rotating the original form through 180° about the c axis[8]. Thus in a virtual change from one twin to another all

$$\delta V_{\alpha\beta} = 0 \tag{12}$$

since the shape of the unit cell as described above is invariant to the 180° rotation about the c axis.

Equations 6, 9, 10, 11 and 12 not only describe the change of shape and bulk of the phase, but also the work done in a virtual change at constant T and stress. A more detailed justification of this is given in ref. 6. In all these equations

$$\operatorname*{Lt}_{\delta N \to 0} (\delta V_{\alpha\beta}/\delta N)_{T, T_{\alpha\beta}} = v_{\alpha\beta} \tag{13}$$

where δN is an increment of mass describing the change, and $v_{\alpha\beta}$ depends on the actual process; in particular in 6 it depends on the direction in the stress field of the surface element at which crystallization or solution is being considered. It is easily seen that when this process is adequately defined, $v_{\alpha\beta}$ is a function of the intensive variables T, $T_{\alpha\beta}$ and the chemical concentrations.

Before concluding this section, the definition of $V_{\alpha\beta}$ may easily be extended so that 2 describes the coordinates of a multiphase system, and it can be an additive quantity in the sense that for the system

$$V_{\alpha\beta} = \sum_i V_{\alpha\beta}^{(i)} \tag{14}$$

where the summation is over the phases. This is so if no slip is allowed between solid phases. For a fluid/solid interface since any displacements **u** are allowed which describe the change of shape of the external surfaces bounding the fluid, we may choose **u** as continuous at the interface, and thus 14 is satisfied.

3. EQUILIBRIUM CONDITIONS VIA THE GIBBS FUNCTION

For a multiphase closed system

$$\mathrm{d}U = (T\,\mathrm{d}S + \sum_i T_{\alpha\beta}^{(i)}\mathrm{d}V_{\alpha\beta}^{(i)}) \tag{15}$$

if T is the same for all phases. Hence if we define a Gibbs function

$$G = U - TS - \sum_i T_{\alpha\beta}^{(i)} V_{\alpha\beta}^{(i)} \tag{16}$$

where U and S are the internal energy and entropy of the whole system, then

$$\mathrm{d}G = -S\,\mathrm{d}T - \sum_i V_{\alpha\beta}^{(i)}\,\mathrm{d}T_{\alpha\beta}^{(i)} \tag{17}$$

$$= 0$$

for

$$\mathrm{d}T = \mathrm{d}T_{\alpha\beta}^{(i)} = 0 \text{ all } i.$$

The equilibrium conditions may easily be obtained by the usual arguments. First for the case of crystallization or solution at a plane surface, consider a virtual change where T, $T_{\alpha\beta}$ for the solid, and P for the solution are all held constant. Then

$$\delta G = \delta N\{u - u_L - T(s - s_L) + P(v - v_L)\} \tag{18}$$
$$= 0$$

on using 6 and the fact that

$$-T_{\alpha\beta}n_\alpha n_\beta = P \tag{19}$$

u, s, v are the specific quantities for the solid, and u_L, s_L, v_L are the partial quantities for the solution. Hence for equilibrium

$$\mu_L = u - Ts + Pv \tag{20}$$

the well-known Gibbs result[11].

For the α–β quartz transition the equilibrium condition for the coexistence of the two forms at the transition surface is given by

$$\Delta u - T\,\Delta s - (T_{xx} + T_{yy})\,\Delta v_{xx} - T_{zz}\,\Delta v_{zz} = 0 \tag{21}$$

where

$$\begin{aligned} \Delta v_{zz} &= v\,\Delta c/c \\ \Delta v_{xx} &= \Delta v_{yy} = v\,\Delta a/a \end{aligned} \tag{22}$$

From 21 and 22 it is easy to prove that

$$\begin{aligned} \partial T_c/\partial P_x &= +v(\Delta a/a)/\Delta s \\ \partial T_c/\partial P_z &= +v(\Delta c/c)/\Delta s \end{aligned} \tag{23}$$

where e.g. $P_x = -T_{xx}$. From x-ray studies[10]

$$\Delta a/a = (2{\cdot}2 \pm 0{\cdot}2)\,\Delta c/c \tag{24}$$

and both quantities in 23 are positive. Since Δs is positive the lefthand expressions in 23 should be and are observed to be positive. Moreover, Coe and Paterson[9] showed experimentally that

$$(\partial T_c/\partial P_x):(\partial T_c/\partial P_z) = 2{\cdot}1 \tag{25}$$

in excellent agreement with 23 and 24.

Finally Thomas and Wooster[8] studied experimentally the de-twinning (Dauphiné) effect of non-hydrostatic stresses on quartz. Since for the twins 12 holds, the condition for coexistence of twins is

$$\Delta f = 0 \tag{26}$$

where f is the specific Helmholtz free energy, and this is consistent with and justifies the empirical principle that de-twinning occurs best when maximal energy is stored.

Thus, in general, equilibrium conditions for chemical and phase changes may be determined by the Gibbs function, the quantities of the form $v_{\alpha\beta}$ of 13, being thermodynamic 'unknowables' just as are the specific and partial volumes in hydrostatics, and must for a particular process be determined experimentally or from a theoretical model.

ACKNOWLEDGEMENTS

The author is grateful to the Conference authorities and to the Council of the University of Canterbury for financial support to enable him to travel from New Zealand to Cardiff.

REFERENCES

1 K. Ito, *J. Fac. Sci. Univ. Tokyo*, **16** (2), 347 (1966).
2 W. B. Kamb, *J. Geol.* **67**, 153 (1959).
3 W. B. Kamb, *J. Geophys. Res.* **66**, 259 (1961).
4 M. Kumazawa, *J. Earth Sci., Univ. Nagoya*, **11**, 145 (1963).
5 A. G. McLellan, *Proc. Roy. Soc. A*, **307**, 1 (1968).
6 A. G. McLellan, *Proc. Roy. Soc. A*, **314**, 443 (1970).
7 J. Verhoogen, *Trans. Am. Geophys. Un.* **32**, 251 (1951).
8 L. A. Thomas and W. A. Wooster, *Proc. Roy. Soc. A*, **208**, 43 (1951).

[9] R. S. Coe and M. S. Paterson, *J. Geophys. Res.* **74**, 4921 (1969).
[10] C. Berger, L. Eyrand, M. Richard and R. Rivière, *Bull. Soc. Chim. France No. 106*, 628 (1966).
[11] J. W. Gibbs, *Scientific Papers 1*. Longmans, Green: London (1906).

SOME ASPECTS OF THE THERMODYNAMIC LIMIT

A. Münster

Institute of Theoretical Physical Chemistry, University of Frankfurt (Main), Graefstrasse 38, 6 Frankfurt (Main), Western Germany

ABSTRACT

The historical development of statistical mechanics over the last hundred years is outlined, culminating in the work of Ruelle and of Fisher in 1963. The 'thermodynamic limit' is defined and conditions on intermolecular potentials and limit theorems are next examined as preliminaries to a detailed consideration of the associated equivalence problem. In conclusion the limits of applicability of thermodynamics are noted.

I. INTRODUCTION

In the early days of statistical mechanics, it was recognized that the laws of thermodynamics could be derived from molecular theory only for systems containing a large number of particles, i.e. for macroscopic systems. This was pointed out by Boltzmann[1] when introducing Stirling's formula, and it has been stated more explicitly by Gibbs[2] in the foreword to his famous treatise on *Elementary Principles in Statistical Mechanics.* He writes: 'The laws of thermodynamics, as empirically determined, express the approximate and probable behaviour of systems of a great number of particles, or, more precisely, they express the laws of mechanics as they appear to beings who have not the fineness of perception to enable them to appreciate quantities of the order of magnitude of those which relate to single particles, and who cannot repeat their experiments often enough to obtain any but the most probable results. The laws of statistical mechanics apply to systems of any number of degrees of freedom and are exact. . . . The laws of thermodynamics may be easily obtained from the principles of statistical mechanics of which they are an incomplete expression'. Gibbs[2] has also shown that in defining analogues of thermodynamic quantities conceptual difficulties can be avoided only if the number of particles is assumed to be very large. This may be illustrated by means of an example. If the statistical analogue of the entropy is defined by the canonical and the grand canonical ensemble respectively, it turns out that these quantities are different from each other. It can be shown, however, that the grand canonical entropy may be expressed as the most probable value of the canonical quantity plus a term depending on the number of particles which becomes completely negligible for macroscopic systems. It is seen that Gibbs always had in mind a macroscopic but finite system. He was in fact not interested in the problem how the laws of thermodynamics could be obtained from statistical mechanics in full mathematical rigour. This, however, is precisely the question on which the more recent development has focused attention and which nowadays is usually termed as 'asymptotic problem of statistical mechanics'.

In this new field the first great success was achieved in 1922 by Darwin and Fowler[3,4] who treated quantum systems of non-interacting particles. The analogous problem in classical statistics was solved by Khinchin[5] whose book was published in the U.S.A. in 1949. In spite of significant differences in the details of the mathematical technique, the underlying basic idea is the same in both methods, namely the use of a generating function which is nothing else but the canonical partition function. This entails that in the final result the thermodynamic relation between the entropy and the Helmholtz free energy appears as the leading term of an asymptotic expansion. For later considerations, it will be useful to write down the essential results in a rather simple form. Let $\Phi(E)$ denote the Gibbs energy function, $Q(\beta)$ with $\beta = 1/kT$ the canonical partition function and $\bar{E}$ the average energy of a system of the canonical ensemble. Then in a first step one obtains for large N, if N is the number of particles,

$$W(E) = \frac{\exp(-\beta E)\exp(\Phi(E))}{Q(\beta)} = \frac{1}{(2\pi B)^{\frac{1}{2}}}\exp\left(-\frac{1}{2}\frac{(E-\bar{E})^2}{B}\right) + O(N^{-1}) \quad (1)$$

where $W(E)$ is the probability density (frequency function) of the energy in the canonical ensemble. It is seen that for very large systems, this probability density tends to a Gaussian. Now let us assume that the average energy $\bar{E}$ equals the microcanonical energy $E^{\star}$. Then, on taking logarithms and dividing by N, we obtain from equation 1

$$N^{-1}\Phi(\bar{E}) = N^{-1}\ln Q(\beta) + N^{-1}\beta\bar{E} + O(N^{-1}\ln N) \quad (2)$$

or, introducing thermodynamic quantities per particle,

$$Ts(e, v) = -f(T, v) + e + O(N^{-1}\ln N) \quad (3)$$

Now, in the case of non-interacting particles, for N, $E \to \infty$ at $v = \text{const.}$ on the RHS the existence of the limit functions is trivial. Thus passing to the limit we may safely conclude that on the LHS the limit function exists as well which immediately leads to the well-known thermodynamic relation between entropy and Helmholtz free energy.

This was the state of affairs in 1949. It is now easy to see that the extension of the aforementioned results to interacting particles and other ensembles will meet with two additional problems. In the first place, the derivation of equation 1 is based on the central limit theorem of probability theory which means that the total energy E is assumed to be a sum of N independent random variables. Obviously this is no longer true in the case of interacting particles. Secondly, for interacting particles the existence of the limit functions on the RHS of equations 2 and 3 is by no means trivial but will depend rather on the details of the intermolecular interaction.

In the following period which starts in 1949 with van Hove's[6] paper on a limit theorem for the canonical ensemble, several attempts were made to solve the aforementioned problems for special cases. Some important results have been obtained but we cannot go into the details here and the interested reader is referred to the literature[7]. We shall turn rather to the most recent development which was induced by the fundamental work of Ruelle[8] and Fisher[9] in 1963.

II. DEFINITIONS. STATEMENT OF THE PROBLEM

In what follows we shall frequently denote extensive parameters by X_i and intensive parameters by P_i. Remembering that in the Gibbs fundamental equation the entropy is a function of extensive state variables only, any Massieu–Planck function Φ_k depending on k intensive parameters appears as a k-fold Legendre transform of the entropy, viz.

$$\Phi_k = S - \sum_{i=1}^{k} P_i X_i \tag{4}$$

satisfying the differential equation

$$\mathrm{d}\Phi_k = -\sum_{i=1}^{k} X_i \mathrm{d}P_i + \sum_{j=k+1}^{r} P_j \mathrm{d}X_j \tag{5}$$

where

$$\frac{\partial \Phi_k}{\partial P_i} = -X_i, \qquad \frac{\partial \Phi_k}{\partial X_j} = P_j \qquad (i \leqslant k < j) \tag{6}$$

Turning to statistical mechanics we first observe that any statistical ensemble depending on k intensive parameters generates the analogue of a Massieu–Planck function Φ_k by the equation

$$\Phi_k = \ln \Xi_k \tag{7}$$

where Ξ_k is the partition function of the ensemble and, for the sake of simplicity, Boltzmann's constant has been put equal to unity. The function Φ_k as defined by equation 7 satisfies a differential equation of the form of equation 5. On the other hand, in the semi-classical approximation we have

$$\exp(\Phi_k) = \int_0^\infty \ldots \int_0^\infty \exp\left(-\sum_{i=1}^{k} P_i X_i\right) \exp(\Phi) \prod \mathrm{d}X_i \tag{8}$$

That is to say that the partition function of the k-ensemble is the k-fold Laplace transform of the microcanonical partition function.

It is easy to see and it can be shown explicitly that for finite systems equations 4 and 8 are not consistent with each other. We therefore consider a sequence of systems characterized by the variable $X_r^{(n)} = nX_r^{(1)}$ where $X_r^{(1)}$ is some fixed reference value and $n \geqslant 1$. Furthermore we introduce the notation

$$\varphi_k^{(n)} = \Phi_k^{(n)}/X_r^{(n)}, \qquad x_j = X_j^{(n)}/X_r^{(n)} \tag{9}$$

By '*thermodynamic limit*' we mean the limit process

$$n \to \infty \quad \text{for} \quad \begin{aligned} P_i &= \text{const.}, \quad (i = 1, 2, \ldots, k) \\ x_j &= \text{const.}, \quad (j = k + 1, \ldots n - 1) \end{aligned} \tag{10}$$

Thus what has been called above the 'asymptotic problem' amounts to a study of the thermodynamic limit comprising the following subproblems:

(a) Existence of the limit $\lim_{n\to\infty} \varphi_k^{(n)}$ and its derivatives (limit theorems)
(b) Consistency of the thermodynamic quantities defined with the aid of various ensembles (equivalence problem)
(c) Thermodynamic stability
(d) Phase transitions.

These questions are closely connected with one another and therefore cannot be discussed independently. Here, however, we are mainly concerned with the equivalence problem. Questions (a) and (c) will be touched upon rather briefly whereas we shall leave out the problem of phase transitions.

III. CONDITIONS ON INTERMOLECULAR POTENTIALS. LIMIT THEOREMS

The existence of the limit functions will depend on assumptions on the intermolecular potentials. For one thing, the forces of attraction could be so strong that the system 'collapses' as the number of particles increases, so that φ_k^{∞} diverges to $+\infty$. On the other hand, the forces of repulsion could decrease so little with increasing separation that φ_k^{∞} diverges to $-\infty$. Following Fisher[9], we therefore make the following assumptions:

(A) Condition of stability

For the potential energy of a system of N interacting particles $U^{(N)}$ there exists a lower bound

$$U^{(N)} \geqslant -Nu_A \tag{11}$$

for all values of the coordinates and all N, where u_A is a fixed natural number. Potentials which satisfy condition A are called stable potentials.

(B1) Condition of weak tempering

Let us imagine that the N particles have been split up into two groups of N_1 and N_2 particles with coordinates $\vec{q}_i$ and $\vec{q}_j'$ respectively. The interaction energy between these groups will be denoted by $U^{(N_1, N_2)}(\vec{q}_i; \vec{q}_j')$. Then for all N_1 and N_2 and some arbitrary fixed R_0 and u_B and $\varepsilon > 0$

$$U^{(N_1, N_2)}(\vec{q}_i; \vec{q}_j') \leqslant \frac{N_1 N_2 u_B}{R^{3+\varepsilon}} \tag{12}$$

if $|\vec{q}_i - \vec{q}_j'| \geqslant R \geqslant R_0$ holds for all i and j and $(N_1 + N_2)/R^{3+\varepsilon}$ is sufficiently small.

(B2) Condition of strong tempering

Under the same assumptions as before

$$U^{(N_1, N_2)}(\vec{q}_i, \vec{q}_j) \leqslant 0 \tag{13}$$

whenever $|\vec{q}_i - \vec{q}_j'| \geqslant R_0$ for all i and j.

On the basis of the assumptions A and Bl, Fisher[9] has proved that for the canonical ensemble the limit φ_k^∞ exists and is a continuous convex and non-decreasing function of the volume per particle v. From the convexity property which is essentially equivalent to thermodynamic stability conditions, existence and properties of the derivatives are obtained by the use of well-known theorems on convex functions[10]. It cannot be shown, however, that we have

$$\frac{\partial \varphi_k^\infty}{\partial v} = \lim_{n \to \infty} \frac{\partial \varphi_k^{(n)}}{\partial v} \tag{14}$$

as required for a complete statistical foundation of thermodynamics. Mathematically this shortcoming arises from the fact that convexity can only be proved for the limit function φ_k^∞.

Analogous results have been obtained for the grand canonical ensemble[9] whereas van der Linden[11] has proved a limit theorem for the microcanonical ensemble on the basis of assumptions A and B2.

IV. EQUIVALENCE PROBLEM

The formal nature of the equivalence problem becomes immediately obvious from the comparison of equations 4 and 8. We have to prove that, at the thermodynamic limit, the Laplace transformation of the statistical partition functions reduces asymptotically to the Legendre transformation of the Massieu–Planck functions. We may still ask, however, for the physical meaning of this problem. There are two possible answers which of course only elucidate two aspects of the same situation. In the first place we may consider that any statistical ensemble is linked conceptually with a particular physical situation of the system of interest. The microcanonical ensemble represents an isolated system, the canonical ensemble a system in contact with a heat bath, and similarly for the other ensembles. The laws of thermodynamics, however, do not depend on a particular physical situation. Thus it must be shown that these 'boundary conditions' become meaningless at the thermodynamic limit. On the other hand we may look on the asymptotic expansion 3. It can be shown that the higher order terms arise from statistical fluctuations. In thermodynamics, however, this concept does not appear at all. We therefore must prove that fluctuations vanish asymptotically at the thermodynamic limit.

The aforementioned physical aspects correspond closely to the mathematical methods which in recent years have been applied to the general treatment of the equivalence problem. In the following we shall describe the essential features of these methods from a more physical point of view, again leaving out the mathematical details.

(i) Method of van der Linden and Mazur

The first method due to van der Linden and Mazur[11, 12] is based first on assumptions A and B2 of Section III and secondly on certain inequalities for the microcanonical ensemble. These essentially state that the phase volume is a never negative and never decreasing convex function of the energy. In a first step it is shown that the limit theorem for the microcanonical ensemble mentioned in Section III generates an analogous limit theorem for the canonical ensemble. We briefly sketch the main idea of the proof. Let us imagine that the system is divided into two subvolumes with numbers of particles and energies N_1, E_1 and $N - N_1, E - E_1$ respectively. Then using a well-known formula for the microcanonical ensemble and the aforementioned inequalities it is easily established that we must have for the phase volume

$$\Omega^\star(E) \geqslant \int_0^E \exp(\Phi_1(E_1))\,\Omega_2^\star(E - E_1)\,dE_1 \tag{15}$$

This property leads in fact to the limit theorem for the microcanonical ensemble. Since the right hand member is a convolution integral, we obtain with the aid of the convolution theorem for the Helmholtz free energy per particle

$$-N\,\mathrm{f}(N) \leqslant -N_1\,\mathrm{f}(N_1) - (N - N_1)\,\mathrm{f}(N - N_1) \tag{16}$$

Functions satisfying an inequality of this form are called subadditive functions. For these we have a limit theorem[13] which combined with assumption A leads to the desired limit theorem for the canonical ensemble.

In dealing with the equivalence problem itself the following ingenious device is used. First we generalize the above division to a division into n subsystems which are assumed to have equal energies $E^\star$, equal numbers of particles $N^\star$ and equal volumes $V^\star$. Then, according to the theory of the microcanonical ensemble and condition B2 it must be true that

$$s_n^\star(E^\star, N^\star, V^\star) \geqslant {}^{(n)}s^\star(E^\star, N^\star, V^\star) \geqslant s^\star(E^\star, N^\star, V^\star) \tag{17}$$

The left hand member of this inequality is simply the entropy of the original system divided by the number of subsystems. In the second member the interaction between the subsystems is neglected but not the energy distribution between them, i.e. they are assumed to be in thermal equilibrium. In the last member the subsystems are considered to be isolated. Thus we have three different physical situations and we shall show that these differences become meaningless at the thermodynamic limit.

The second member of expression 17 refers to an 'ideal system'. Hence, considering the subsystems as 'particles', we may apply Khinchin's argument. Passing to the limit $n \to \infty$ for fixed $N^\star$ [which does not affect the last member of 17] we then obtain from equation 3 and the limit theorem for the microcanonical ensemble

$$s^\infty(e, v) \geqslant -\beta\,\mathrm{f}(\beta, v, N) + \beta\bar{e}(\beta, v, N) \geqslant s(e, v, N) \tag{18}$$

where $N^\star$ has been replaced by N and quantities per particle have been introduced. Next we study the thermodynamic limit of a subsystem ($N \to \infty$, $\beta = \text{const.}$, $v = \text{const.}$). Under this process the first member becomes simply $s^\infty[e(\beta, v), v]$ and the last member converges to the same limit.

Hence, making use of the limit theorem for the canonical ensemble, we obtain

$$s^{\infty}[e(\beta, v), v] = -\beta f^{\infty}(\beta, v) + \beta e^{\infty}(\beta, v) \tag{19}$$

which is a special case of the thermodynamic relation 4. This argument can be carried through for all conceivable ensembles although the details become more complicated. Under stronger assumptions about the intermolecular forces (two-body central forces satisfying an additional condition) van der Linden[14] has also proved equation 14.

(ii) Method of Still, Haubold and Münster

The other approach due to Still, Haubold and Münster[15] is essentially a generalization of Khinchin's method to interacting particles and any conceivable ensemble. In comparison with the first method, assumptions about intermolecular forces here enter only via limit theorems but they are not used explicitly. Furthermore thermodynamic relations are obtained as leading terms of asymptotic expansions. The underlying assumptions are essentially the existence of a limit theorem for the function $\varphi_{l+1}(P)$ and the exclusion of phase transition points.

As explained in section I, we have first to prove that the frequency function $W(y)$ with

$$y = (X - \overline{X})/B^{\frac{1}{2}} \tag{20}$$

tends to a Gaussian at the thermodynamic limit. This problem is of considerable interest in itself since in many applications the frequency function is assumed to be a Gaussian. Now it can be shown without difficulty that the sequence of characteristic functions

$$\psi^{(n)}(t) = \int_{-\infty}^{+\infty} e^{ity}\, W^{(n)}(y)\, dy \tag{21}$$

for $n \to \infty$ converges to a Gaussian in any finite t-interval and, moreover, that the moments of the frequency function converge to the moments of a Gaussian with dispersion one. From this result, however, we cannot conclude that the functions

$$W^{(n)}(y) = \frac{1}{2\pi}\int_{-\infty}^{+\infty} e^{-ity}\, \psi^{(n)}(t)\, dt \tag{22}$$

converge to a Gaussian. The reason is that convergence of the $\psi^{(n)}(t)$ has been proved only for finite t-intervals. Thus in performing the integral of equation 22 we must have some knowledge about the 'tail' of the integrand. As shown by Mazur and van der Linden[16] this knowledge is indeed available for the canonical ensemble but unfortunately this is not true in the general case. To overcome this difficulty we replace the original frequency function by a 'smoothed' frequency function $W_s^{(n)}(y)$ which is obtained from the former by convolution with an appropriate smoothing function $S^{(n)}(y)$ in such a way that the smoothed frequency function remains normalized and non-negative. Thus we define

$$W_s^{(n)}(y) = \int_{-\infty}^{+\infty} W^{(n)}(y')\, S^{(n)}(y - y')\, dy' \tag{23}$$

It is easy to see that the smoothing procedure is nothing else but a local averaging. This, however, means that the characteristic function is damped for large values of t. Since we know $S^{(n)}(y)$ explicitly this gives us the required knowledge about the tail. Therefore we are now able to show that for large n we have

$$W_s^{(n)}(X) = \frac{1}{(2\pi B)^{\frac{1}{2}}} \exp\left(-\frac{1}{2}\frac{(X-\bar{X})^2}{B}\right)[1 + O(n^{-1}\ln^2 n)] \tag{24}$$

which is the generalization of equation 1.

Although 'smoothing' or 'coarse graining' is nothing new in statistical mechanics we still wish to give some justification for the above procedure. This is achieved by showing rigorously that the functions $W^{(n)}(y)$ and $W_s^{(n)}(y)$, at the thermodynamic limit, cannot be distinguished by any conceivable macroscopic measurement. From equation 24 we obtain by a straight-forward argument

$$\varphi_l^n = \varphi_{l+1}^n + P_{l+1}\bar{x}_{l+1} + O(n^{-1}\ln^2 n) \tag{25}$$

which on passing to the limit $n \to \infty$ yields again the thermodynamic relation 4 and at the same time proves the existence of φ_l^∞. Formally equation 25 was proved only for smoothed functions. But it can be shown rigorously that $\varphi_{l+1.s}^{(n)} = \varphi_{l+1}^{(n)}$ for all n. Then it follows from the uniqueness of the Legendre transformation, that there is one and only one function φ_l^∞ satisfying the thermodynamic equation 4. Hence we are left with two possibilities. Either the 'true' limit function φ_l^∞ does not exist or does not satisfy equation 4. Then 'smoothing' is a necessary step in the foundation of thermodynamics. Or the 'true' limit function φ_l^∞ satisfies equation 4 as well, then it cannot be distinguished from the smoothed function. Indeed if we assume existence of a limit theorem for φ_l (which has been avoided so far) then we can prove that $\varphi_l^\infty = \varphi_{ls}^\infty$ almost everywhere.

Similar results are obtained for the derivatives of φ_l^∞. In particular it can be shown that differentiation and passage to the thermodynamic limit commute. From this thermodynamic stability conditions are obtained in the general form first given by Schottky, Ulich and Wagner[17] and it is shown that they are identical with conditions for the statistical fluctuations.

V. LIMITS OF APPLICABILITY OF THERMODYNAMICS

Up to now we have exclusively treated the formal derivation of thermodynamics from statistical mechanics which necessarily requires the introduction of the concept of an infinitely large system. However, experimental physics always deals with finite systems, so that we can expect this to impose certain bounds on the applicability of the thermodynamic formalism. These are determined by the neglected terms in equations 3 and 25 which have their origin in statistical fluctuations. Thus the domain of validity of thermodynamics is determined by the condition that fluctuations are negligible with respect to the accuracy of measurements. In most cases this will be true for macroscopic systems. Planck[18], however, has already discussed an interesting case where thermodynamics breaks down even for finite macroscopic systems.

Consider a Debye crystal, say a cube of edge length 1 cm ($N \approx 10^{21}$) with characteristic temperature $\Theta_D = 10^2\,°K$ which has been cooled to a temperature of $10^{-5}\,°K$. Then from a general formula derived by Münster[19] we have

$$\frac{1}{T} = \left(\frac{\partial S}{\partial E}\right)_0 + \frac{1}{2}\bar{E}_D^2\left(\frac{\partial^3 S}{\partial E^3}\right)_0 \frac{\overline{(E - \bar{E}_D)^2}}{\bar{E}_D^2} \tag{26}$$

with

$$\bar{E}_D = \frac{3\pi^4}{5} Nk \frac{T^4}{\Theta_D^3}, \; \bar{E} = \bar{E}_D + \frac{9}{8} Nk\Theta_D \tag{27}$$

Numerical calculation shows that both terms of the RHS of equation 26 are of the same order of magnitude $10^5[°K]^{-1}$. Thus, as stated already by Planck[18], near the absolute zero the concepts of entropy and absolute temperature are no longer uniquely definable because fluctuations of energy become important. It is obvious that this has some bearing on the interpretation of Nernst's heat theorem[20].

REFERENCES

1. L. Boltzmann, *S.B. Akad. Wiss. Wien*, **76**, 373 (1877) [*Wissensch. Abh.*, Vol. II, p 177. New York (1968)].
2. J. W. Gibbs, *Elementary Principles in Statistical Mechanics*. New Haven (1902) [Collected works, Vol. II, p VIII, 203, New Haven (1948)].
3. C. S. Darwin and R. H. Fowler, *Phil. Mag.* **44**, 450 and 823 (1922); **45**, 1 (1923).
4. R. H. Fowler, *Statistical Mechanics*, 2nd ed. Cambridge University Press: London (1936).
5. A. J. Khinchin, *Mathematical Foundation of Statistical Mechanics*. New York (1949).
6. L. van Hove, *Physica*, **15**, 951 (1949).
7. A. Münster, *Statistical Thermodynamics*, Vol. I. Springer: Berlin and New York (1969).
8. D. Ruelle, *Helv. Phys. Acta*, **36**, 183 (1963).
9. M. E. Fisher, *Arch. Ration. Mech. Anal.* **17**, 377 (1964).
10. G. H. Hardy, J. E. Littlewood and G. Pólya, *Inequalities*, 2nd ed., Cambridge University Press: London (1952).
11. J. van der Linden, *Physica*, **32**, 642 (1966).
12. J. van der Linden and P. Mazur, *Physica*, **36**, 491 (1967).
13. E. Hille and R. S. Phillips, *Functional Analysis and Semi-groups*. Providence (1957).
14. J. van der Linden, *Physica*, **38**, 173 (1968).
15. E. Still, K. Haubold and A. Münster, *Z. Naturforschung*, **24a**, 201 and 412 (1969).
16. P. Mazur and J. van der Linden, *J. Math. Phys.* **4**, 271 (1963).
17. W. Schottky, H. Ulich and C. Wagner, *Thermodynamik*. Springer: Berlin (1929).
18. M. Planck, *Theorie der Wärmestrahlung*, 4te Aufl., p 218. Springer: Berlin (1921).
19. A. Münster, *Z. Phys.* **136**, 179 (1953).
20. M. J. Klein, *The Laws of Thermodynamics*. (*R.C. Scu. Intern. Fis. 'Enrico Fermi'*, Corso X) Bologna (1960).

DETERMINATION OF THE FREE VOLUME AND THE ENTROPY BY A MONTE CARLO METHOD

EVELINE M. GOSLING AND K. SINGER

Department of Chemistry, Royal Holloway College, Englefield Green, Surrey

ABSTRACT

The configurational integral can in principle, though not in practice, be determined from the ratio of accepted to rejected moves ($\langle a_N(v)\rangle$) in the Monte Carlo method of Metropolis *et al.*, if the moves consist of simultaneous random displacements of N particles uniformly distributed over the sample volume v. In the usual Monte Carlo process only one particle at a time is displaced within a small volume. If this volume is suitably chosen (v_D), the acceptance ratio ($\langle a_1(v_D)\rangle$) determines the mean free volume per particle ($v_f = v_D\langle a_1(v_D)\rangle$). Evidence is presented supporting the approximate validity of the relationship $\langle a_N(v)\rangle = \langle a_1(v_D)\rangle^N$ which permits the evaluation of the configurational integral. The entropies calculated in this way for a system of 108 Lennard-Jones particles with parameters corresponding to argon, are in good agreement with the experimental values for solid and liquid argon. The results indicate that the full amount of the communal entropy appears on fusion.

In the well-known Monte Carlo (MC) method of Metropolis *et al.*[1] certain many-dimensional integrals containing the Boltzmann factor as weighting function are evaluated by means of the following procedure: configurations of a system of N particles are generated by small random displacements of single particles; if $\mathbf{r}_\alpha^N$ is the configuration after α moves and the next trial move leads to $\mathbf{r}_{\alpha'}^N$, then this move is rejected (by setting $\mathbf{r}_{\alpha+1}^N = \mathbf{r}_\alpha^N$) if and only if the difference between the potential energies $\Phi(\mathbf{r}_{\alpha'}^N) - \Phi(\mathbf{r}_\alpha^N) = \Delta_\alpha\Phi > 0$ and $\exp(-\Delta_\alpha\Phi/kT) < \eta$, η being a number chosen at random between 0 and 1; otherwise the move is accepted (by setting $\mathbf{r}_{\alpha+1}^N = \mathbf{r}_{\alpha'}^N$).

If in this process configurations were generated by simultaneous displacements of all N particles to random positions uniformly distributed over the sample volume v, the configurational integral could in principle be evaluated from the mean acceptance ratio for the trial moves

$$\frac{\int_0^L \cdots \int_0^L \exp\left[-\{\Phi(\mathbf{r}^N) - \Phi_{\text{min.}}\}/kT\right]\, d\mathbf{r}^N}{\int_0^L \cdots \int_0^L d\mathbf{r}^N} = \frac{Q'(NvT)\exp(\Phi_{\text{min.}}/kT)}{v^N}$$

$$= \lim_{M\to\infty} \sum_{\alpha=1}^{M} \frac{\exp\left[-\{\Phi(\mathbf{r}_\alpha^N) - \Phi_{\text{min.}}\}/kT\right]}{M} = \frac{\text{no. of accepted moves}}{\text{no. of attempted moves}}$$

$$= \langle a_N(v)\rangle \qquad (1)$$

Table 1. Comparison of the statistics of consecutive acceptances (runs) of MC moves with the corresponding statistics for independent events of equal probability, calculated with the formula[2] $f(M) = M(1-p)\,p^a/(1-p^a)$, with $p = \langle a_1(v_d)\rangle$, $v_d = (4\pi/3)\,R_d^3$

	$T = 50{\cdot}0$		$V = 24{\cdot}24$		$N = 108$	
a	2	8	14	20	26	32
$f_{obs.}$	51375	7922	2619	1032	403	175
$f_{calc.}$	51376	7900	2619	1010	410	170
	$T = 97{\cdot}0$		$V = 28{\cdot}48$		$N = 108$	
a	2	11	20	29	38	47
$f_{obs.}$	54540	5881	1776	668	241	103
$f_{calc.}$	54540	5906	1808	654	250	97
	$T = 97{\cdot}0$		$V = 28{\cdot}48$		$N = 108$	
a	2	11	20	29	38	47
$f_{obs.}$	69528	6461	1632	488	129	38
$f_{calc.}$	69482	6439	1626	473	143	44
	$T = 97{\cdot}0$		$V = 28{\cdot}48$		$N = 108$	
a	2	4	6	8	10	12
$f_{obs.}$	36428	9089	2898	943	315	95
$f_{calc.}$	36371	9157	2840	931	311	104
	$T = 97{\cdot}0$		$V = 28{\cdot}48$		$N = 108$	
a	2	4	6	8	10	12
$f_{obs.}$	32407	7427	2068	607	175	53
$f_{calc.}$	32517	7379	2031	585	171	50
	$T = 97{\cdot}0$		$V = 28{\cdot}48$		$N = 108$	
a	2	4	6	8	10	12
$f_{obs.}$	28700	5757	1421	333	91	27
$f_{calc.}$	28544	5711	1360	336	84	21
	$T = 97{\cdot}0$		$V = 28{\cdot}48$		$N = 108$	
a	2	4	6	8	10	12
$f_{obs.}$	25043	4419	938	207	36	13
$f_{calc.}$	25165	4446	919	196	42	9
	$T = 97{\cdot}0$		$V = 28{\cdot}48$		$N = 108$	
a	1	2	3	4		
$f_{obs.}$	2842	57	0	0		
$f_{calc.}$	2842	60	1	0		
	$T = 97{\cdot}0$		$V = 28{\cdot}48$		$N = 256$	
a	1	2	3	4		
$f_{obs.}$	6402	131	0	0		
$f_{calc.}$	6402	134	3	0		
	$T = 298{\cdot}2$		$V = 69{\cdot}98$		$N = 108$	
a	2	24	46	68	90	112
$f_{obs.}$	60098	3117	980	342	143	62
$f_{calc.}$	60082	3109	948	355	142	58
	$T = 298{\cdot}2$		$V = 69{\cdot}98$		$N = 108$	
a	1	2	3	4	5	6
$f_{obs.}$	24269	3460	567	94	17	5
$f_{calc.}$	24269	3467	564	94	16	3

T is temperature in °K.
V is volume in cm^3 $mole^{-1}$.
N is number of particles.
R_d is range of displacements in Å.
M is number of attempted displacements.
a is number of displacements accepted consecutively (length of run).
$f_{obs.}$ and $f_{calc.}$ are the observed and calculated frequences of the number of displacements accepted consecutively.

$M = 12800$			$R_d = 0{\cdot}065$		
38	44	50	56	62	68
80	32	14	7	3	0
71	30	12	5	2	1
$M = 127600$			$R_d = 0{\cdot}069$		
56	65	74	83	92	101
47	24	9	6	2	1
38	15	6	2	1	0
$M = 169600$			$R_d = 0{\cdot}086$		
56	65	74	83	92	
12	4	2	1	0	
13	4	1	0	0	
$M = 170800$			$R_d = 0{\cdot}310$		
14	16	18	20	22	
29	15	5	1	0	
35	12	4	1	0	
$M = 170800$			$R_d = 0{\cdot}344$		
14	16				
16	3				
15	4				
$M = 171200$			$R_d = 0{\cdot}379$		
14	16	18	20		
7	3	3	0		
5	1	0	0		
$M = 171600$			$R_d = 0{\cdot}413$		
14	16	18	20		
5	3	1	0		
2	0	0	0		
$M = 132800$			$R_d = 2{\cdot}07$		
$M = 299008$			$R_d = 2{\cdot}07$		
$M = 127600$			$R_d = 0{\cdot}116$		
134	156	178	200	222	244
31	11	8	5	2	1
24	10	4	2	1	0
$M = 145600$			$R_d = R_D = 11{\cdot}62$		
7	8	9			
1	1	0			
0	0	0			

($\Phi_{\text{min.}}$ is the lowest potential energy attained, and the angular bracket indicates an ensemble average; $v = L^3$).

This method is impracticable because the acceptance ratio $\langle a_N(v)\rangle$ is too small and the numerical labour too great. Both difficulties are overcome in the usual process by the displacement of one particle at a time within a small volume $v_d(\ll v)$, centred on the current position of the particle. The mean acceptance ratio, $\langle a_1(v_d)\rangle$, has no physical significance here. When the range of attempted displacements is increased, the product $v_d\langle a_1(v_d)\rangle$ increases up to a limiting value v_d^* such that for $v_D \geqslant v_d^*$ the probability of acceptance of moves ending outside v_D it is virtually zero. For solids and liquids not too near the critical point, $v_D < v/N$; for fluids of lower density even large displacements are accepted, i.e. $v_D = v$. The acceptance ratio $\langle a_1(v_D)\rangle$ is the MC estimate of the integral

$$\langle \int \exp(-(\Phi^1(\mathbf{r}_1) - \Phi^1_{\text{min.}})/kT)\,\mathrm{d}\mathbf{r}_1\rangle_{N-1}/v_D \tag{2}$$

where $\Phi^1(\mathbf{r}_1)$ and $\Phi^1_{\text{min.}}$ are the energy of particle 1 and its minimum value within v_D, and the subscript $N-1$ indicates an average over all configurations of particles 2 to N. The numerator of expression 2 can be interpreted as defining the free volume, v_f.

The evaluation of the configurational integral is based on the proposition that, to a sufficient approximation,

$$\langle a_N(v)\rangle = \langle a_1(v)\rangle^N \tag{3}$$

First, it will be noted that the acceptance ratio for displacements over the whole volume v can be determined from

$$v_D\langle a_1(v_D)\rangle = v\langle a_1(v)\rangle \tag{4}$$

(for dense systems $\langle a_1(v)\rangle$ is too small to be obtained directly.) Secondly, since the MC evaluation of a many-dimensional integral can be effected either by sampling the different dimensions simultaneously or consecutively (in cycles)

$$\langle a_N(v)\rangle = \langle a_1(v)^N\rangle \tag{5}$$

i.e. the mean probability of acceptance of a simultaneous displacement of N particles equals the joint probability of acceptance of N successive moves of one particle at a time. Thirdly, numerical experiments indicate that

$$\langle a_1(v)^N\rangle = \langle a_1(v)\rangle^N \tag{6}$$

The statistics of runs of acceptances in the MC process have been compared with the known statistics of runs of independent events[2]. The results are summarized in *Table 1*. To obtain runs of length N (= 108), it is necessary to reduce the range of displacements, R_d, drastically. Alternatively, when R_d equals the radius of the spherical volume v_D, only short runs are obtained. Under both these conditions the observed statistics agree closely with the statistics of independent events. Although this evidence for the validity of equation 6 is not entirely conclusive, it does indicate that the two sides of 6

Table 2. Data for the calculation of entropies and comparison with the entropy of argon

No.	V $cm^3 mol^{-1}$	T °K	M 10^3	$\langle a_1(v_D)\rangle$	$N_0 v_f$ $cm^3 mol^{-1}$	R_D Å	$\frac{\langle \Phi \rangle - \Phi_{min.}}{NkT}$ A	B	S_{MC} J. °K^{-1} mol^{-1} A	B	S_{argon}
1. a, f	24·24 (c)	50·0	266	0·0240	0·092	1·18	0·166	0·535	22·6	24·6	23·3 d
2. b, f	24·24 (c)	86·3	176	0·00885	0·167	1·67	0·124	—	33·4	—	—
3. b, f	24·51 (c)	86·3	266	0·00891	0·171	1·97	0·147	0·757	34·5	38·2	38·8 d
4. b, g	28·48 (l)	97·0	398	0·0210	0·467	2·07	0·204	0·613	52·3	55·7	55·9 e
5. c, g	28·48 (l)	97·0	298	0·0214	0·476	2·07	0·180	0·631	51·4	55·8	55·9 e
6. b, g	29·69 (l)	97·0	266	0·0265	0·615	2·10	0·216	—	54·4	—	58·4 e
7. a, g	29·69 (l)	97·0	133	0·0281	0·652	2·10	0·229	0·595	55·0	57·6	58·4 e
8. b, g	28·48 (l)	108·0	266	0·0224	0·499	2·07	0·245	0·570	54·3	57·0	58·5 e
9. a, g	28·48 (l)	136·0	266	0·0232	0·516	2·27	0·196	0·545	59·1	62·0	63·3 e
10. b, g	38·01 (l)	136·0	180	0·0804	3·04	2·46	0·180	0·475	71·7	74·1	74·6 e
11. b, h	69·98 (g)	173·2	146	0·121	8·47	11·62	0·245	—	83·7	—	91·0 d
12. b, h	69·98 (g)	298·2	297	0·1634	12·35	11·62	0·138	0·306	92·2	93·6	100·6 d

a: $N = 108$, LJ parameters $\varepsilon/k = 117{\cdot}2$, $\sigma = 3{\cdot}405$ Å (ref. 3).
b: $N = 108$, LJ parameters $\varepsilon/k = 119{\cdot}8$, $\sigma = 3{\cdot}405$ Å (ref. 7).
c: $N = 256$, LJ parameters $\varepsilon/k = 119{\cdot}8$, $\sigma = 3{\cdot}405$ Å (ref. 7).
d: interpolated from data in ref. 8.
e: interpolated from data in ref. 9.
f: $C = 1$ g: $C = N^N/N!$ h: $C = 1/N!$
A: method A; B: method B.

differ by less than an order of magnitude. A small error would have little effect on the entropy: since

$$S/Nk = (1/N)\log(v^N\langle a_1(v)^N\rangle) + \text{other terms},$$

a discrepancy by a factor ten between the two sides of 6 would (with $N = 108$) lead to an error of less than three per cent in the configurational entropy. The configurational partition function can therefore be written in the form

$$Q = C\exp(-\Phi_{\text{min.}}/kT)(v\langle a_1(v)\rangle)^N = C\exp(-\Phi_{\text{min.}}/kT)\,v_f^N \qquad (7)$$

The value of C depends on the nature of the system: for localized particles (in solids), $C = 1$; for dense fluids, when $v_D < v/N$, $C = N^N/N!$; for fluids of lower density, when $v_D = v$, $C = 1/N!$. There is a (not very extensive) region below the critical temperature where $v/N < v_D < v$, when the present method is not applicable.

Entropies have been calculated with the formula

$$S/Nk = (\langle\Phi\rangle - \Phi_{\text{min.}})/NkT + 1{\cdot}5\{1 + \log(2\pi mkT/h^2)\} + (1/N)\log C + \log v_f \qquad (9)$$

for a model consisting of 108 (or 256) Lennard–Jones (12–6) particles with parameters corresponding to argon. As a first approximation $\Phi_{\text{min.}}$ was equated to the lowest total potential energy attained in the MC experiment (method A). *Table 2* shows that this leads to entropies which are consistently by about eight per cent lower than the experimental values. This procedure is also unsatisfactory because $(\langle\Phi\rangle - \Phi_{\text{min.}})/NkT$ estimated in this manner varies as $N^{-1/2}$ instead of being an intensive quantity. This objection does not hold if $\Phi_{\text{min.}}$ is set equal to $(N/2)\langle\Phi^1_{\text{min.}}\rangle$ where $\langle\Phi^1_{\text{min.}}\rangle$ is the mean value of the minimum of the potential energy of a particle in the field of the other particles. This step may be further justified as follows. Equation 2 can be written in the form

$$\langle a_1(v)\rangle = \langle \textstyle\int F_1(\mathbf{r}^N)\,d\mathbf{r}_1\rangle_{N-1}/v$$

$$\equiv \textstyle\int..\int (F_N/F_1)\,d\mathbf{r}_2..d\mathbf{r}_N \int F_1\,d\mathbf{r}_1 / \int..\int (F_N/F_1)\,d\mathbf{r}_2..d\mathbf{r}_N \int d\mathbf{r}_1$$

where $F_N = \exp(-(\Phi(\mathbf{r}^N) - \Phi_{\text{min.}})/kT)$. If equations 5 and 6 are to be valid, one must equate F_1 to

$$\exp(-(\Phi^1(\mathbf{r}^N) - \Phi^1_{\text{min.}}(\mathbf{r}_2,..,\mathbf{r}_N))/2kT)$$

and

$$\Phi_{\text{min.}} \text{ to } N\langle\Phi^1_{\text{min.}}\rangle/2,$$

where

$$\Phi^1(\mathbf{r}^N) = \sum_{j=2}^{N} \varphi(|\mathbf{r}_j - \mathbf{r}_1|)$$

and

$$\Phi^1_{\text{min.}}(\mathbf{r}_2,...,\mathbf{r}_N) = \sum_{j=2}^{N} \varphi(|\mathbf{r}_i - \mathbf{r}_{1(\text{min.})}(\mathbf{r}_2,...,\mathbf{r}_N)|).$$

$\langle \Phi^1_{min.} \rangle$ is determined from separate MC data. The values so obtained are in satisfactory agreement with the experimental entropies for solid and liquid argon (*Table 2*, method B). The persistent discrepancy for the gaseous V, T point is almost certainly due to the difficulty of determining $\langle \Phi^1_{min.} \rangle$ in this case.

In spite of this uncertainty, which calls for further investigation, the present methods has some advantages compared with the MC method employed in the impressive work of Hansen and Verlet[4]. The statistical convergence of the acceptance ratio is excellent and the statistical error in the entropies after 2 to 3×10^5 moves is of the order of one per cent. This compares very favourably with the accuracy of the MC estimate of pressure which is required in methods based on the integration of $p\,dv$ along isotherms[4]. Moreover, whereas the evaluation by the latter method of the free energy and entropy at one V,T point requires MC pressure data for $\simeq 12$ V,T points, the proposed method requires data for one V,T point only. A similar saving of computational labour applies to MC calculations of the free energy of mixtures: these have been evaluated by integration with respect to continuously varying parameters along an isotherm[5]. The most promising field of application of the present method is perhaps the calculation of the free volume and the entropy of different molecular and ionic species in mixtures.

REFERENCES

1 N. Metropolis, A. W. Rosenbluth, M. N. Rosenbluth, A. H. Teller and E. Teller, *J. Chem. Phys.* **21**, 1087 (1953).
2 W. Feller, *Probability Theory and its Applications*, p 266. Wiley: New York (1950).
3 I. R. McDonald and K. Singer, *J. Chem. Phys.* **50**, 2308 (1969).
4 J.-P. Hansen and L. Verlet, *Phys. Rev.* **184**, 151 (1969).
5 K. Singer, *Chem. Phys. Letters*, **3**, 164 (1969).
6 I. R. McDonald and K. Singer, *J. Chem. Phys.* **47**, 4766 (1967).
7 A. Michels, H. Wijker and H. Wijker, *Physica*, **15**, 627 (1949).
8 F. Din, *Thermodynamic Functions of Gases* (edited by F. Din), Vol. II, pp 181–185. Butterworth: London (1956).
9 R. K. Crawford and W. B. Daniels, *J. Chem. Phys.* **50**, 3171 (1969).

THE THERMODYNAMIC DESCRIPTION OF PHASE TRANSITIONS*

MICHAEL H. COOPERSMITH

Department of Physics and Center for Advanced Studies, University of Virginia, McCormick Road, Charlottesville, Va 22903, U.S.A.

ABSTRACT

The historical fact of the ability of a system to avoid showing a phase transition when taken along a suitable thermodynamic path is examined and a way offered of looking at the 'critical point' problem in keeping with various prevailing viewpoints. The semi-phenomenological view is given, followed by exact results, perturbation expansions, a description of equilibrium phenomena and illustrations which involve the Riemann surface but omit the property of convexity.

1. INTRODUCTION

Despite the recent heroic attempts to construct microscopic models whose exact solutions show phase transitions, there remains a peculiar gap between the equations of state which arise from these models and the experimentally defined relationships among the thermodynamic parameters. This gap is all the more striking due to the fact that good quantitative agreement with experiment can be obtained for a wide range of parameters but there is a decided lack of qualitative agreement near just those values of the parameters which appear to be most interesting. These are the values near a critical point of the system.

That the neighbourhood of a critical point is indeed the most interesting place to look at a phase transition is not only a latter day phenomenon but also an historical fact. In looking over the literature pertaining to changes of state, one is immediately struck by the concern of early workers in the field with the fact that, above certain critical values of the parameters, no change of state of the system can be observed. Thus, as early as 1822, one Baron de la Tour[1] observed that various gases including carbon dioxide and water vapour exist in only one phase when the temperature is above the value now commonly known as the critical temperature. Later experiments confirmed these results more quantitatively[2] but de la Tour's work had already uncovered the basic phenomenon described by van der Waals[3] in his now classic equation of state. This 'continuity of liquid and vapour states' (as van der Walls referred to it) or the ability of a system to avoid showing a phase transition when taken along a suitable thermodynamic path represents the most outstanding feature of the critical region and may, in fact, be thought of as the essence of a critical point. We offer here a way of

* As the author was unable to be present at the Conference, this paper was read by title only.

looking at the problem which ties together various prevailing viewpoints. Much of this material is excerpted from an article of the same title by the author[4].

II. THE SEMI-PHENOMENOLOGICAL VIEW

The idea of a semi-phenomenological explanation for the continuity of states originated, as already mentioned, with van der Waals, was continued by Weiss[5] for an explanation of the Curie point of ferromagnetism, and reached its culmination in the generalization of the van der Waals equation of state proposed by Widom[6] a few years ago. The prefix 'semi' is justified because one can concoct microscopic models which yield the van der Waals and Weiss equations of state in some approximation. These models have been discussed extensively in the literature[7] and will not be taken up here. We shall only note that the physical basis of these equations of state comes about from the competition between energy and entropy in the Helmholtz free energy, $F = E - TS$, the one arising from the weak, long range attraction between the particles and the other from the strong, short range repulsion (generally taken as a hard core). Widom's homogeneous equations of state do not have such a simple explanation[8] but if we accept the idea of homogeneity as at least an approximation to some general physical principle, then these equations too may be described as semi-phenomenological.

In order to put this viewpoint in its proper perspective, we refer to an analysis of free energies (i.e. thermodynamic potentials) which show the phenomenon of the critical point as given by the author[9]. This treatment relies on the sole assumption of analytical continuation of the thermodynamic potential in both its variables. The main results are:

(1) In the neighbourhood of the Curie (critical) point, the Weiss theory free energy $F(H, T)$ can be represented by an algebraic function of third degree. The chemical potential corresponding to the van der Waals equation of state in the neighbourhood of the critical point has the same form when regarded as a function of reduced pressure and temperature after a transformation which makes the reduced vapour pressure equal to zero for all temperatures. (We use the word 'reduced' to denote the difference between a variable and its critical value.)

(2) Magnetic free energies which are algebraic functions homogeneous in the applied field and reduced temperature form a special case of the analogue of Widom's homogeneous functions.

The assumption that a thermodynamic potential is an algebraic function of its natural variables is clearly an approximation since (a) the critical exponents must be rational and (b) no logarithmic behaviour can be obtained. However, we shall see that algebraic functions act as a framework upon which to hang the more general description which will include logarithmic behaviour as well as other singularities.

III. EXACT RESULTS

It seems increasingly clear that a simple closed analytical expression for the full ferromagnetic problem with applied field or the related problem of a fluid is by its very nature impossible to obtain. The reason boils down to a question of representation of a function. In ref. 9 it is seen that even the simplest algebraic functions which show the characteristic behaviour of a critical point cannot be represented in closed form but are defined as implicit functions by a polynomial equation. The representation is then realized by expanding the individual terms in the polynomial about any point of interest, thus obtaining a representation of the function about that point. This representation is in the form of an ordinary Taylor expansion if the point is non-singular and a Taylor expansion in the variable $\zeta = z^{1/n}$ if the point is a singular one. Here z is the original variable, n is the degree of the defining polynomial (more accurately, the order of the singular point) and the expansion may start with a negative but finite power of ζ. In the neighbourhood of a critical point, we are dealing with a function of two variables and the points of interest lie on either the coexistence curve or the critical isotherm. The critical point itself is excluded. For the magnetic case, the expansions are of the form

$$F(H, T) = \sum_{n=0}^{\infty} a_n(T - T_c)\, H^n \tag{1}$$

for the first case and

$$F(H, T) = \sum_{n=0}^{\infty} b_n(H)\,(T - T_c)^n \tag{2}$$

for the second. The coefficients a and b are themselves expandable in the form required for a singular (branch) point in one variable, so that we have

$$a_n(z) = \sum_{i=r_n}^{\infty} A_{ni} z^{i/t_n} \tag{3}$$

and

$$b_n(z) = \sum_{i=s_n}^{\infty} B_{ni} z^{i/u_n} \tag{4}$$

where r and s are positive or negative integers and t and u are the orders of the branch points. The numbers r_n/t_n and s_n/u_n (referred to as α_n and β_n in ref. 9) are the leading exponents and are commonly known as the critical exponents for the nth derivative of the free energy with respect to H and T respectively. For the two-dimensional Ising model, only the asymptotic forms of the leading terms of $a_0^{\pm}(T - T_c)$ and $a_1^{-}(T - T_c)$ have been found analytically. (The plus and minus signs refer to whether T is greater than or less than T_c, that is, whether we are above or below the critical point.) Onsager's well known solution[10] gives $a_0^{+}(T - T_c)$ and $a_0^{-}(T - T_c)$ both asymptotically equal to $(T - T_c)^2 \ln(|T - T_c|)$ while the published solution of Yang[11] for the spontaneous magnetization gives the asymptotic form of $a_1^{-}(T - T_c)$ as $(T_c - T)^{\frac{1}{8}}$. None of the other as and bs are known even asymptotically†. Nevertheless, the mean field theories together with some recent work on perturbation expansions to be discussed in the next section contain what we believe to be the essential features of phase transitions.

† Various numerical calculations for the a_ns and b_ns up to $n = 8$ have been made. See refs. 12 and 13.

IV. PERTURBATION EXPANSIONS

The expansion about the ideal gas pressure with the fugacity as the variable has been returned to recently by a number of authors, among them Andreev[14], Langer[15] and Fisher[16]. It is shown that if one considers an expansion in the fugacity of real droplets (as opposed to mathematical clusters), one is led to a series which represents a function which has an essential singularity on the real axis for all temperatures. Thus, the conjecture[17] that there be a singularity at the condensation point (as opposed to the critical point) has been verified in a physically appealing approximation. Furthermore, Fisher[16] has also analysed a one-dimensional model which shows these properties.

We consider the Riemann surface of an algebraic function of the type proposed in ref. 9. On a neighbourhood of this Riemann surface not containing a branch point, one may define a single valued function[18]. We now imagine a function defined on the Riemann surface which has an essential singularity at the crunode (see ref. 9) and is linear in a neighbourhood of the crunode. If the essential singularity be chosen so that its value and the values of all its derivatives approaching from the real axis are zero (the most common example is e^{-1/z^2}) then we will have a function of the type which has been found for the droplet model[16]. A specific example is the following. Let the algebraic function of two variables be denoted by $f(z, T - T_c)$ where z is the variable appropriate to the physical system (reduced chemical potential for a fluid, magnetic field for a ferromagnet). Let $g(z)$ be a function which has an essential singularity of the type previously mentioned. The function $f[g(z) + z, T - T_c]$ then has all the properties ascribed by Fisher to the grand potential of his one-dimensional model. Furthermore, the line of essential singularities is continued above the transition temperature as in that model.

V. DESCRIPTION OF EQUILIBRIUM PHENOMENA

In the preceding sections, we have attempted to show the relationship between the currently accepted views of critical phenomena and the picture presented by the author in ref. 9. In that paper, explicit use was made of algebraic functions as a representation of that thermodynamic potential whose natural variables are intensive. But we have already seen in Section II that algebraic functions cannot provide all the observed behaviour. We now pose the following question. What class of functions can exhibit all the known behaviour and yet is elementary in the sense that it has already been studied?

Before discussing this problem we might remark that there is a natural analogue of this situation in the description of decay of physical states. That problem might be thought of in the same light as the present one. That is, we have available an elementary function (the decreasing exponential) whose properties are obviously well known and which describes the qualitative physical behaviour. Nevertheless, as in the present case, the connection between the underlying microscopic dynamics and the observed macroscopic behaviour has never been completely established†.

What then are the elementary functions describing critical phenomena? We have already seen that algebraic functions alone are capable of describing many aspects of the two phase region. The main feature of algebraic functions of two variables which is most useful for the present situation is the breaking up of the Riemann surface into disjoint pieces when one of the variables has the particular value which corresponds physically to the critical value of the thermodynamic variable. This feature may be kept by working with functions defined on the Riemann surface of the appropriate algebraic function. Since the only missing aspect of the equilibrium situation is the possible logarithmic behaviour of the coefficients $a_n(T - T_c)$ defined in equation 1, we need only assume that the functions defined on the Riemann surface are Abelian integrals or products of analytical functions and Abelian integrals. Such functions are elementary in the sense defined above. Since an Abelian integral is defined as the integral of a rational function of an algebraic function and its variable on the Riemann surface of that algebraic function, then its singularities include poles and logarithmic singularities, but no essential singularities‡.

With these functions, we may now have $a_0(T - T_c) \sim (T - T_c) \ln (T - T_c)$ as in the two-dimensional Ising model. Also, since the singularities of an Abelian integral (in one variable) are isolated, we preserve the property of analytical continuation in two variables from one side of the two phase region to the other. The same procedure may be used to define a function with a non-algebraic branch point on the Riemann surface of an algebraic function. This leads to irrational critical exponents. Finally, as in Section IV, we can add an essential singularity at the crunode by forming the function $F[g(z) + z, T - T_c]$ where $F(x, y)$ is an Abelian integral defined on the Riemann surface of an algebraic function of x and y and $g(z)$ is a function with an essential singularity of the type $\exp(-1/z^2)$ at the origin.

VI. ILLUSTRATIONS

The following free energies serve to illustrate the method of description of the preceding section. Because of space limitations, we shall not analyse the Riemann surfaces but merely write down the defining equations for the algebraic functions and indicate briefly how the analysis should be done.

The first example is the familiar molecular field theory with the reduced thermodynamic potential f given as a function of the applied field H and reduced temperature $T - T_c = t$. Since f is known to be homogeneous and the exponent δ is equal to three, we are dealing with a third order algebraic function. Its defining equation is

---

† An example of this is the decay of an arbitrary state to its thermodynamic equilibrium value. While various approximation schemes enable one to calculate the relaxation time (i.e. the normalization of the exponent governing the time dependence), its actual existence has never been rigorously demonstrated.

‡ Since we will not make any computational use of Abelian integrals here, we merely catalogue some of their properties. The interested reader is referred to G. A. Bliss, ref. 18 for the precise definition of Abelian integrals as well as theorems relating to them.

$$P(f; H, t) = f^3 + \tfrac{3}{2}t^2 f^2 + (\tfrac{9}{16}t^4 + \tfrac{27}{8}H^2 t)f + \tfrac{81}{64}H^4 + \tfrac{9}{32}H^2 t^3 = 0 \quad (5)$$

Computation of the discriminant of P (the remainder when P is divided by $\partial P/\partial f$) shows that the branch points are located at $H = \pm\tfrac{2}{3}(-t)^{\frac{3}{2}}$ in agreement with the positions of the end points of the loops in the molecular field theory. The critical exponents are obtained by inserting the expansions of equations 1 and 2 into the defining equation 5 and equating coefficients of equal powers of H or t. We find easily:

$$\begin{aligned} a_0^- &= -\tfrac{3}{4}(-t)^2 & & \\ a_1^- &= \pm 3^{\frac{1}{2}}(-t)^{\frac{1}{2}} & a_0^+ &= 0 \\ a_2^- &= -\tfrac{1}{4}(-t)^{-1} & a_1^+ &= 0 \\ b_0 &= -(\tfrac{3}{4})^{\frac{4}{3}}H^{\frac{4}{3}} & a_2^+ &= -\tfrac{1}{2}t^{-1} \\ b_1 &= (\tfrac{3}{2})^{\frac{2}{3}}H^{\frac{2}{3}} & & \\ b_2 &= -\tfrac{1}{2} & & \end{aligned} \quad (6)$$

where the plus and minus signs on the *a*s refer to the sign of t, that is whether we are above or below the critical point. For $H = 0$, factorization of P into $f(f + \tfrac{3}{4}t^2)^2$ indicates that the Riemann surface of f becomes disjoint on the coexistence curve and that there is a crunode at $H = 0$.

The second example is the somewhat less physical three-dimensional ideal Bose gas. Following Gunton and Buckingham, we look at the free energy for fixed density and variable density of particles in the zero momentum state. Within a phase factor, the order parameter or Bose moment ψ is the square root of the density of zero momentum particles and the conjugate variable ζ is the Bose field. The reduced thermodynamic potential f as a function of ζ and t is homogeneous and the exponent δ is equal to five so we are now dealing with a fifth order algebraic function. Its defining equation is

$$\begin{aligned} P(f; \zeta, t) = f^5 - \tfrac{1}{6}t^3 f^4 + \tfrac{25}{12}t\zeta^2 f^3 - \tfrac{1}{3}t^4\zeta^2 f^2 + \tfrac{125}{108}t^2\zeta^4 f + (\tfrac{5}{6})^5\zeta^6 \\ - \tfrac{1}{6}t^5\zeta^4 = 0 \end{aligned} \quad (7)$$

The root of the discriminant locate the branch points at $\zeta = \pm(16/5^{\frac{5}{2}})(-t)^{\frac{5}{2}}$ and $\zeta = 0$. This sticking of two of the branch points at $\zeta = 0$ shows that the expansion of equation 1 must be changed to

$$f(\zeta, t) = \sum_{n=0}^{\infty} a_{n/2}^-(-t)\,\zeta^{n/2} \quad (8)$$

for $t < 0$. The *a*s and *b*s are found to be:

$$\begin{aligned} a_0^- &= 0 & a_0^+ &= \tfrac{1}{6}t^3 \\ a_{\frac{1}{2}}^- &= 0 & a_1^+ &= 0 \\ a_1^- &= \pm(-t)^{\frac{1}{2}} & a_2^+ &= -\tfrac{1}{2}t^2 \\ a_{\frac{3}{2}}^- &= \pm\tfrac{1}{3}(-t)^{-\frac{3}{4}} & b_0 &= -\tfrac{5}{6}\zeta^{\frac{6}{5}} \\ a_2^- &= \tfrac{1}{8}(-t)^{-2} & b_1 &= \tfrac{1}{2}\zeta^{\frac{4}{5}} \\ & & b_2 &= \tfrac{1}{10}\zeta^{\frac{2}{5}} \end{aligned} \quad (9)$$

For $\zeta = 0$, factorization of P into $f^4(f - \frac{1}{6}t^3)$ again shows that there is a crunode at $\zeta = 0$. However, the fact that $a_{\frac{3}{2}}^-$ exists means that the analogue of the susceptibility is infinite at $\zeta = 0$ for all $t < 0$.

The branch points in the neighbourhood of the origin illustrate another aspect of the critical region which has become popular recently, mainly with the experimentalists. Since the two physical branches of f which form the crunode for $t < 0$ must be connected to the other non-physical branches, there must be a branch on each of them which approaches the origin of the H-plane as $t \to 0$. The trajectory of these branch points in the $t - H$ manifold is the spinodal line originally referred to by van der Waals[19]. Since the spinodal line lies on a non-physical (perhaps, metastable) sheet, it is unreachable by an equilibrium experiment. Nevertheless, by measuring a thermodynamic quantity along, say, an isotherm close to the critical isotherm, one may approach the branch point on the spinodal line and thus measure the associated exponent. The most obvious thing to measure along an isotherm is the magnetization as a function of H[20, 21]. In the molecular field theory, this behaves as

$$M \sim [H - H(t)]^{\frac{1}{2}} \tag{10}$$

where $H(t)$ is the trajectory of branch points. One may also approach the spinodal line along other curves; typically, for a liquid, C_V may be measured along a non-critical isochore.

VII. CONCLUSION

We have attempted to take some of the mystery out of the mathematical description of thermodynamic potentials near a critical point by showing that there are elementary functions which possess all the properties normally ascribed to these potentials in the vicinity of such a point. For the thermodynamic potential whose natural variables are naturally intensive, these properties may be summarized as follows. The thermodynamic potential is denoted by F and its reduced natural variables by z and t.

(1) F is a single valued function of z and t with a singularity in z and t separately at $z = t = 0$ (critical point).

(2) $(\partial F/\partial z)_t$ has a discontinuity along the line $z = 0$ for $t < 0$ (coexistence curve).

(3) F has no singularity in t along the line $t = 0$ and $z \neq 0$ (critical isotherm) nor anywhere else in a neighbourhood $(0 < z < z_0,\ 0 < t < t_0)$ of the critical point.

The elementary analytical function which has the above properties is a single valued analytic function defined on the Riemann surface of an algebraic function. The algebraic function $\mathrm{f}(z, t)$ has the following characteristics.

(1) f has a branch point in z and t separately at $z = t = 0$.

(2) The two smallest real branches of f form a crunode (crossing point) in the z-plane along the line $z = 0$ for $t < 0$.

(3) The sheets of f contain branch points whose trajectories in the $z - t$ manifold pass through the origin $(z = t = 0)$. That part of the Riemann surface of the true thermodynamic potential which is relevant in the neighbourhood of the critical point is given by the Riemann surface of f.

None of the other sheets of the true thermodynamic potential affect the critical exponents.

This last characteristic of f is illustrated by the molecular field theory. The equation of state $\tanh[(H + M)/kT] = M$ has an infinite number of solutions (sheets of M as a function of H) but only the three sheets which contain the two branch points (cusps) which come together at the origin determine the critical behaviour. Thus, the relevant part of the Riemann surface for the molecular field is given by the Riemann surface of the third order algebraic function of Section VI. This Riemann surface is seen to be connected in the H-plane for $t \neq 0$ (the defining polynomial is irreducible in H for $t \neq 0$) but becomes disjoint when $t = 0$ (the polynomial factorizes when $t = 0$).

The property of convexity of the thermodynamic potential has been left out of the above list since it plays no direct role in determining the qualitative properties of the Riemann surface. It is, rather, a feature which must be taken into account when constructing specific examples of thermodynamic potentials and will, of course, play a role in determining the relationships among the critical exponents, the so-called scaling laws.

REFERENCES

1. C. de la Tour, *Ann. Chim. (Phys.)*, [2], **21**, 127 (1822).
2. T. Andrews, *Trans. Roy. Soc. A*, **178**, 45 (1887).
3. J. D. van der Waals, *Dissertation* (Leiden), unpublished.
4. M. H. Coopersmith, *Advanc. Chem. Phys.* In press.
5. P. Weiss, *J. Phys., France*, **6**, 667 (1907).
6. B. Widom, *J. Chem. Phys.* **43**, 3898 (1965).
7. M. Kac, G. E. Uhlenback and P. C. Hemmer, *J. Math. Phys.* **4**, 216 (1963).
8. L. P. Kadanoff, *Physics (N.Y.)*, **2**, 263 (1966).
9. M. H. Coopersmith, *Phys. Rev.* **172**, 230 (1968).
10. L. Onsager, *Phys. Rev.* **65**, 117 (1944).
11. C. N. Yang, *Phys. Rev.* **85**, 808 (1952).
12. C. Domb and D. L. Hunter, *Proc. Phys. Soc.* **86**, 1147 (1965).
13. G. A. Baker, H. E. Gilbert, J. Eve and G. S. Rushbrooke, *Physics Letters*, **22**, 269 (1966).
14. A. F. Andreev, *Soviet Phys. JETP*, **18**, 1415 (1964).
15. J. S. Langer, *Ann. Phys. (N.Y.)*, **41**, 108 (1967).
16. M. E. Fisher, *Physics (N.Y.)*, **3**, 255 (1967).
17. S. F. Streeter and J. E. Mayer, *J. Chem. Phys.* **7**, 1025 (1939).
18. G. A. Bliss, *Algebraic Functions*, p 29. Dover: New York (1966).
19. J. S. Rowlinson, *Liquids and Liquid Mixtures*, p 205. Butterworths: London (1959).
20. J. T. Ho and J. D. Litster, *Phys. Rev. Letters*, **22**, 603 (1969).
21. M. H. Coopersmith, *Phys. Letters*, **30A**, 192 (1969).

C. THERMO-MECHANICS

SOME RECENT RESULTS IN THE THEORY OF FADING MEMORY

BERNARD D. COLEMAN

Mellon Institute and Department of Mathematics, Carnegie–Mellon University, Pittsburgh, Pa 15213, U.S.A.

ABSTRACT

An outline is given of the phenomenological theory of fading memory recently explored by V. J. Mizel and the author. The theory provides a general framework in which one can derive the restrictions which the second law of thermodynamics places on the constitutive equations of materials with memory.

1. INTRODUCTION

In theories of the dynamical behaviour of continua, there are several ways of describing the dissipative effects which, in addition to heat conduction, accompany deformation. The oldest way is to employ a viscous stress which depends on the rate of strain, as is done in the theory of Navier–Stokes fluids. In another description of dissipation, one postulates the existence of internal state variables which influence the stress and obey differential equations in which the strain appears. A third approach is to assume that the entire past history of the strain influences the stress in a manner compatible with a general postulate of smoothness or 'principle of fading memory'.

Experience in high-polymer physics shows that the mechanical behaviour of many materials, including polymer melts and solutions, as well as amorphous, crosslinked solids and semi-crystalline plastics, is more easily described within the theory of materials with fading memory than by theories of the viscous-stress type, which do not account for gradual stress-relaxation, or by theories which rest on a finite number of internal state variables and which, therefore, give rise to discrete relaxation spectra when linearized.

Some years ago, Walter Noll and I proposed a systematic procedure for rendering explicit the restrictions which the second law places on constitutive relations[1]. The procedure was easily applied in theories of materials of the viscous-stress type[1, 2] and in theories which employ evolution equations for internal state variables[5]; these applications did not yield results which a physicist would consider surprising and were presented as attempts at clarification, with the emphasis laid upon logical relations. Implementation of the procedure in the theory of materials with memory was a different matter, however, for it there led to conclusions[4] which, although not anticipated by other arguments, have recently been shown to have important bearing on wave propagation[5] and dynamical stability[6, 7]. Here I should like to discuss the restrictions which the second law places on the response functionals of materials with memory. Although it is possible to develop analogous theories for materials with 'permanent memory'†, I emphasize materials which possess 'fading memory' in the sense that configurations experienced in the recent past have a stronger influence on the present values of the stress and free energy than configurations experienced in the distant past.

2. PROCESSES, CONSTITUTIVE ASSUMPTIONS, AND THE SECOND LAW

Let a fixed reference configuration $\mathscr{R}$ be assigned for the body $\mathscr{B}$ under consideration, and identify each of the material points X of $\mathscr{B}$ with the place ξ in space that X occupies when $\mathscr{B}$ has the configuration $\mathscr{R}$. A *thermodynamic process* of $\mathscr{B}$ is a collection of functions of ξ and time compatible with the laws of balance of momentum and energy. For the materials covered by the present theory, each process consists of eight functions: (1) the *motion* $\boldsymbol{\chi}$, with $\mathbf{x} = \boldsymbol{\chi}(\xi, t)$ called the *position* at time t of the material point located at ξ in $\mathscr{R}$, (2) the local absolute *temperature* θ, which is assumed to be positive, (3) the symmetric *stress tensor* $\mathbf{T}$ of Cauchy, (4) the specific *internal energy* ε, per unit mass, (5) the specific *entropy* η, per unit mass, (6) the *heat flux vector* $\mathbf{q}$, (7) the *body force* $\mathbf{b}$, per unit mass (exerted on $\mathscr{B}$ at $\mathbf{x} = \boldsymbol{\chi}(\xi, t)$ by the 'external world', i.e. by other bodies which do not intersect $\mathscr{B}$), and (8) the *rate of heat supply* r (i.e. the radiation energy, per unit mass and unit time, absorbed by $\mathscr{B}$ at $\mathbf{x} = \boldsymbol{\chi}(\xi, t)$, and furnished by the 'external world'). The first six of these functions determine the process, for once $\boldsymbol{\chi}$, θ, $\mathbf{T}$, ε, η, and $\mathbf{q}$ have been specified for all ξ and t, the functions $\mathbf{b}$ and r are determined[1] by the requirement that the process shall obey the laws of balance of momentum and energy, which state that, for each part $\mathscr{P}$ of $\mathscr{B}$ and each time t,

$$\frac{\mathrm{d}}{\mathrm{d}t}\int_{\mathscr{P}} \dot{\mathbf{x}}\,\mathrm{d}m = \int_{\mathscr{P}} \mathbf{b}\,\mathrm{d}m + \int_{\partial\mathscr{P}} \mathbf{Tn}\,\mathrm{d}a \tag{2.1}$$

and

$$\frac{\mathrm{d}}{\mathrm{d}t}\int_{\mathscr{P}} (\varepsilon + \tfrac{1}{2}\,\dot{\mathbf{x}}\cdot\dot{\mathbf{x}})\,\mathrm{d}m = \int_{\partial\mathscr{P}} (\dot{\mathbf{x}}\cdot\mathbf{b} + r)\,\mathrm{d}m + \int_{\partial\mathscr{P}} (\dot{\mathbf{x}}\cdot\mathbf{Tn}) - \mathbf{q}\cdot\mathbf{n})\,\mathrm{d}a \tag{2.2}$$

In these equations $\mathrm{d}m$ is the element of mass in the body, $\partial\mathscr{P}$ is the surface of $\mathscr{P}$ in the configuration at time t, $\mathrm{d}a$ is the element of surface area, $\mathbf{n}$ is the exterior unit normal vector to $\partial\mathscr{P}$, and the superposed dots denote material time-derivatives.

The specific free energy‡ ψ is defined by

$$\psi = \varepsilon - \theta\eta \tag{2.3}$$

† See, for examples, Owen's discussion of the thermodynamics of materials with elastic range[8], Owen and Williams's theory of rate-independent materials[9], and a recent essay[10], in which Owen and I generalize the present treatment.

‡ Also called the 'Helmholtz free energy per unit mass'.

The *deformation gradient* $\mathbf{F}$ is the gradient of $\boldsymbol{\chi}(\boldsymbol{\xi}, t)$ with respect to $\boldsymbol{\xi}$:

$$\mathbf{F} = \mathbf{F}(\boldsymbol{\xi}, t) = \nabla_{\xi}\boldsymbol{\chi}(\boldsymbol{\xi}, t) \tag{2.4}$$

It is assumed that $\mathbf{F}$ is non-singular; hence

$$\det \mathbf{F} \neq 0 \tag{2.5}$$

The Piola–Kirchhoff tensor, $\mathbf{S} = \mathbf{S}(\boldsymbol{\xi}, t)$, is defined by

$$\mathbf{S} = (1/\rho)\,\mathbf{T}\mathbf{F}^{T^{-1}}, \qquad \rho\mathbf{S}\mathbf{F}^{T} = \mathbf{T} \tag{2.6}$$

with ρ the mass density. I denote by $\mathbf{g}$ the spatial gradient of the temperature, i.e. the gradient of θ considered as a function of the present position $\mathbf{x} = \boldsymbol{\chi}(\boldsymbol{\xi}, t)$:

$$\mathbf{g} = \nabla_{\mathbf{x}}\theta(\boldsymbol{\chi}^{-1}(\mathbf{x}, t), t) \quad \text{or} \quad \mathbf{F}^{T}\mathbf{g} = \nabla_{\xi}\theta(\boldsymbol{\xi}, t) \tag{2.7}$$

Now, let $\mathbf{F}(\tau)$ and $\theta(\tau)$ be the deformation gradient and temperature at time τ at a fixed material point X. The functions $\mathbf{F}^t$ and θ^t, defined by

$$\mathbf{F}^t(s) = \mathbf{F}(t-s), \quad \theta^t(s) = \theta(t-s), \quad 0 \leqslant s < \infty \tag{2.8}$$

are called the *history up to t of the deformation gradient at* X and the *history up to t of the temperature at* X; $\mathbf{F}^t$ maps $[0, \infty)$ into the set of non-singular tensors, while θ^t maps $[0, \infty)$ into the set of positive numbers.

Each material is characterized by constitutive relations which limit the class of processes possible in a body comprised of the material. In the thermodynamics of materials with memory‡, a *simple material* is one for which the free energy, the stress, the entropy, and the heat flux are determined when the history of the deformation gradient, the history of the temperature, and the present value of the temperature gradient are specified. Thus, at each material point of a simple material there hold equations of the form:

$$\left.\begin{aligned} \psi(t) &= \mathfrak{p}(\mathbf{F}^t, \theta^t; \mathbf{g}(t)) \\ \eta(t) &= \mathfrak{h}(\mathbf{F}^t, \theta^t; \mathbf{g}(t)) \\ \mathbf{S}(t) &= \mathfrak{s}(\mathbf{F}^t, \theta^t; \mathbf{g}(t)) \\ \mathbf{q}(t) &= \mathfrak{q}(\mathbf{F}^t, \theta^t; \mathbf{g}(t)) \end{aligned}\right\} \tag{2.9}$$

It is assumed that the four functions $\mathfrak{p}$, $\mathfrak{h}$, $\mathfrak{s}$ and $\mathfrak{q}$ are given at each material point; these functions, called 'response functionals' or 'constitutive functionals' depend, of course, on the choice of the reference configuration§. A process is said to be *admissible* in the simple material if, in addition to obeying the balance laws 2.1 and 2.2, it obeys the constitutive relations 2.9.

If one regards $\mathbf{q}/\theta$ to be a vectorial flux of entropy and r/θ to be a scalar supply of entropy, then it is natural to define the *rate of production of entropy* in a part $\mathscr{P}$ of $\mathscr{B}$ to be

$$\Gamma(\mathscr{P}, t) = \frac{\mathrm{d}}{\mathrm{d}t}\int_{\mathscr{P}} \eta\, \mathrm{d}m - \left[\int_{\mathscr{P}} \frac{r}{\theta}\,\mathrm{d}m - \int_{\partial\mathscr{P}} \mathbf{q}\cdot\mathbf{n}\,\mathrm{d}a\right] \tag{2.10}$$

‡ Cf. Ref. 4.

§ Cf. Noll[11].

The *Clausius–Duhem inequality*[12] is the assertion that

$$\Gamma(\mathscr{P}, t) \geqslant 0 \tag{2.11}$$

In our paper[1] of 1963, Noll and I pointed out that in many branches of continuum physics the second law of thermodynamics can be given a precise mathematical meaning if it is interpreted to be the following principle.

Dissipation Principle. *For every admissible thermodynamic process in a body* $\mathscr{B}$, *the Clausius–Duhem inequality* 2.11 *must hold at all times* t *and in all parts* $\mathscr{P}$ *of* $\mathscr{B}$.

It is clear that this principle implies that response functionals cannot be chosen arbitrarily. In Section 4 I shall list the restrictions which the principle places on $\mathfrak{p}$, $\mathfrak{h}$, $\mathfrak{s}$ and $\mathfrak{q}$ in 2.9 when these functions obey the postulate of regularity called the 'principle of fading memory'‡. First, however, I should like to outline a recently developed axiomatic approach[13] to the theory of fading memory.

3. ON THE THEORY OF FADING MEMORY

Let us follow a procedure employed in ref. 4, and use Greek majuscules, such as Λ, to denote ordered pairs $(\mathbf{L}, \lambda)$, with $\mathbf{L}$ a tensor and λ a scalar. The definitions

$$\left.\begin{aligned} \alpha\Lambda_1 + \beta\Lambda_2 = \alpha(\mathbf{L}_1,\lambda_1) + \beta(\mathbf{L}_2,\lambda_2) = (\alpha\mathbf{L}_1 + \beta\mathbf{L}_2, \alpha\lambda_1 + \beta\lambda_2), \\ \Lambda_1\cdot\Lambda_2 = \operatorname{tr}(\mathbf{L}_1\mathbf{L}_2^T) + \lambda_1\lambda_2 \end{aligned}\right\} \tag{3.1}$$

make the set of all such ordered pairs a 10-dimensional vector space $\mathscr{V}_{(10)}$ with norm

$$|\Lambda| = \sqrt{\Lambda\cdot\Lambda} = \sqrt{\operatorname{tr}(\mathbf{L}\mathbf{L}^T) + \lambda^2} \tag{3.2}$$

The elements of $\mathscr{V}_{(10)}$ of the type

$$\boldsymbol{\Gamma} = (\mathbf{F}, \theta) \tag{3.3}$$

with $\mathbf{F}$ a non-singular tensor and θ a positive number, form a cone $\mathscr{C}$ in $\mathscr{V}_{(10)}$§. At a given material point in a process, the *total history up to t*, i.e. the history up to t of the deformation gradient and temperature, is the function $\boldsymbol{\Gamma}^t = (\mathbf{F}^t, \theta^t)$, mapping $[0, \infty)$ into $\mathscr{C}$:

$$\boldsymbol{\Gamma}^t(s) = (\mathbf{F}^t(s), \theta^t(s)) \quad \text{for } 0 \leqslant s < \infty. \tag{3.4}$$

The ordered pair $\boldsymbol{\Sigma}$, defined by

$$\boldsymbol{\Sigma} = (\mathbf{S}, -\eta) = \left(\frac{1}{\rho}\mathbf{T}\mathbf{F}^{T^{-1}}, -\eta\right) \in \mathscr{V}_{(10)}, \tag{3.5}$$

‡ For earlier studies of the principle of fading memory, see refs. 14–18. Ref. 19 surveys work done up to 1965.

§ A subset $\mathfrak{A}$ of a vector space is called a *cone* if $\mathbf{u} \in \mathfrak{A}$ and b > 0 imply $b \in \mathfrak{A}$

is called the *stress-entropy vector*[4]. If one writes simply ψ for $\psi(t)$, $\mathbf{g}$ for $\mathbf{g}(t)$, and $\boldsymbol{\Sigma}$ for $\boldsymbol{\Sigma}(t)$, the constitutive equations 2.9 become, in the present notation,

$$\left.\begin{aligned} \psi &= \mathfrak{p}(\boldsymbol{\Gamma}^t; \mathbf{g}) \\ \boldsymbol{\Sigma} &= \mathfrak{S}(\boldsymbol{\Gamma}^t; \mathbf{g}) \\ \mathbf{q} &= \mathfrak{q}(\boldsymbol{\Gamma}^t; \mathbf{g}) \end{aligned}\right\} \tag{3.6}$$

where the response functional $\mathfrak{S}$ has the 'components'

$$\mathfrak{S} = (\mathfrak{s}, -\mathfrak{h}) \tag{3.7}$$

It is frequently possible to prove theorems in a branch of continuum physics without specifying the form of response functionals, but usually one must assume something about their smoothness. For this reason several topologies have been proposed as appropriate for sets of histories[14–18, 4, 8, 9, 10].

Let us suppose that the histories $\boldsymbol{\Lambda}^t$ of interest form a cone $\mathfrak{C}$ in a Banach function-space $\mathfrak{B}$. Certain basic, but usually tacit, assumptions of physical theories place limitations on the choice of the function space $\mathfrak{B}$ and its norm $\|\cdot\|$. I list below three of these requirements‡.

(**1**) Given an arbitrary history $\boldsymbol{\Gamma}^t$ in the domain D of a constitutive functional§ and a positive number σ, one expects to find in D the history $\boldsymbol{\Gamma}^{t+\sigma}$ for processes in which $\boldsymbol{\Gamma} = (\mathbf{F}, \theta)$ (at some fixed material point) has the history $\boldsymbol{\Gamma}^t$ up to time t and is constant throughout the interval $[t, t+\sigma]$. The history $\boldsymbol{\Gamma}^{t+\sigma}$ in such a process is called the 'static continuation of $\boldsymbol{\Gamma}^t$ by the amount σ'. The static continuation of a history should be well defined even if one identifies the history with the set of functions at zero distance from it in $\mathfrak{B}$.

(**2**) If the history $\boldsymbol{\Gamma}^t$ of $\boldsymbol{\Gamma}$ up to time t is in the domain D, then one expects to find in D the histories $\boldsymbol{\Gamma}^{t-\sigma}$ of $\boldsymbol{\Gamma}$ up to previous times $t-\sigma, \sigma \geqslant 0$. These earlier histories are called 'σ-sections of $\boldsymbol{\Gamma}^t$'.

(**3**) Since it should be possible to evaluate response functionals at 'equilibrium states', one expects D to contain constant histories of the form $\boldsymbol{\Gamma}^t(s) \equiv \boldsymbol{\Omega},\ 0 \leqslant s < \infty$.

Victor Mizel and I have found some apparently useful implications of these elementary physical requirements, and I summarize below some of our results[13].

Let μ be an *influence measure*; that is, a non-trivial‖, sigma-finite, positive, regular Borel measure on $[0, \infty)$ and let $\mathscr{S}$ be the set of all μ-measurable functions ϕ mapping $[0, \infty)$ into $[0, \infty)$. Let v be a function on $\mathscr{S}$ such that for all ϕ (or ϕ_i) in $\mathscr{S}$:

(i) $0 \leqslant v(\phi) \leqslant \infty$, and $v(\phi) = 0$ if and only if $\phi(s) = 0$ μ-a.e.¶;

(ii) $v(\phi_1 + \phi_2) \leqslant v(\phi_1) + v(\phi_2)$, and $v(a\phi) = av(\phi)$ for all numbers $a \geqslant 0$;

‡ Cf. Coleman and Mizel[18, 13].

§ D is here the domain for a fixed value of $\mathbf{g}$.

‖ i.e. not identically zero.

¶ i.e. for all s in $[0, \infty)$ except for a set Z with $\mu(Z) = 0$.

(iii) if $\phi_1(s) \leqslant \phi_2(s)$ μ-a.e., then $\nu(\phi_1) \leqslant \nu(\phi_2)$;
(iv) there is at least one function ψ in $\mathscr{s}$ with $0 < \nu(\psi) < \infty$;
(v) if $\psi, \phi_1, \phi_2, \ldots$ are in $\mathscr{s}$ and if $\phi_n(s) \uparrow \psi(s)$ μ-a.e., then $\nu(\phi_n) \uparrow \nu(\psi)$.

Such a function ν is called a non-trivial function norm, relative to μ, with the sequential Fatou property‡.

Let $\overline{V}$ be the set of μ-measurable functions mapping $[0, \infty)$ into $\mathscr{V}_{(10)}$, and let $\|\cdot\|$ be the function on $\overline{V}$ defined by

$$\|\boldsymbol{\Phi}\| = \nu(|\boldsymbol{\Phi}|) \tag{3.8}$$

for each $\boldsymbol{\Phi}$ in $\overline{V}$. I write V for the set of all $\boldsymbol{\Phi}$ in $\overline{V}$ with $\|\boldsymbol{\Phi}\| < \infty$. Each function $\boldsymbol{\Phi}$ in V is called a *history*, and its independent variable (usually denoted by s) is called the *elapsed time*. The value $\boldsymbol{\Phi}(0)$ of $\boldsymbol{\Phi}$ at $s = 0$ is called the *present value* of $\boldsymbol{\Phi}$, and the *past values* are those for which $0 < s < \infty$. The function space $\mathfrak{V}$ obtained by calling two functions $\boldsymbol{\Phi}_1$ and $\boldsymbol{\Phi}_2$ in V the same whenever $\|\boldsymbol{\Phi}_1 - \boldsymbol{\Phi}_2\| = 0$ is easily shown to be a Banach space; it is called a *history space* or, at length, a *Banach function space formed from histories with values in* $\mathscr{V}_{(10)}$.

Let C be the class of functions $\boldsymbol{\Phi}$ in V such that $\boldsymbol{\Phi}(s)$ is in $\mathscr{C}$ for all $s \geqslant 0$, and let $\mathfrak{C}$ be the set of equivalence classes obtained by calling the same those elements $\boldsymbol{\Phi}_1, \boldsymbol{\Phi}_2$ of C for which $\|\boldsymbol{\Phi}_1 - \boldsymbol{\Phi}_2\| = 0$. Clearly, C is a cone in V, and $\mathfrak{C}$ is a cone contained in the Banach function-space $\mathfrak{V}$. Let $\mathfrak{C}$ be the domain D of definition of the response functionals in (3.6)[20]§.

If $\boldsymbol{\Psi}$ is a function on $[0, \infty)$ and σ a positive number, then the *static continuation* of $\boldsymbol{\Psi}$ by the amount σ is the function $\boldsymbol{\Psi}^{(\sigma)}$ on $[0, \infty)$ defined by[16]

$$\boldsymbol{\Psi}^{(\sigma)}(s) = \begin{cases} \boldsymbol{\Psi}(0), & 0 \leqslant s < \sigma \\ \boldsymbol{\Psi}(s-\sigma), & \sigma < s < \infty \end{cases} \tag{3.9}$$

and the *σ-section* of $\boldsymbol{\Psi}$ is the function $\boldsymbol{\Psi}_{(\sigma)}$ on $[0, \infty)$ given by[18]

$$\boldsymbol{\Psi}_{(\sigma)}(s) = \boldsymbol{\Psi}(s + \sigma), \qquad 0 \leqslant s < \infty \tag{3.10}$$

If $\boldsymbol{\Psi}$ is the history up to t of $\boldsymbol{\Gamma} = (\mathbf{F}, \theta)$ (at a fixed material point X in some particular process), then $\boldsymbol{\Psi}_{(\sigma)}$ is the history of $\boldsymbol{\Gamma}$ up to $t - \sigma$, while $\boldsymbol{\Psi}^{(\sigma)}$ gives the history of $\boldsymbol{\Gamma}$ up to $t + \sigma$ assuming that $\boldsymbol{\Gamma}$ is held constant from t to $t + \sigma$. The physical requirements (**1**) and (**2**) stated above are made precise by laying down the following two postulates[18, 13].

Postulate 1. *If a given function $\boldsymbol{\Phi}$ is in C, then all its static continuations $\boldsymbol{\Phi}^{(\sigma)}$, $\sigma \geqslant 0$, are also in C. Furthermore, if $\boldsymbol{\Phi}$ and $\boldsymbol{\Psi}$ in C are such that $\|\phi - \psi\| = 0$, then $\|\boldsymbol{\Phi}^{(\sigma)} - \boldsymbol{\Psi}^{(\sigma)}\| = 0$ for all $\sigma \geqslant 0$.*

Postulate 2. *If $\boldsymbol{\Phi}$ is in C, then so also are all its σ-sections, $\boldsymbol{\Phi}_{(\sigma)}$, $\sigma \geqslant 0$.*

Employing Postulate 1, one can easily prove the following theorem which shows that the present value $\boldsymbol{\Phi}(0)$ of a history $\boldsymbol{\Phi}$ has a special status, in the

‡ Cf. Luxemburg and Zaanen[21] and the literature quoted by them.

§ When the dependence on **g** is under discussion, the domain is taken to be the set $\mathfrak{C} \times \mathscr{V}_{(3)}$, which forms a cone in $\mathfrak{V} \oplus \mathscr{V}_{(3)}$.

sense that the norm $\|\boldsymbol{\Phi}\| = \nu(|\boldsymbol{\Phi}|)$ places greater emphasis on $\boldsymbol{\Phi}(0)$ than on any individual past value.

Theorem 1‡. *The influence measure μ must have an atom at $s = 0$ and be absolutely continuous on $(0, \infty)$ with respect to Lebesgue measure.*

Postulates 1 and 2, together, yield

Theorem 2§. *Either $\mu((0, \infty)) = 0$, or Lebesgue measure is absolutely continuous on $(0, \infty)$ with respect to μ.*

Thus the μ-measure of the singleton $\{0\}$ is not zero, and if $\mu((0, \infty))$ is not zero, then an arbitrary subset of $(0, \infty)$ has zero μ-measure if and only if it has zero Lebesgue measure. So as to have a non-trivial theory, *let us assume that $\mu((0, \infty))$ is not zero*. Since the measure μ is employed here only to render precise the expression 'μ-a.e.' in the axioms (i), (iii), (iv) and (v) for ν, Theorems 1 and 2 imply that one can here replace μ with the Borel measure on $[0, \infty)$ that assigns the value 1 to the singleton $\{0\}$ and equals Lebesgue measure when restricted to Borel subsets of $(0, \infty)$.

If $\boldsymbol{\Phi}$ is a function in V, the restriction of $\boldsymbol{\Phi}$ to $(0, \infty)$ is called the *past history* of $\boldsymbol{\Phi}$ and is denoted by $\boldsymbol{\Phi}_r$. Let V_r be the set of all functions $\boldsymbol{\Phi}_r$ obtained by restricting members of V to $(0, \infty)$, and define $\|\cdot\|_r$ on V_r by

$$\|\boldsymbol{\Phi}_r\|_r = \|\boldsymbol{\Phi}\chi_{(0,\infty)}\| \tag{3.11}$$

with $\chi_{(0,\infty)}$ the characteristic function‖ of $(0, \infty)$. The *space of past histories* is the function space $\mathfrak{B}_r$ obtained by calling the same those elements $\boldsymbol{\Phi}_r$, $\boldsymbol{\Psi}_r$ of V_r for which $\|\boldsymbol{\Phi}_r - \boldsymbol{\Psi}_r\|_r = 0$. It is easily verified that $\|\cdot\|_r$ is a norm on $\mathfrak{B}_r$ and that $\mathfrak{B}_r$ is a Banach space. I write C_r for the set of functions in V_r with values in $\mathscr{C}$ and $\mathfrak{C}_r$ for the corresponding cone in $\mathfrak{B}_r$.

An immediate consequence of Theorems 1 and 2 is

Theorem 3 ¶. *$\mathfrak{B} = \mathscr{V}_{(10)} \oplus \mathfrak{B}_r$, and the norm $\|\cdot\|$ on $\mathfrak{B}$ is equivalent to the norm $\|\cdot\|'$ defined by*

$$\|\boldsymbol{\Phi}\|' = |\boldsymbol{\Phi}(0)| + \|\boldsymbol{\Phi}_r\|_r \tag{3.12}$$

Here $|\cdot|$ is the original norm 3.2 on $\mathscr{V}_{(10)}$, $\|\cdot\|$ is the norm on $\mathfrak{B}$ defined in 3.8, and $\|\cdot\|_r$ is the norm on $\mathfrak{B}_r$ defined in 3.11. The equivalence of $\|\cdot\|'$ and $\|\cdot\|$ means that there exist two positive numbers c_1 and c_2 such that

$$c_1\|\boldsymbol{\Phi}\| \leqslant \|\boldsymbol{\Phi}\|' \leqslant c_2\|\boldsymbol{\Phi}\|$$

for all $\boldsymbol{\Phi}$ in $\mathfrak{B}$. It follows from Theorem 3 that, even after the functions in V are grouped together to form the equivalence classes comprising $\mathfrak{B}$, each history $\boldsymbol{\Phi}$ has a well-defined present value $\boldsymbol{\Phi}(0)$.

If $\boldsymbol{\Omega}$ is a vector, $\boldsymbol{\Omega}$‡ denotes the constant function on $[0, \infty)$ with value $\boldsymbol{\Omega}$:

$$\boldsymbol{\Omega}^{\ddagger}(s) = \boldsymbol{\Omega}, \qquad 0 \leqslant s < \infty \tag{3.13}$$

‡ Ref. 13, Thm 2.1.

§ Ref. 13, Thm 2.2.

‖ i.e. $\chi_{(0,\infty)}$ has domain $[0, \infty)$ and is such that $\chi_{(0,\infty)}(s) = 1$ for $s \in (0, \infty)$, while $\chi_{(0,\infty)}(0) = 0$.

¶ Ref. 13, Thm 3.1.

The following postulate embodies the third of the physical requirements listed above.

Postulate 3. *The space* $\mathfrak{V}$ *contains non-trivial constant functions. That is, for each vector* $\boldsymbol{\Omega}$ *in* $\mathscr{C}$, *the function* $\boldsymbol{\Omega}^{\dagger}$ *is in* C.

It follows from this assumption that given any functional $\mathfrak{f}$ on $\mathfrak{C}$, one can define a function $\mathfrak{f}^{\circ}$ on $\mathscr{C}$ by the formula

$$\mathfrak{f}^{\circ}(\boldsymbol{\Omega}) = \mathfrak{f}(\boldsymbol{\Omega}^{\dagger}) \qquad \text{for all} \quad \boldsymbol{\Omega} \in \mathscr{C} \tag{3.14}$$

$\mathfrak{f}^{\circ}$ is called the *equilibrium response function corresponding to* $\mathfrak{f}$.[4] If $\mathfrak{f}$ is a continuous functional on $\mathfrak{C}$, then $\mathfrak{f}^{\circ}$ is continuous on $\mathscr{C}$.

The norm $\|\cdot\|$ on $\mathfrak{V}$ is said to have the *relaxation property*‡, if, for each function $\boldsymbol{\Phi}$ in V,

$$\lim_{\sigma \to \infty} \|\boldsymbol{\Phi}^{(\sigma)} - \boldsymbol{\Phi}(0)^{\dagger}\| = 0 \tag{3.15}$$

where, in accord with 3.13, $\boldsymbol{\Phi}(0)^{\dagger}$ is the constant function on $[0, \infty)$ with value $\boldsymbol{\Phi}(0)$. Clearly, $\|\cdot\|$ has the relaxation property if and only if 3.15 holds for each $\boldsymbol{\Phi}$ in C. Hence the assumption of the relaxation property is equivalent to the assertion that every continuous functional $\mathfrak{f}$ on $\mathfrak{C}$ obeys the relation

$$\lim_{\sigma \to \infty} \mathfrak{f}(\boldsymbol{\Phi}^{(\sigma)}) = \mathfrak{f}(\boldsymbol{\Phi}(0)^{\dagger}) = \mathfrak{f}^{\circ}(\boldsymbol{\Phi}(0)) \tag{3.16}$$

for each $\boldsymbol{\Phi}$ in $\mathfrak{C}$; that is, in the limit of large σ, the response $\mathfrak{f}(\boldsymbol{\Phi}^{(\sigma)})$ to the static continuation $\boldsymbol{\Phi}^{(\sigma)}$ of an arbitrary history $\boldsymbol{\Phi}$ depends only on the present value of $\boldsymbol{\Phi}$ and is given by the equilibrium response function defined in 3.14.

Postulate 4. *The norm* $\|\cdot\|$ *has the relaxation property.*

Postulates 1 to 4 yield

Theorem 4. *Let* $\Lambda_1(\cdot)$ *and* $\Lambda_2(\cdot)$ *be functions mapping* $(-\infty, \infty)$ *into* $\mathscr{V}_{(10)}$ *such that, for each t, the histories* Λ_1^t *and* Λ_2^t *are in* V. *If* $\lim_{t \to \infty} |\Lambda_1(t) - \Lambda_2(t)| = 0$, *then* $\lim_{t \to \infty} \|\Lambda_1^t - \Lambda_2^t\| = 0$.

A function $\boldsymbol{\Phi}$ in C is called a *tame history* if:

(α) $\boldsymbol{\Phi}$ is differentiable in the classical sense at $s = 0$; that is

$$\dot{\boldsymbol{\Phi}}(0) \stackrel{\text{def}}{=} -\frac{\mathrm{d}}{\mathrm{d}s}\boldsymbol{\Phi}(s)\bigg|_{s=0} = \lim_{s \to 0^+} \frac{\boldsymbol{\Phi}(0) - \boldsymbol{\Phi}(s)}{s} \tag{3.17}$$

exists.

(β) $\boldsymbol{\Phi}_r$, the past history of $\boldsymbol{\Phi}$, is an absolutely continuous function on $(0, \infty)$.

(γ) $\mathfrak{V}$ contains an element $\dot{\boldsymbol{\Phi}}$, called the *time-derivative* of $\boldsymbol{\Phi}$, which obeys the equation

$$\dot{\boldsymbol{\Phi}}(s) = -\frac{\mathrm{d}}{\mathrm{d}s}\boldsymbol{\Phi}(s), \qquad \mu\text{-a.e.} \tag{3.18}$$

For technical reasons, one assumes

‡ Ref. 18; see also refs. 16, 17, 4, 13 and 20.

Postulate 5. *Tame histories with time-derivatives of compact support are dense in $\mathfrak{C}$. That is, given any $\boldsymbol{\Psi}$ in $\mathfrak{C}$ and any $\delta > 0$, there exists a tame history $\boldsymbol{\Phi}$ in $\mathfrak{C}$ such that $\| \boldsymbol{\Psi} - \boldsymbol{\Phi} \| < \delta$ and $\dot{\boldsymbol{\Phi}}(s) = \mathbf{0}$ for all s outside a closed bounded set in $[0, \infty)$.*

It follows from Postulate 5 that $\mathfrak{B}$ is separable, that continuous functions of compact support are dense in $\mathfrak{B}$, and that $\mathfrak{B}$ has the following *dominated-convergence property*‡ familiar in the theory of Lebesgue spaces: *If $\boldsymbol{\Psi}$* belongs to $\mathfrak{B}$ and if $\boldsymbol{\Phi}_j$ is a sequence of elements of $\mathfrak{B}$ with $|\boldsymbol{\Phi}_j(s)| \leqslant |\boldsymbol{\Psi}(s)|$, μ-a.e., such that $\boldsymbol{\Phi}_j(s) \to \mathbf{0}$, μ-a.e., *then* $\|\boldsymbol{\Phi}_j\| \to 0$.

Let $\mathfrak{f}$ be a continuous function mapping $\mathfrak{C}$ into a metric space. It follows from Theorem 3 that $\mathfrak{f}$ can be regarded equally well as a function of ordered pairs $(\boldsymbol{\Phi}(0), \boldsymbol{\Phi}_r)$ with $\boldsymbol{\Phi}(0)$ in $\mathscr{C}$ and $\boldsymbol{\Phi}_r$ in $\mathfrak{C}_r$, i.e.

$$\mathfrak{f}(\boldsymbol{\Phi}) = \mathfrak{f}(\boldsymbol{\Phi}(0); \boldsymbol{\Phi}_r) \tag{3.19}$$

and the continuity of $\mathfrak{f}$ over $\mathfrak{C}$ implies that $\mathfrak{f}(\boldsymbol{\Phi}(0); \boldsymbol{\Phi}_r)$ is jointly continuous in the two variables, $\boldsymbol{\Phi}(0)$ in $\mathscr{C}$, and $\boldsymbol{\Phi}_r$ in $\mathfrak{C}_r$. Now, $\mathfrak{f}$ can be used to define a functional transformation mapping functions $\boldsymbol{\Lambda}(\cdot)$ on $(-\infty, \infty)$ into functions $\phi(\cdot)$ on $(-\infty, \infty)$, by setting

$$\phi(t) = \mathfrak{f}(\boldsymbol{\Lambda}^t) \tag{3.20}$$

for each $t \in (-\infty, \infty)$, where $\boldsymbol{\Lambda}^t(s) = \boldsymbol{\Lambda}(t-s)$, $s \geqslant 0$. Employing Postulates 1, 2, 3 and 5, one can prove that the functional transformation, $\boldsymbol{\Lambda}(\cdot) \rightsquigarrow \phi(\cdot)$, preserves regularity in the following sense.

Theorem 6§. *Let $\mathfrak{f}$ be a continuous function on $\mathfrak{C}$ with values in a metric space, and suppose that $\boldsymbol{\Lambda}(\cdot)$ is a function on $(-\infty, \infty)$ with $\boldsymbol{\Lambda}^t$ in $\mathfrak{C}$ for each t. If $\boldsymbol{\Lambda}(\cdot)$ is a regulated function, i.e. a function for which the limits $\lim_{\tau \to t^+} \boldsymbol{\Lambda}(\tau)$ and $\lim_{\tau \to t^-} \boldsymbol{\Lambda}(\tau)$ exist for each t, then $\phi(\cdot)$, given by 3.20, is also a regulated function. Furthermore, $\phi(\cdot)$ can suffer discontinuities at only those times t_i at which $\boldsymbol{\Lambda}(\cdot)$ is discontinuous; at all other times $\phi(\cdot)$ is continuous.*

(To obtain this result one first shows that the mapping $t \rightsquigarrow \boldsymbol{\Lambda}_r^t \in \mathfrak{C}_r$ is continuous, for all t, even for those at which $\boldsymbol{\Lambda}(\cdot)$ experiences a discontinuity.)

Let $\mathfrak{U}$ be a cone in a Banach space $\mathfrak{B}$, and let $\mathfrak{D}$ be the subspace of $\mathfrak{B}$ spanned by $\mathfrak{U}$. A real-valued function $\mathfrak{g}$ defined on $\mathfrak{U}$ is said to be *continuously Fréchet-differentiable on* $\mathfrak{U}$‖ if, for each $\boldsymbol{\phi}$ in $\mathfrak{U}$ and every $\boldsymbol{\xi}$ in $\mathfrak{B}$ with $\boldsymbol{\phi} + \boldsymbol{\xi}$ in $\mathfrak{U}$,

$$\mathfrak{g}(\boldsymbol{\phi} + \boldsymbol{\xi}) = \mathfrak{g}(\boldsymbol{\phi}) + \mathrm{d}\mathfrak{g}(\boldsymbol{\phi}|\boldsymbol{\xi}) + o(\|\boldsymbol{\xi}\|) \tag{3.21}$$

where $\mathrm{d}\mathfrak{g}(\cdot|\cdot)$ is defined and continuous on $\mathfrak{U} \times \mathfrak{D}$ and is such that $\mathrm{d}\mathfrak{g}(\boldsymbol{\phi}|\boldsymbol{\xi})$ is a linear function of $\boldsymbol{\xi}$ for each $\boldsymbol{\phi}$. The linear functional $\mathrm{d}\mathfrak{g}(\boldsymbol{\phi}|\cdot)$ is called the *Fréchet derivative of* $\mathfrak{g}$ *at* $\boldsymbol{\phi}$.

An argument given in ref. 20 here yields

‡ For theorems of this type, see Luxemburg and Zaanen[21], Thm 2.2, and Luxemburg[22], Thm 46.2. See also ref. 13, Remarks 3.1 and 3.2.

§ Ref. 13, Remark 3.3; see also ref. 18, Remark 5.1.

‖ Of course the definition can be employed for other types of subsets of $\mathfrak{B}$, such as open subsets.

Theorem 7‡. (Chain Rule.) *If* $\mathfrak{g}$ *is a real-valued continuously Fréchet-differentiable function on* $\mathfrak{C}$, *then, for each tame history* $\boldsymbol{\Phi}$ *in* $\mathfrak{C}$, *the derivative*

$$\dot{\mathfrak{g}} \overset{\text{def}}{=} \lim_{\sigma \to 0^+} \frac{\mathfrak{g}(\boldsymbol{\Phi}) - \mathfrak{g}(\boldsymbol{\Phi}_{(\sigma)})}{\sigma} \tag{3.22}$$

exists and is given by

$$\dot{\mathfrak{g}} = d\mathfrak{g}(\boldsymbol{\Phi}|\dot{\boldsymbol{\Phi}}) \tag{3.23}$$

with $\dot{\boldsymbol{\Phi}}$ *the time derivative defined in* 3.18.

Suppose $\mathfrak{g}$ is continuously Fréchet-differentiable on $\mathfrak{C}$, and recall that $\mathfrak{g}(\boldsymbol{\Phi})$ can be written

$$\mathfrak{g}(\boldsymbol{\Phi}) = \mathfrak{g}(\boldsymbol{\Phi}(0); \boldsymbol{\Phi}_r) \tag{3.24}$$

where $\boldsymbol{\Phi}(0)$, in $\mathscr{C}$, is the present value of $\boldsymbol{\Phi}$, and $\boldsymbol{\Phi}_r$, in $\mathfrak{C}_r$, is the restriction of $\boldsymbol{\phi}$ to $(0, \infty)$. The assumed differentiability of $\mathfrak{g}$ on $\mathfrak{C}$ implies the existence, for each $\boldsymbol{\Phi}$, of the *instantaneous derivative*[4] $D\mathfrak{g}(\boldsymbol{\Phi})$ and the *past-history derivative* $\delta\mathfrak{g}(\boldsymbol{\Phi}|\cdot)$, which are determined by the equations

$$\mathfrak{g}(\boldsymbol{\Phi}(0) + \boldsymbol{\Omega}; \boldsymbol{\Phi}_r) = \mathfrak{g}(\boldsymbol{\Phi}(0); \boldsymbol{\Phi}_r) + D\mathfrak{g}(\boldsymbol{\Phi})\cdot\boldsymbol{\Omega} + o(|\boldsymbol{\Omega}|) \tag{3.25}$$

and

$$\mathfrak{g}(\boldsymbol{\Phi}(0); \boldsymbol{\Phi}_r + \boldsymbol{\Psi}_r) = \mathfrak{g}(\boldsymbol{\Phi}(0); \boldsymbol{\Phi}_r) + \delta\mathfrak{g}(\boldsymbol{\Phi}|\boldsymbol{\Psi}_r) + o(\|\boldsymbol{\Psi}_r\|_r); \tag{3.26}$$

3.25 holds for all $\boldsymbol{\Omega}$ in $\mathscr{V}_{(10)}$ with $\boldsymbol{\Phi}(0) + \boldsymbol{\Omega}$ in $\mathscr{C}$, while 3.26 holds for all $\boldsymbol{\Psi}_r$ in $\mathfrak{V}_r$ with $\boldsymbol{\Phi}_r + \boldsymbol{\Psi}_r$ in $\mathfrak{C}_r$. For each $\boldsymbol{\Phi}$ in $\mathfrak{C}$, the value $D\mathfrak{g}(\boldsymbol{\Phi})$ of $D\mathfrak{g}$ is a vector in $\mathscr{V}_{(10)}$, and $\delta\mathfrak{g}(\boldsymbol{\Phi}|\cdot)$ is a linear function on $\mathfrak{V}_r$. The functionals $D\mathfrak{g}$ and $\delta\mathfrak{g}$ determine $d\mathfrak{g}$ through the relation

$$d\mathfrak{g}(\boldsymbol{\Phi}|\boldsymbol{\Psi}) = D\mathfrak{g}(\boldsymbol{\Phi})\cdot\boldsymbol{\Psi}(0) + \delta\mathfrak{g}(\boldsymbol{\Phi}|\boldsymbol{\Psi}_r) \tag{3.27}$$

and one can write the chain rule 3.23 in the form

$$\dot{\mathfrak{g}} = D\mathfrak{g}(\boldsymbol{\Phi})\cdot\dot{\boldsymbol{\Phi}}(0) + \delta\mathfrak{g}(\boldsymbol{\Phi}|\dot{\boldsymbol{\Phi}}_r) \tag{3.28}$$

with $\dot{\boldsymbol{\Phi}}(0)$ the present value, and $\dot{\boldsymbol{\Phi}}_r$ the past history, of $\dot{\boldsymbol{\Phi}}$.

There is now assembled here apparatus sufficient for a precise statement of the principle of fading memory as used in the thermodynamics of simple materials.

Postulate of Smoothness for Response Functionals. *For each simple material there exists a history space* $\mathfrak{V}$, *formed as described above, such that*

(1) $\mathfrak{C}$, *the cone in* $\mathfrak{V}$ *corresponding to functions mapping* $[0, \infty)$ *into* $\mathscr{C}$, *obeys Postulates 1–5;*

(2) *the functionals* $\mathfrak{p}$, $\mathfrak{S}$ *and* $\mathfrak{q}$ *in* 3.6 *are defined and continuous on* $\mathfrak{C} \times \mathscr{V}_{(3)}$§;

(3) *the functional* $\mathfrak{p}$ *is continuously Fréchet-differentiable on* $\mathfrak{C} \times \mathscr{V}_{(3)}$‖.

‡ Ref. 20, Remark 1 and Appendix II; see also refs. 4 and 23. The proof given in ref. 20 does not require Postulate 4.

§ In 3.6, $\Gamma^t \in \mathfrak{C}$, while $\mathbf{g} \in \mathscr{V}_{(3)}$.

‖ $\mathfrak{C} \times \mathscr{V}_{(3)}$ is here considered a cone in $\mathfrak{V} \oplus \mathscr{V}_{(3)}$.

4. CONSEQUENCES OF THE SECOND LAW

It is easily shown that under appropriate assumptions of regularity for the dependence of χ and θ upon ξ and t, it follows from the balance laws 2.1 and 2.2 that the Clausius–Duhem inequality 2.11 can be written in the form[4]

$$\dot{\psi} \leqslant \boldsymbol{\Sigma} \cdot \dot{\boldsymbol{\Gamma}} - \frac{1}{\rho\theta} \mathbf{q} \cdot \mathbf{g} \tag{4.1}$$

Working with this local form of the inequality one can prove the following theorem which gives the restrictions which the second law places on the response functionals $\mathfrak{p}$, $\mathfrak{h}$, $\mathfrak{s}$ and $\mathfrak{q}$ in 2.9 [or, equivalently, $\mathfrak{p}$, $\mathfrak{S}$ and $\mathfrak{q}$ in 3.6].

Theorem 8‡. *It follows from the Dissipation Principle and the Postulate of Smoothness for Response Functionals that*

(i) *the functionals* $\mathfrak{p}$ *and* $\mathfrak{S}$ *are independent of* $\mathbf{g}$; i.e.

$$\psi(t) = \mathfrak{p}(\boldsymbol{\Gamma}^t), \qquad \boldsymbol{\Sigma}(t) = \mathfrak{S}(\boldsymbol{\Gamma}^t) \tag{4.2}$$

whenever $\boldsymbol{\Gamma}^t$ *is in* $\mathfrak{C}$;

(ii) *the functional* $\mathfrak{S}$ *is determined by the functional* $\mathfrak{p}$ *through the 'generalized stress relation'*

$$\mathfrak{S} = \mathrm{D}\mathfrak{p}, \tag{4.3}$$

i.e.

$$\boldsymbol{\Sigma}(t) = \mathrm{D}\mathfrak{p}(\boldsymbol{\Gamma}^t), \tag{4.4}$$

whenever $\boldsymbol{\Gamma}^t$ *is in* $\mathfrak{C}$;

(iii) *for each tame history* $\boldsymbol{\Gamma}^t$ *in* $\mathfrak{C}$

$$\delta\mathfrak{p}(\boldsymbol{\Gamma}^t | \dot{\boldsymbol{\Gamma}}^t) \leqslant 0 \tag{4.5}$$

and

$$\mathfrak{q}(\boldsymbol{\Gamma}^t;\mathbf{g}) \cdot \mathbf{g} \leqslant -\rho\theta\delta\mathfrak{p}(\boldsymbol{\Gamma}^t;\mathbf{g}) \tag{4.6}$$

Furthermore, (i), (ii) *and* (iii), *when taken together, give not only a necessary, but also a sufficient condition on* $\mathfrak{p}$, $\mathfrak{S}$ *and* $\mathfrak{q}$ *for* (4.1) *to hold for all* $\mathbf{g}$ *in* $\mathscr{V}_{(10)}$ *and all tame histories* $\boldsymbol{\Gamma}^t$ *in* $\mathfrak{C}$.

When I gave this theorem in my essay[4] of 1964, I proved it using a form of the principle of fading memory less general than that described here§. In the present terminology one can say that I employed the Postulate of Smoothness stated at the end of Section 3, but used for $\mathfrak{B}$ a special type of history space, namely a Hilbert space $\mathfrak{H}$ formed from functions $\boldsymbol{\Phi}$, mapping $[0, \infty)$ in $\mathscr{V}_{(10)}$, for which

$$\|\boldsymbol{\Phi}\|^2 \stackrel{\text{def}}{=} |\boldsymbol{\Phi}(0)|^2 + \int_0^\infty |\boldsymbol{\Phi}(s)|^2 k(s)\,\mathrm{d}s \tag{4.7}$$

‡ Ref. 4, Thm 1, p 19; see also ref. 20, Thm 1.

§ The form of the principle used in ref. 4 was drawn from earlier work done in collaboration with Walter Noll[14, 15, 16].

is finite; $k(s)$ was a fixed, positive, monotone-decreasing function, assumed to be summable on $(0, \infty)$, and called the 'influence function'‡. Later, Victor Mizel and I[20] observed that Theorem 8 remains valid in the present more general theory.

The conclusions (i), (ii) and (iii) of Theorem 8 have some interesting consequences which I list below.

Theorem 9§. *Of all total histories ending with a given value of* $\boldsymbol{\Gamma} = (\mathbf{F}, \theta)$, *that corresponding to constant values of* $\boldsymbol{\Gamma}$ *for all times has the least free energy; i.e. for each* $\boldsymbol{\Gamma}^t$ *in* $\mathfrak{C}$

$$\mathfrak{p}^{\circ}(\boldsymbol{\Gamma}^t(0)) \leqslant \mathfrak{p}(\boldsymbol{\Gamma}^t) \tag{4.8}$$ ||

Theorem 10¶. *If* $\boldsymbol{\Gamma}$ *is a vector in* $\mathscr{C}$, *and if* $\boldsymbol{\Gamma}^{\dagger}$ *is the constant function defined by* $\boldsymbol{\Gamma}^{\dagger}(s) \equiv \boldsymbol{\Gamma}$, *then for all functions* $\boldsymbol{\Phi}_r$ *in* $\mathfrak{B}_r$,

$$\delta\mathfrak{p}(\boldsymbol{\Gamma}^{\dagger} | \boldsymbol{\Phi}_r) = 0 \tag{4.9}$$

Theorem 11*. *The equilibrium response functions corresponding to* $\mathfrak{p}$ *and* $\mathfrak{S}$ *obey the classical relation*

$$\mathfrak{S}^{\circ}(\boldsymbol{\Gamma}) = \nabla\mathfrak{p}^{\circ}(\boldsymbol{\Gamma}) \quad \text{for all } \boldsymbol{\Gamma} \text{ in } \mathscr{C} \tag{4.10}$$

where $\nabla\mathfrak{p}^{\circ}$ *is the ordinary gradient of* $\mathfrak{p}^{\circ}$.

Although the proof of Theorem 8 does not employ Postulate 4, the proofs of Theorems 9, 10 and 11 do**.

ACKNOWLEDGEMENT

I am grateful to the U.S. National Science Foundation for making possible my attendance at the Conference and to the U.S. Air Force Office of Scientific Research for supporting parts of the research reported here.

REFERENCES

1 B. D. Coleman and W. Noll, *Arch. Ration. Mech. Anal.* **13**, 167 (1963).
2 B. D. Coleman and V. J. Mizel, *J. Chem. Phys.* **40**, 1116 (1964).
3 B. D. Coleman and M. E. Gurtin, *J. Chem. Phys.* **47**, 597 (1967).
4 B. D. Coleman, *Arch. Ration. Mech. Anal.* **17**, 1, 230 (1964).
5 B. D. Coleman and M. E. Gurtin, *Arch. Ration. Mech. Anal.* **19**, 266, 317 (1965).
6 B. D. Coleman and V. J. Mizel, *Arch. Ration. Mech. Anal.* **29**, 105 (1968); **30**, 173 (1968).
7 B. D. Coleman and E. H. Dill, *Arch. Ration. Mech. Anal.* **30**, 197 (1968).
8 D. R. Owen, *Arch. Ration. Mech. Anal.* **31**, 91 (1968).

---

† Cf refs. 14 16. No further assumptions on k are needed for the main theorems of ref. 4. See ref. 18 for an axiomatic approach to history spaces obeying 4.7.

§ Ref. 4, Thm 3, p 26; see also ref. 20, Thm 2.

|| See equation 3.14 for the definition of $\mathfrak{p}^{\circ}$, the equilibrium response function corresponding to $\mathfrak{p}$.

¶ Ref. 4, Corollary to Thm 3, p 26; ref. 18, Thm 3.

* Ref. 4, Remark 11, p 27; see also ref. 20, Thm 4.

** For further discussion of this point, see ref. 10 and the papers referred to therein.

[9] D. R. Owen and W. O. Williams, *Arch. Ration. Mech. Anal.* **33**, 288 (1969).
[10] B. D. Coleman and D. R. Owen, *Arch. Ration. Mech. Anal.* **36**, 245 (1970).
[11] W. Noll, *Arch. Ration. Mech. Anal.* **2**, 197 (1958); **27**, 1 (1967).
[12] C. Truesdell and R. A. Toupin, *The Classical Field Theories*, in the *Encyclopedia of Physics*, Vol. III/1, Springer: Heidelberg (1960).
[13] B. D. Coleman and V. J. Mizel, *Arch. Ration. Mech. Anal.* **29**, 18 (1968).
[14] B. D. Coleman and W. Noll, *Arch. Ration. Mech. Anal.* **6**, 355 (1960).
[15] B. D. Coleman and W. Noll, *Rev. Mod. Phys.* **33**, 239 (1961); errata **36**, 1103 (1964).
[16] B. D. Coleman and W. Noll, *Proc. Internat. Symp. Second-order Effects, Haifa*, 530 (1962).
[17] C.-C. Wang, *Arch. Ration. Mech. Anal.* **18**, 117 (1965).
[18] B. D. Coleman and V. J. Mizel, *Arch. Ration. Mech. Anal.* **23**, 87 (1966).
[19] C. Truesdell and W. Noll, *The Non-linear Field Theories of Mechanics*, in the *Encyclopedia of Physics*, Vol. III/3, Springer: Heidelberg (1965).
[20] B. D. Coleman and V. J. Mizel, *Arch. Ration. Mech. Anal.* **27**, 255 (1967).
[21] W. A. J. Luxemburg and A. C. Zaanen, *Math. Annalen*, **149**, 150 (1963); **162**, 337 (1966).
[22] W. A. J. Luxemburg, *Indag. Math.* **27**, 229 (1965).
[23] V. J. Mizel and C.-C. Wang, *Arch. Ration. Mech. Anal.* **23**, 124 (1966).

A NEW SYSTEMATIC APPROACH TO NON-EQUILIBRIUM THERMODYNAMICS

INGO MÜLLER

Institut für theoretische Physik, RWTH Aachen, 51 Aachen

ABSTRACT

The theory presented here is based on the following definitions and postulates:

(i) A process is defined as a set of functions $\rho(\mathbf{x}, t)$, $\chi(\mathbf{x}, t)$, $\vartheta(\mathbf{x}, t)$ representing the density, the motion and the (empirical) temperature.

(ii) Constitutive equations are formulated for the stress tensor $\mathbf{t}$, the internal energy ε and the heat flux $\mathbf{q}$, such that a set $\mathbf{t}$, ε, $\mathbf{q}$ belongs to every process.

(iii) A thermodynamic process is defined as a process which is a solution of the equations of balance for the mass, the momentum and the energy.

(iv) (Entropy Principle) In a body $\mathscr{B}$ there exists a scalar extensive quantity which cannot decrease in any thermodynamic process, if its flux through the surface of $\mathscr{B}$ vanishes, and whose density η and flux $\boldsymbol{\Phi}$ are given by constitutive relations. This quantity is called entropy.

The ideas (i)–(iv) can be put to use to derive restrictive conditions on the constitutive relations for $\mathbf{t}$, ε, $\mathbf{q}$, η and $\boldsymbol{\Phi}$, and these restrictions are valid for arbitrary non-equilibrium processes.

In order to obtain restrictions on $\mathbf{t}$, ε and $\mathbf{q}$ alone one needs further information about thermodynamic processes in a body and essentially the only processes about which useful information exists are uniform equilibrium processes. We know from experience that:

(v) There exists a uniform equilibrium in a body on whose boundary the fluxes of mass and energy vanish.

(vi) If such a body $\mathscr{B}_T$ consists of two subbodies $\mathscr{B}^\alpha$ ($\alpha = 1, 2$) of different materials, there exists a uniform equilibrium in each of the subbodies and their temperatures are equal.

(vii) If E_T, H_T and V_T are internal energy, entropy and volume of $\mathscr{B}_T$, and E^α, H^α and V^α the internal energies, entropies and volumes of $\mathscr{B}^\alpha$, then the following relations hold:

$$\left(\frac{\partial E_T}{\partial V^1}\right)_{H_T V^2} = \left(\frac{\partial E^1}{\partial V^1}\right)_{H^1} \text{ and } \left(\frac{\partial E_T}{\partial V^2}\right)_{H_T V^1} = \left(\frac{\partial E^2}{\partial V^2}\right)_{H^2}$$

From (i)–(vii) the existence of an absolute temperature can be proved for uniform equilibrium processes, but not more generally.

The theory based on (i)–(vii) avoids any specific assumptions on the entropy and the entropy flux, in particular it avoids the customary assumptions that the flux and supply of entropy are equal to the quotients of flux and supply of internal energy and absolute temperature.

If the principles (i)–(vii) are applied to a simple heat-conducting fluid, one may obtain, apart from all the familiar results for the material, a hyperbolic equation of heat conduction; and this is an immediate consequence of avoiding specific assumptions concerning the entropy and its flux.

---

1. INTRODUCTION

An evaluation of the literature on thermodynamics shows that writers in this field make some specific assumptions which are motivated by thermostatics, whose validity or range of validity, however, is never assessed, except by intuition. Such specific assumptions are:

(i) there exists an absolute temperature in non-equilibrium,

(ii) the flux of entropy is equal to the heat flux divided by the absolute temperature,

(iii) the supply of entropy is equal to the supply of internal energy divided by the absolute temperature.

The entropy inequality based on these assumptions is used as a restrictive condition on the constitutive equations for stress, heat flux, internal energy and entropy[1].

After recalling basic concepts of thermodynamics and introducing the notation in Section 2, I propose (in Section 3) an entropy principle, which is free of the specific assumptions (i) to (iii) above, and obtain restrictions on constitutive equations for arbitrary non-equilibrium processes. The method for deriving these restrictions is illustrated for simple heat conducting fluids, because these materials furnish the simplest non-trivial examples. In Section 4 I show that these restrictions, if supplemented by axioms on the behaviour of fluids in equilibrium, lead to all the restrictive conditions of linear irreversible thermodynamics and, moreover, allow for a finite speed of the propagation of heat.

In a recently published paper[2] I have developed the method, which is presented here, for a relativistic fluid.

2. THERMODYNAMIC PROCESSES

The objective of thermodynamics is to determine the future development of the density ρ, the motion χ_k and the (empirical) temperature ϑ of a body $\mathscr{B}$. For this one employs the equations of balance of mass, momentum and energy:

$$\begin{aligned} &(\partial \rho/\partial t) = (\rho v_k)_{,k} = 0 \\ &(\partial \rho v_i/\partial t) + (\rho v_i v_k - t_{ik})_{,k} = 0 \\ &(\partial \rho \varepsilon/\partial t) + (\rho \varepsilon v_k + q_k)_{,k} = t_{ik} v_{i,k}{}^{*} \end{aligned} \tag{2.1}$$

together with the constitutive relations for the stress t_{ik}, the heat flux q_i and the specific internal energy ε for a simple heat-conducting fluid:

$$\begin{aligned} t_{ik} &= t_{ik}(\rho, \vartheta, \vartheta_i, \dot{\vartheta}) \\ q_i &= q_i(\rho, \vartheta, \vartheta_i, \dot{\vartheta}) \qquad \dot{\vartheta} \equiv (\partial \vartheta/\partial t) + v_k \vartheta_k \\ \varepsilon &= \varepsilon(\rho, \vartheta, \vartheta_i, \dot{\vartheta}) \end{aligned} \tag{2.2}$$

---

* Throughout this paper I shall use the customary cartesian tensor notation.

The equations 2.1 form a complete set of differential equations for the determination of the unknown fields ρ, χ_k and ϑ from initial and boundary conditions.

In writing the constitutive relations 2.2 I have made use of the principle of equipresence[3] and I also wish to use the principle of objectivity[4] according to which the constitutive functions are isotropic tensor, vector or scalar functions with respect to the Euclidean group[5] of transformations*. These same principles will be applied to the constitutive relations for the entropy flux Φ_i and the specific entropy η [see equations (3.2)].

A solution of the set 2.1 and 2.2 of differential equations will be called a *thermodynamic process*.

3. THE ENTROPY PRINCIPLE

The entropy principle assumed here reads: In every body there exists an extensive quantity, the entropy, with a non-negative production:

$$\frac{\partial \rho\eta}{\partial t} + (\rho\eta\vartheta_k + \Phi_k)_{,k} \geqslant 0 \tag{3.1}$$

where Φ_i and η are given by constitutive relations:

$$\begin{aligned} \eta &= \eta(\rho, \vartheta, \vartheta_i, \dot{\vartheta}) \\ \Phi_k &= \Phi_k(\rho, \vartheta, \vartheta_i, \dot{\vartheta}) \end{aligned} \tag{3.2}$$

and the *entropy inequality* 3.1 holds for every thermodynamic process. This entropy principle will now be employed to derive restrictive conditions on the constitutive functions for t_{ik}, q_i, ε, η and Φ_i, η.

I shall make use of a theorem in the theory of partial differential equations[6] which states that, in general, there exists an analytical solution of a system like 2.1, 2.2 for arbitrary analytical initial conditions. If the above assertion about the entropy is true, the entropy inequality, which may be written in the form

$$\begin{aligned} &\left(\eta + \rho\frac{\partial \eta}{\partial \rho}\right)\frac{\partial \rho}{\partial t} + \rho\frac{\partial \eta}{\partial \dot{\vartheta}}\vartheta_{,i}\frac{\partial v_i}{\partial t} + \rho\frac{\partial \eta}{\partial \dot{\vartheta}}\frac{\partial^2 \vartheta}{\partial t^2} + \left[\left(\eta + \rho\frac{\partial \eta}{\partial \rho}\right)v_i + \frac{\partial \Phi_k}{\partial \rho}\right]\frac{\partial \rho}{\partial x_i} \\ &+ \left[\rho\frac{\partial \eta}{\partial \vartheta_{,i}} + 2\rho\frac{\partial \eta}{\partial \dot{\vartheta}}v_i + \frac{\partial \Phi_i}{\partial \dot{\vartheta}}\right]\frac{\partial \vartheta_{,i}}{\partial t} + \left[\left(\frac{\partial \eta}{\partial \vartheta_{,k}} + \frac{\partial \eta}{\partial \dot{\vartheta}}v_k\right)\vartheta_i + \frac{\partial \Phi_i}{\partial \vartheta_{,k}}\right. \\ &\left. + \frac{\partial \Phi_i}{\partial \dot{\vartheta}}v_k\right]\vartheta_{,ik} + \left[\rho\eta\delta_{ik} + \rho\frac{\partial \eta}{\partial \dot{\vartheta}}v_i\vartheta_{,k} + \frac{\partial \Phi_k}{\partial \dot{\vartheta}}\vartheta_{,i}\right]v_{i,k} + \rho\frac{\partial \eta}{\partial \vartheta}\frac{\partial \vartheta}{\partial t} \\ &\qquad + \left[\rho v_i\frac{\partial \eta}{\partial \vartheta} + \frac{\partial \Phi_i}{\partial \vartheta}\right]\vartheta_{,i} \geqslant 0 \end{aligned} \tag{3.3}$$

must hold for all such solutions of initial value problems and, in particular, for arbitrary values of

$$\rho_{,i}, \left(\frac{\partial \vartheta}{\partial t}\right)_{,i}, \vartheta_{,ik}, v_{i,k} \tag{3.4}$$

at one point at the initial time.

---

* Tacit use has already been made in 2.2 of the principle of objectivity, because only objective quantities were admitted in the set of variables.

Of course, 3.3 contains also the derivatives $\partial\rho/\partial t$, $\partial\vartheta_i/\partial t$ and $\partial^2\vartheta/\partial t^2$, but these are related to the quantities 3.4 by the system 2.1 and 2.2 of differential equations:

$$
\begin{aligned}
\frac{\partial\rho}{\partial t} &= -(\rho v_k)_{,k} \\
\frac{\partial\vartheta_i}{\partial t} &= -v_k v_{i,k} + \frac{1}{\rho} t_{ik,k} \\
\frac{\partial^2\vartheta}{\partial t^2} &= \left(v_k v_{i,k} - \frac{1}{\rho} t_{ik,k}\right)\vartheta_{,i} + \frac{1}{(\partial\varepsilon_{,k}/\partial\vartheta)}\left(-\vartheta_k\varepsilon_{,k} + \frac{1}{\rho} t_{ik} v_{i,k} - \frac{1}{\rho} q_{k,k}\right. \\
&\qquad \left. + \frac{\partial\varepsilon}{\partial\rho}(\rho v_k)_{,k} - \frac{\partial\varepsilon}{\partial\vartheta}\frac{\partial\vartheta}{\partial t} - \frac{\partial\varepsilon}{\partial\vartheta_{,k}}\frac{\partial\vartheta_{,k}}{\partial t}\right)
\end{aligned}
\tag{3.5}
$$

I insert 2.2 in 3.5, carry out all differentiations and then eliminate $\partial\rho/\partial t$, $\partial\vartheta_i/\partial t$ and $\partial^2\vartheta/\partial t^2$ from 3.3 and 3.5 thus obtaining an inequality which is explicitly linear in the derivatives 3.4; these, however, are arbitrary (except for the obvious symmetry of $\vartheta_{,ik}$) and therefore the inequality could easily be violated unless their factors vanish. Therefore the entropy principle requires the following restrictions on the constitutive relations 2.2 and 3.2:

$$
\begin{aligned}
&\frac{\partial\Phi_i}{\partial\rho} + \Lambda\frac{\partial q_i}{\partial\rho} = 0 \\
&\frac{\partial\Phi_i}{\partial\vartheta_{,k}} + \frac{\partial\Phi_k}{\partial\vartheta_{,i}} + \Lambda\left(\frac{\partial q_i}{\partial\vartheta_{,k}} + \frac{\partial q_k}{\partial\vartheta_{,i}}\right) = 0 \\
&\frac{\partial\Phi_i}{\partial\vartheta} + \Lambda\frac{\partial q_i}{\partial\vartheta} + \rho\left(\frac{\partial\eta}{\partial\vartheta_{,k}} + \Lambda\frac{\partial\varepsilon_{,i}}{\partial\vartheta_{,k}}\right) = 0
\end{aligned}
\tag{3.6}
$$

$$
\begin{aligned}
&-\rho^2\left(\frac{\partial\eta}{\partial\rho} + \Lambda\frac{\partial\varepsilon}{\partial\rho}\right)\delta_{ik} - \Lambda t_{ik} + \left(\frac{\partial\Phi_k}{\partial\vartheta} + \Lambda\frac{\partial q_k}{\partial\vartheta}\right)\vartheta_{,i} = 0 \\
&\left(\frac{\partial\eta}{\partial\vartheta} + \Lambda\frac{\partial\varepsilon}{\partial\vartheta}\right)\dot{\vartheta} + \frac{1}{\rho}\left(\frac{\partial\Phi_k}{\partial\vartheta} + \Lambda\frac{\partial q_k}{\partial\vartheta}\right)\vartheta_{,k} \geqslant 0
\end{aligned}
\tag{3.7}
$$

where Λ is defined by

$$
\Lambda = -\left(\frac{\partial\eta}{\partial\vartheta}\right)\left(\frac{\partial\varepsilon}{\partial\vartheta}\right)^{-1}
\tag{3.8}
$$

The conditions 3.6, 3.7 hold for *all thermodynamic processes* in a simple heat conducting fluid, they represent relations between the constitutive functions t_{ik}, q_k, ε, η, Φ_i, and, if η and Φ_i were known explicitly, we should have here what we are interested in, namely restrictions on the constitutive functions t_{ik}, q_k and ε alone. However, η and Φ_i are not known explicitly and it is therefore necessary that the conditions 3.6, 3.7 be supplemented by axioms which express experience with processes in simple heat conducting fluids. Unfortunately observational methods are such that useful information can only be had for time-independent processes and indeed only for particularly simple time-independent processes.

4. PROPERTIES IN UNIFORM EQUILIBRIUM

Let us first consider a body $\mathscr{B}$, consisting of a simple heat-conducting fluid, on whose boundary the normal components of the heat flux and of the velocity vanish (in the appropriate frame). Experience then leads to the formulation of the following axiom:

(a) There exists a uniform and time-independent process in $\mathscr{B}$, which I shall call a uniform equilibrium or, briefly, an equilibrium for the purposes of this paper.

Let us now consider a body $\mathscr{B}_T$ consisting of two subbodies $\mathscr{B}^\alpha(\alpha = 1, 2)$ of different simple heat-conducting fluids and let the normal components of the heat flux and the velocity vanish on the boundary of $\mathscr{B}_T$, whereas only the vanishing of the normal velocity is required on the boundaries of $\mathscr{B}^\alpha$.

Experience shows that:

(b) there exists an equilibrium in the bodies $\mathscr{B}^\alpha$ and the temperature is uniform throughout $\mathscr{B}_T$.

With (a) and (b) the energy E_T and the entropy H_T of $\mathscr{B}_T$ in equilibrium may depend on the temperature ϑ and the volumes V^1 and V^2, whereas the energies E^1, E^2 and the entropies H^1, H^2 may depend on ϑ, V^1 (for $\alpha = 1$) and ϑ, V^2 (for $\alpha = 2$). Therefore E_T can be written as $E_T(H_T, V^1, V^2)$ and E^α as $E^\alpha(\vartheta, V^\alpha)$ and for these functions the following relations are postulated:

(c) $$\left(\frac{\partial E^T}{\partial V^1}\right)_{H_T, V^2} = \left(\frac{\partial E^1}{\partial V^1}\right)_{H^1} \text{ and } \left(\frac{\partial E_T}{\partial V^2}\right)_{H_T, V^1} = \left(\frac{\partial E^2}{\partial V^2}\right)_{H^2}$$

This last postulate is usually motivated by the form of the energy equation in a 'quasistatic and adiabatic process'.

I shall now use these axioms to obtain information about $\Lambda|_E$ (i.e. Λ in equilibrium).

In equilibrium the entropy production Σ, which is the LHS of 3.7, assumes its minimum value namely zero and a necessary condition for this is

$$\left.\frac{\partial \Sigma}{\partial \vartheta}\right|_E = 0 \text{ or } \frac{\partial \eta|_E}{\partial \vartheta} = -\Lambda|_E \frac{\partial \varepsilon|_E}{\partial \vartheta} \tag{4.1}$$

All conditions 3.6 are identically satisfied in equilibrium except the last one which shows that $t_{ik}|_E$ has the form $t_{ik}|_E = -p|_E \delta_{ik}$ and that

$$\frac{\partial \eta|_E}{\partial \rho} = -\Lambda|_E \left(\frac{\partial \varepsilon|_E}{\partial \rho} - \frac{p|_E}{\rho^2}\right) \tag{4.2}$$

$p|_E$ is called the pressure in equilibrium.

The equations 4.1 and 4.2 can be combined to

$$\mathrm{d}\eta|_E = -\Lambda|_E \frac{\partial \varepsilon|_E}{\partial \vartheta} \mathrm{d}\vartheta - \Lambda|_E \left(\frac{\partial \varepsilon|_E}{\partial \rho} - \frac{p|_E}{\rho^2}\right) \mathrm{d}\rho \tag{4.3}$$

or, when $\eta|_E$ is considered as a function of $\varepsilon|_E$ and ρ

$$\mathrm{d}\eta|_E = -\Lambda|_E \left(\mathrm{d}\varepsilon|_E + p|_E \,\mathrm{d}\left(\frac{1}{\rho}\right)\right) \tag{4.4}$$

or, finally, by multiplying by the mass M of $\mathscr{B}$ and introducing the internal energy E and entropy H of $\mathscr{B}$ in equilibrium:

$$\mathrm{d}H = -\Lambda|_E(\mathrm{d}E + p\,\mathrm{d}V) \tag{4.5}$$

Let us now consider a process between two equilibria characterized by (E, V) and $(E + \mathrm{d}E, V + \mathrm{d}V)$ such that $\mathrm{d}E + p\,\mathrm{d}V \neq 0$; if such a process occurs without exchange of entropy through the surface of $\mathscr{B}$, $\mathrm{d}H$ is positive and hence, from 4.5, $\Lambda|_E$ cannot be zero. If it is further assumed that $\mathrm{d}H$ is finite, then $\Lambda|_E$ must also be finite and, since the initial equilibrium (E, V) is arbitrary, $\Lambda|_E$ must either be a positive valued or a negative valued function.

Let us now consider the situation to which axiom (b) above applies: Both subbodies $\mathscr{B}^\alpha$ are in equilibrium, so that 4.5 holds for both:

$$\mathrm{d}E^\alpha = -\frac{1}{\Lambda^\alpha|_E}\mathrm{d}H^\alpha - p^\alpha\,\mathrm{d}V^\alpha \tag{4.6}$$

and thus we have for the energy $E = E^1 + E^2$ of $\mathscr{B}_T$

$$\mathrm{d}E_T = -\frac{1}{\Lambda^1|_E}\mathrm{d}H^1 - \frac{1}{\Lambda^2|_E}\mathrm{d}H^2 - p^1\,\mathrm{d}V^1 - p^2\,\mathrm{d}V^2 \tag{4.7}$$

This seems to suggest that E_T depends on the four variables H^1, H^2, V^1, V^2; however, E_T as well as H_T, the entropy of $\mathscr{B}_T$, depend on the common temperature ϑ of $\mathscr{B}^\alpha (\alpha = 1, 2)$ and on V^1, V^2 so that we may write

$$\mathrm{d}E_T(H_T, V^I, V^{II}) = \frac{\partial E_T(H_T, V^1, V^2)}{\partial H_T}\mathrm{d}H_T + \frac{\partial E_T(H_T, V^1, V^2)}{\partial V^1}\mathrm{d}V^1 + \frac{\partial E_T(H_T, V^1, V^2)}{\partial V^2}\mathrm{d}V^2 \tag{4.8}$$

and with $\mathrm{d}H_T = \mathrm{d}H^1 + \mathrm{d}H^2$ and axiom (c) we conclude by comparing 4.8 and 4.7:

$$\frac{\partial E_T(H_T, V^1, V^2)}{\partial H_T} = -\frac{1}{\Lambda^\alpha|_E} \quad (\alpha = 1, 2) \tag{4.9}$$

The LHS of 4.9 may be considered to be a function of ϑ, V^1, V^2 whereas the RHS could only depend on ϑ, V^1 (for $\alpha = 1$) or on ϑ, V^2 (for $\alpha = 2$). Hence $\Lambda^\alpha|_E$ cannot depend on V^α at all and we have

$$\Lambda^1|_E = \Lambda^2|_E \equiv \Lambda|_E \equiv -\frac{1}{T(\vartheta)} \tag{4.10}$$

or, in words, $\Lambda|_E$ is a universal function of the temperature alone, its negative reciprocal T is called the *absolute temperature.*

Since $\Lambda|_E$ is finite and non-vanishing, so is T, and by one single experiment with any simple heat-conducting fluid we may determine its sign: *The* *absolute temperature T is positive.* Combining 4.10 and 4.5 we obtain the Gibbs equation

$$\mathrm{d}H = (1/T)(\mathrm{d}E + p\,\mathrm{d}V) \tag{4.11}$$

5. CONDITIONS ON LINEAR CONSTITUTIVE RELATIONS

We may be inclined to consider constitutive relations which are linear in $\vartheta_{,i}$ and $\dot{\vartheta}$ as good approximations to the exact relations in the 'neighbourhood' of uniform equilibrium. The most general representations of such linear constitutive relations are:

$$\begin{aligned} t_{ik}^L &= -(p|_E + p_{\dot{\vartheta}}|_E\dot{\vartheta})\,\delta_{ik} \\ q_k^L &= -\kappa|_E\vartheta_{,k} \\ \varepsilon^L &= \varepsilon|_E + \varepsilon_{\dot{\vartheta}}|_E\dot{\vartheta} \\ \Phi_i^L &= \varphi|_E\vartheta_{,i} \\ \eta^L &= \eta|_E + \eta_{\dot{\vartheta}}|_E\dot{\vartheta} \end{aligned} \tag{5.1}$$

where all coefficients $p|_E$, $p_{\dot{\vartheta}}|_E$, $\kappa|_E$, $\varepsilon|_E$, $\varepsilon_{\dot{\vartheta}}|_E$, $\varphi|_E$, $\eta|_E$ and $\eta_{\dot{\vartheta}}|_E$ may depend on ρ and ϑ.

If the relations 5.1 and the linear representation of Λ (see 4.10)

$$\Lambda = -\frac{1}{T} + \Lambda_{\dot{\vartheta}}|_E\dot{\vartheta} \tag{5.2}$$

are introduced in the equations 3.6, 3.8, they yield the following restrictions on the coefficients:

$$\frac{\partial\eta|_E}{\partial\vartheta} = \frac{1}{T}\frac{\partial\varepsilon|_E}{\partial\vartheta}, \qquad \eta_{\dot{\vartheta}}|_E = \frac{1}{T}\varepsilon_{\dot{\vartheta}}|_E, \qquad p_{\dot{\vartheta}}|_E = -\rho^2 T\Lambda_{\dot{\vartheta}}|_E\left(\frac{\partial\varepsilon|_E}{\partial\rho} - \frac{p|_E}{\rho^2}\right)$$

$$\frac{\partial\eta|_E}{\partial\rho} = \frac{1}{T}\left(\frac{\partial\varepsilon|_E}{\partial\rho} - \frac{p|_E}{\rho^2}\right), \quad \varphi|_E = -\frac{\kappa|_E}{T}, \tag{5.3}$$

While $5.3_{1,2}$ have already been discussed in Section 4, 5.3_4 states a new result: The linear entropy flux is equal to the linear heat flux divided by the absolute temperature:

$$\Phi_i^L = q_i^L/T \tag{5.4}$$

This relation, however, does not hold in a non-linear theory in general, although that is commonly assumed in thermodynamic theories.

If the linear relations 5.1 are introduced into the residual inequality 3.7 we obtain after some straightforward calculation and by use of the previous results 5.3

$$\varepsilon_{\dot{\vartheta}}|_E \leqslant \frac{T^2}{\mathrm{d}T/\mathrm{d}\vartheta}\Lambda_{\dot{\vartheta}}|_E\frac{\partial\varepsilon|_E}{\partial\vartheta}, \quad \kappa|_E\frac{\mathrm{d}T}{\mathrm{d}\vartheta} \geqslant 0 \tag{5.5}$$

If, as is usually done, the empirical temperature ϑ is chosen to be identical to the absolute temperature, 5.5_2 states the well known result that the heat conductivity is non-negative.

6. THE LINEAR EQUATION OF HEAT CONDUCTION

We might be content in this presentation of a new approach to the old subject of non-equilibrium thermodynamics to derive assumptions and results of the established theories on that subject and show their limitations. In this sense, if there were no other result than the derivation of the Gibbs equation 4.11, of the inequality 5.5 on κ, and of the *linear* relation 5.4 between the fluxes of entropy and heat, we should still have been able to compose a list of all the assumptions and axioms that lead to these results in a manner which allows us to appreciate their validity.

There is, however, a bonus, even for such a singularly simple material as the simple heat-conducting fluid considered here: When we introduce $5.1_{2,3}$ into 2.1_3 and consider a process with ρ time-independent and uniform and $\vartheta_{,i} \equiv 0$, we find in a linear theory the following form of the linear differential equation of heat conduction

$$\left(\frac{\partial \varepsilon|_E}{\partial \vartheta}\right)\frac{\partial \vartheta}{\partial t} + \varepsilon_{\dot{\vartheta}}|_E \frac{\partial^2 \vartheta}{\partial t^2} = \kappa|_E \Delta\vartheta$$

Provided that $\varepsilon_{\dot{\vartheta}}|_E > 0$, this is a hyperbolic differential equation which implies a finite speed for the propagation of disturbances in temperature; and the fact that the theory presented here allows for this finite speed gives it a marked advantage over other thermodynamic theories which predict infinite speed. Of course, I cannot prove here that indeed $\varepsilon_{\dot{\vartheta}}|_E > 0$ holds; what we *can* conclude from 5.5_1 is that $\varepsilon_{\dot{\vartheta}}|_E$ *may* be positive.

REFERENCES

1. B. D. Coleman and W. Noll, *Arch. Ration. Mech. Anal.* **13**, 231 (1963).
2. I. Müller, *Arch. Ration. Mech. Anal.* **34**, 259 (1969).
3. C. Truesdell and R. Toupin, *Handbuch der Physik*, **Bd III/1**, 704. Springer: Berlin (1960).
4. W. Noll, *Arch. Ration. Mech. Anal.* **2**, 197 (1968).
5. C. Truesdell and R. Toupin, *Handbuch der Physik*, **Bd III/1**, 453. Springer: Berlin (1960).
6. R. Courant and D. Hilbert, *Methods of Mathematical Physics*, Vol. II, p 39ff. Interscience: New York (1962).

ON THE VALIDITY OF THE CLAUSIUS–DUHEM INEQUALITY

J. U. KELLER

Institut für Theoretische Physik, Technische Hochschule Aachen, Templergraben 55, D-51Aachen, West Germany

ABSTRACT

The author discusses the applicability of various theories for describing processes in continuous matter and deals in some detail with the entropy-free thermodynamics of irreversible processes, basing his discourse on the work of J. Meixner in this field.

Today there are essentially three different phenomenological theories available for describing processes in continuous matter:

(1) The classical Thermodynamics of Irreversible Processes (TIP)[1].

(2) The Non-linear Field-Theories of Mechanics and their thermodynamic extensions (NFT)[2].

(3) The Entropy-free Thermodynamics of Processes (ETIP)[3,4].

These theories have in common the conservation laws of continuous matter and the laws of thermostatics. They differ essentially in the arguments on which the constitutive equations (CE) are based. Classical TIP starts from a generalized Clausius–Duhem Inequality (CDI) for the well-known (specific) thermostatic entropy s_{st}

$$\rho \dot{s}_{st} + \operatorname{div} \mathbf{J}_{st} \geqslant 0 \tag{1}$$

The quantity s_{st} is well defined by the second law of thermodynamics and is a function $s_{st} = s_{st}(u, \rho)$ for fluids and a function $s_{st} = s_{st}(u, F_{ij})$ for solid systems. Here u means the (specific) internal energy, $\rho \ldots$ density, $F_{ij} \ldots$ deformation gradient, $\dot{s}_{st} \ldots$ substantial time derivative of s_{st} and $\mathbf{J}_{st} \ldots$ current of s_{st}. In the case of thermodynamic systems without diffusion one has

$$\mathbf{J}_{st} = \mathbf{q}/T \tag{2}$$

with $\mathbf{q} \ldots$ heat flux and $T \ldots$ absolute temperature. The inequality 1 has been proved valid in quite a few applications[1]. Nevertheless it must be emphasized that in TIP its validity is postulated and cannot be concluded from the second law (ref. 1, p 424; ref. 4, p 33 etc.).

To get CE in NFT one starts with a generalized CDI

$$\rho \dot{s}_{N} + \operatorname{div} \mathbf{J}_{N} \geqslant 0 \tag{3}$$

for a hypothetical (specific) Non-equilibrium Entropy (NEE), s_N. Here $\dot{s}_N$ means the substantial time derivative and $\boldsymbol{J}_N$ the current of s_N. Unfortunately in NFT neither s_N nor $\boldsymbol{J}_N$ is defined uniquely. Therefore today the physical significance of these quantities is not clear. Beside this, in NFT nothing is said about the physical meaning of the inequality 3, the validity of which is merely postulated.

Now within the framework of a phenomenological theory the concept of NEE s_N is at least questionable. The reason is that s_N cannot be defined uniquely. This has been shown by J. Meixner[3].

Therefore one may ask whether thermodynamics of processes can be developed without postulating the inequality 1 and without using the concept of NEE or the inequality 3. This is indeed possible. J. Meixner[4] has developed a theory of processes in continuous matter without using the inequalities 1 or 3 and without using the concept of NEE. This theory will therefore be called ETIP. Instead of the generalized CDI 1 or 3 in ETIP one has the Fundamental Inequality (FI) [ref. 4, p 88, (6)] which is mainly a consequence of the integral form of the second part of the second law

$$S_{st}(B) - S_{st}(A) \geqslant \int_A^B \frac{\delta Q}{T} \tag{4}$$

Here A and B denote two *equilibrium* states where B is posterior to A and δQ is the heat the system has been supplied with at a temperature T. This inequality has been proved by R. Clausius[5]. In contrast to this, the differential form of the second part of the second law

$$\mathrm{d}S_N \geqslant \delta Q/T \tag{5}$$

is a postulate and cannot be concluded from expression 4 or any other equivalent form of the second law.

Now we shall write down the FI for a fluid or solid system which consists of one component and one phase only and which is not affected by any body forces. Arbitrary time-dependent forces acting on the surface of the system may be prescribed by boundary conditions.

The system can exchange heat and mechanical work through its surface with its surroundings.

In the interior of the system heat conduction, internal friction and a process which will be called 'internal energy exchange' may appear. Following J. Meixner[4], for each element of such a system and each real process which starts at $t = -\infty$ in an equilibrium state, the FI

$$\int_{-\infty}^{t_0} \left\{ \left(\frac{1}{T_{st}} - \frac{1}{T} \right) \dot{u} + \frac{1}{\rho} \left(\frac{P_{st\,ik}}{T_{st}} - \frac{P_{ik}}{T} \right) \eta_{ik} + \frac{1}{\rho} q_i \partial_i \frac{1}{T} \right\} \mathrm{d}t \geqslant 0 \tag{6}$$

holds. Here T_{st} means the thermostatic temperature, T ... thermodynamic temperature, $P_{st\,ik}$ ... thermostatic pressure tensor, P_{ik} ... thermodynamic pressure tensor, v_i ... velocity, $\partial_i = \partial/\partial x_i$, $2\eta_{ik} = \partial_i v_k + \partial_k v_i$ $(i, k = 1, 2, 3)$, $F_{ij} = (\partial/\partial X_j)x_i(X_k, t)$ ... deformation gradient, $x_i(X_k, t)$ ... trajectory of a mass element $\dot{} = (\partial/\partial t) + v_i(\partial/\partial x_i)$ ... substantial time derivative, X_k ... initial position of the mass element.

All quantities are functions of position x_i and time t. In all quantities appearing in expression 6 the dependence on position x_i may be thought to be changed into a dependence on time by use of the trajectory $x_i(X_k, t)$ of the element. In other words, expression 6 holds for a fixed material element along its path. The following relations hold

$$\mathrm{d}s_{st}(u_1 F_{ij}) = \frac{1}{T_{st}}\,\mathrm{d}u + \frac{1}{\rho}\sum_{ikl}(F^{-1})_{ik}\,(P_{st\,kl}/T_{st})\,\mathrm{d}F_{ei} \tag{7}$$

$$2\eta_{ik} = \sum_{l}\{\dot{F}_{il}(F^{-1})_{lk} + \dot{F}_{kl}(F^{-1})_{li}\} \tag{8}$$

$$\rho(t) = \rho(-\infty)\,(\mathrm{Det}\,|F_{ij}(t)|)^{-1} \tag{9}$$

The quantities T_{st}, $P_{st\,ik}$ are defined by equation 7. The thermodynamic temperature T and pressure tensor P_{ik} describe the temperature and the stresses which are actually realized in the system at time t and position x_i. In general T will differ from T_{st} in the same way that P_{ik} will differ from $P_{st\,ik}$.

The FI holds for all times t_0, arbitrary mass elements and all real processes which start in an equilibrium state at $t = -\infty$, i.e. all systems of continuous functions

$$\{u(t'), F_{ij}(t'), q_i(t'), \rho(-\infty), \qquad -\infty < t' \leqslant t\} \tag{10}$$

The integrand of the FI is, apart from a factor ρ, the 'production σ_{st} of thermostatic entropy', i.e.

$$\rho\dot{s}_{st} + \partial_i(q_i/T) = \sigma_{st} \tag{11}$$

$$\sigma_{st} = \left(\frac{1}{T_{st}} - \frac{1}{T}\right)\rho\dot{u} + \left(\frac{P_{st\,ik}}{T_{st}} - \frac{P_{ik}}{T}\right)\eta_{ik} + q_i\partial_i(1/T) \tag{12}$$

σ_{st} need not be positive only but also may assume negative values! In view of the generalized CDI 1, 3, the FI 6 and the relation 11, the following questions arise:

(1) Do materials exist for which one may deduce from 4 or 6 respectively the validity of the CDI 1 for s_{st}? In other words: For what kinds of materials may the thermostatic entropy s_{st} also be interpreted as NEE?

(2) Assume that question 1 is answered in the negative for the material under consideration. Is it possible to define at least a certain NEE in such a manner that both the FI 6 and the CDI 3 for s_N hold?

Here we shall only consider question 1. In the following we give the answer to this question for the system mentioned above (one component, one phase, no body forces) and for some special classes of materials. Some results concerning question 2 will be the subject of another publication.

Now the answer to question 1 depends mainly on the structure of the CE of the material under consideration. Therefore we must make a few remarks on these equations.

One usually assumes (ref. 2, p 56 etc.; ref. 4, p 91) that the present state of a certain element of mass is uniquely determined by the 'history' of the element. Here we confine ourselves to so-called 'simple materials' which are characterized by the fact that their state at time t is completely specified by the histories of u, q_i and F_{ik}. All other quantities, especially the coefficients of u, q_i, η_{ik} which appear in the FI 6,

$$\frac{1}{T_{st}} - \frac{1}{T}, \frac{P_{st\,ik}}{T_{st}} - \frac{P_{ik}}{T}, \partial_i \frac{1}{T} \tag{13}$$

must be functionals of the history 10. These functionals must be single-valued, continuous and invariant against translation of time. Therefore the CE may be written as:

$$\begin{aligned} &\frac{1}{T_{st}} - \frac{1}{T} = \mathscr{F}_0\{\cdot/.\} && \text{(a)} \\ &\partial_i \frac{1}{T} = \mathscr{F}_{1,i}\{\cdot/.\} \qquad i, k = 1, 2, 3 && \text{(b)} \\ &\frac{P_{st\,ik}}{T_{st}} - \frac{P_{ik}}{T} = \mathscr{F}_{2,ik}\{\cdot/.\} && \text{(c)} \end{aligned} \tag{14}$$

Here the bracket $\{\cdot/.\}$ is given by 10. The equations 14b and 14c are generalized laws of heat conduction and internal friction respectively. Equation 14a describes the so-called 'internal energy exchange'. Example: Relaxation of temperature in a homogeneous mixture of ortho- and para-hydrogen. The functionals $\mathscr{F}_\alpha(\alpha = 0, 1, 2)$ must vanish in thermostatic equilibrium. Therefore, it is presumed that for all $t \geqslant t_0$

$$\dot{u}(t) = 0, \quad q_i(t) = 0, \quad \dot{F}_{ij}(t) = 0$$

the relation

$$\lim_{t\to\infty} \mathscr{F}_\alpha\{u(t') \ldots \rho(-\infty); -\infty < t' \leqslant t\} = 0, \quad \alpha = 0, 1, 2 \tag{15}$$

holds. Moreover, the functions $\mathscr{F}_\alpha$ must obey the FI 6 and the principle of material frame indifference (ref. 2, p 44). For literature, see ref. 4, pp 93, 43 respectively.

Two important classes of materials are those of differential type and of integral type (ref. 2, p 93, 98; ref. 4, p 95). The following statements hold*:

(1) For simple thermodynamic materials of the differential type with arbitrary complexity $r \geqslant 1$, one can conclude under certain conditions from the FI 6 that the CDI 1 holds for s_{st}. To be more precise we formulate a theorem:

C_1: The material is simple, of the differential type and has arbitrary complexity $r \geqslant 1$. That is, the functionals are restrained to single-valued and continuous (generally non-linear) functions.

$$\mathscr{F}_\alpha\{\cdot/.\} = \Phi_\alpha^{(r)}(-), \quad \alpha = 0, 1, 2 \tag{16}$$

$$\text{with } (-) = (\ldots \dot{u}^{(k)}(t) \ldots, \ldots \dot{q}_i^{(l)}(t) \ldots, \ldots \dot{F}_{ij}^{(k)}(t) \ldots, \rho(-\infty), k = 0 \ldots r, l = 0 \ldots r-1) \tag{17}$$

C_2: The equilibrium condition 15. In view of 16 and 17 this condition reads

$$\Phi_\alpha^{(r)}(u, 0, 0 \ldots, 0 \ldots 0, F_{ij}, 0 \ldots, \rho(-\infty)) = 0$$

* C... condition, pre-requisite, T... theorem, P... proof.

C_3: The differences between T_{st} and T, $P_{st\,ik}$ and P_{ik} and the temperature gradient $\partial_i T$, which will occur in the material during the process, exist for all finite and infinite values of the *higher* time derivatives of u, q_i, F_{ij} namely $\dot{u}^{(k)}$, $\dot{q}_i^{(k-1)}$, $\dot{F}_{ij}^{(k)}$, $k = 2 \ldots r$. In other words: Let $\dot{u}^{(k)} = a_k$, $\dot{q}_i^{(k-1)} = b_{ik}$, $\dot{F}_{ij}^{(k)} = c_{ijk}$ be arbitrary constants, then

$$\lim_{\tau \to \infty} \Phi_\alpha^{(r)}(u, \dot{u} \ldots \tau a_k \ldots, q_i \ldots \tau b_{ik} \ldots, F_{ij}, \dot{F}_{ij} \ldots \tau c_{ijk} \ldots)$$

$= \hat{\Phi}_\alpha^{(r)}(u, \dot{u}, q_i, F_{ij}, \dot{F}_{ij}) \ldots$ exists and is a single-valued and continuous function of its arguments.

C4: The Fundamental Inequality 6 with 14 and 16, i.e.

$$\int_{-\infty}^{t_0} \{\dot{u}\Phi_0^{(r)} + \frac{1}{\rho}\eta_{ik}\Phi_{2,ik}^{(r)} + \frac{1}{\rho}q_i\Phi_{1,i}^{(r)}\}\, dt \geqslant 0$$

for all histories 10 the functions of which are differentiable r times and vanish or converge to a constant value for $t' \to -\infty$ faster than t'^n with $n = 1, 2, \ldots$ arbitrary.

T: $\sigma_{st} = \rho\dot{u}\Phi_0^{(r)} + \eta_{ik}\Phi_{z,ik}^{(r)} + q_i\Phi_{1,i}^{(r)} \geqslant 0$

When $r = 1$ the condition C_3 is obeyed trivially because the functions $\Phi_\alpha^{(1)}$ ($\alpha = 0, 1, 2$) do not depend on the higher time derivatives $\ddot{u} \ldots$ etc. In this case the CE 14 with 16 are, disregarding the distinction between T_{st} and T, the CE of TIP (ref. 1, p 425). Therefore one can say that, as a consequence of the above theorem, in TIP the CDI 1 for s_{st} does indeed hold. Moreover, one can say that in fact the CDI 1 for s_{st} holds for a much larger class of materials than those used in TIP, namely for materials which are of the differential type with arbitrary complexity $r \geqslant 1$ and which obey C_3.

(2) For simple thermodynamic materials of the integral type of order 1 and complexity 1 which obey a certain principle of fading memory the CDI 1 cannot hold. This is a consequence of the following theorem:

C_1: The material is simple, of the integral type*, of order 1 and complexity 1. That is, the functionals are restricted to

$$\mathscr{F}_\alpha\{\dot{\%}\} = \Lambda_\alpha^{(1)}(—) + \int_{-\infty}^{t} dt'\, \kappa_\alpha^{(1)}(|) \quad \alpha = 0, 1, 2 \tag{18}$$

with

$$\begin{aligned} (—) &= [u(t), \dot{u}(t), q_i(t), F_{ij}(t), \dot{F}_{ij}(t), \rho(-\infty)] \\ (|) &= [u(t'), \dot{u}(t'), q_i(t'), F_{ij}(t'), \dot{F}_{ij}(t'), t - t', \rho(-\infty)] \end{aligned} \tag{19}$$

and $\{\dot{\%}\}$ given by 10.

Further, $\Lambda_\alpha^{(1)}$ and $\kappa_\alpha^{(1)}$ are single-valued, finite and continuous functions of their arguments. Moreover, $\kappa_\alpha^{(1)}$ is integrable in $-\infty < t' \leqslant t$ for all histories 10 the functions of which vanish or converge to a constant value for $t' \to -\infty$ more rapidly than t'^n with $n = 1, 2 \ldots$ arbitrary. The integral

* Our definition of materials of the integral type of order 1 is somewhat more general than the definition given by Truesdell and Noll in ref. 2. (We don't presume the integral kernels $\kappa_\alpha^{(1)}$ to be resolvable in factors which depend on different arguments.)

in equation 18 must converge for $t' \to -\infty$ uniformly with respect to t for all $t \geqslant t_0$.

C_2: The equilibrium condition 15. We confine attention here to materials 18 for which the following conditions hold:

$$\Lambda_\alpha^{(1)}(u, 0, 0, F_{ij}, 0, \rho(-\infty)) = 0 \qquad \text{(a)}$$

$$\kappa_\alpha^{(1)}(u, 0, 0, F_{ij}, 0, t - t', \rho(-\infty)) = 0 \qquad \text{(b)} \qquad (20)$$

$$\lim_{t \to \infty} (t - t')\, \kappa_\alpha^{(1)}(u, \dot{u}, q, F_{ij}, \dot{F}_{ij}, t - t', \rho(-\infty)) = 0 \qquad \text{(c)}$$

This limit may be approached uniformly with respect to t' for all $t' \leqslant t_0$.

The assumption 20c means that the 'memory' of the material decreases for $t' \to -\infty$ faster than t'^{-1}. It can easily be seen that 20 is sufficient, though not necessary, for 15 to hold.

C_3: The CDI 1 with 11, 12, 14 and 18, i.e.

$$\dot{u}(t)\left\{\Lambda_0^{(1)} + \int_{-\infty}^{t} \mathrm{d}t'\kappa_0^{(1)}\right\} + \frac{1}{\rho}\eta_{ik}\left\{\Lambda_{2ik}^{(1)} + \int_{-\infty}^{t} \mathrm{d}t'\kappa_{2,ik}^{(1)}\right\}$$
$$+ \frac{1}{\rho} q_i \left\{\Lambda_{1,i}^{(1)} + \int_{-\infty}^{t} \mathrm{d}t'\kappa_{1,i}^{(1)}\right\} \geqslant 0$$

holds for all t and all histories 10, the functions of which are at least differentiable.

T: $\kappa_\alpha^{(1)}(1) = 0$.

P: See ref. 7.

Therefore when the memory functions $\kappa_\alpha^{(1)}$ in 18 are not identically zero, one can conclude that the CDI 1 for s_{st} cannot be valid for all times t.

(3) For simple thermodynamic materials of the integral type of order 1 and complexity $r = 2$, i.e. the functions $\Lambda_\alpha^{(2)}$ $\kappa_\alpha^{(2)}$ may depend on $(u, \dot{u}, \ddot{u}, q, \dot{q}, F_{ij}, \dot{F}_{ij}, \ddot{F}_{ij})$, the CDI 1 may or may not hold[7].

The question whether CDI 1 holds for simple materials of integral type of higher order and higher complexity is postponed to future investigations.

REFERENCES

[1] J. Meixner and H. G. Reik, *Handbuch der Physik*, **Bd III**/2, pp 413–523. S. Flügge (Ed.). Springer: Berlin (1959).
S. R. de Groot and P. Mazur, *Non-equilibrium Thermodynamics*. North Holland Pub. Co.: Amsterdam (1962).

[2] C. Truesdell and W. Noll, *Handbuch der Physik*, **Bd III**/3, pp 1–602, S. Flügge (Ed.). Springer: Berlin (1965).
B. D. Coleman, *Arch. Ration. Mech. Anal.* **17**, 1–230 (1964).

[3] J. Meixner, *J. Appl. Mech.* Ser. E **33**, 481 (1966).
J. Meixner, *Rheologica Acta*, **7**, 8 (1968).

[4] J. Meixner, *Z. Phys.* **219**, 79 (1969).
J. Meixner, *Arch. Ration. Mech. Anal.* **33**, 33 (1969).

[5] R. Clausius, *Die mechanische Wärmetheorie*, Vol. I, 2nd ed., pp 206, 224. Braunschweig (1876).

[6] J. U. Keller, *Physica*, **46**, 451 (1970).

[7] J. U. Keller, *Z. Naturforschung*, **24a**, 1989 (1969).

ON THE RATIONAL FOUNDATIONS OF THE NON-EQUILIBRIUM THERMODYNAMICS OF ORDINARY FLUID MIXTURES

JOHN L. BARTELT* AND FREDERICK H. HORNE

Department of Chemistry, Michigan State University, East Lansing, Michigan

ABSTRACT

Despite the arbitrary and ambiguous character of certain fundamental parts of the usual theory of non-equilibrium thermodynamics of mixtures, it has been successful in describing a variety of experimental situations. The goals of the research partially reported in this paper have been (1) to strengthen the foundations of the usual, practical theory and (2) to extend its usefulness to experimental situations hitherto regarded as too complicated to analyse. The principles and methods of rational mechanics are used to deduce balance equations which take full account of the kinetic energy of each component and of the partial stress tensor of each component. Similar equations have been obtained previously by similar techniques. The new results reported here stem from a particular choice of independent variables; namely, for a ν-component, non-reacting mixture of fluids, the independent variables are the ν component densities and their gradients, the temperature and its gradient, the $\nu - 1$ independent diffusive velocities, and the ν symmetric and $\nu - 1$ antisymmetric parts of the gradients of component velocity. For the special but very important case of an ordinary fluid mixture, whose constitutive relations are linear in the diffusion velocities and the independent gradients, the source term in the entropy balance equation is a bi-linear form in the diffusion velocities and the independent gradients. This bi-linear form for the entropy source term, which has eluded previous workers, has two immediate consequences: (1) Coupled with the second law, it leads to unambiguous conditions for equilibrium; namely, for equilibrium with respect to all processes in a ν-component, non-reacting fluid mixture, it is necessary and sufficient that the following shall all vanish: the temperature gradient, the diffusion velocities, and the symmetric and antisymmetric parts of the gradients of component velocity. (2) It permits full comparison with the practical theory, wherein the entropy source plays a central role. It also opens new experimental avenues since it contains terms which arise from component kinetic energy and component stress tensor.

I. INTRODUCTION

The 'thermodynamics of irreversible processes'[1] has enjoyed a history of practical success in describing a diversity of transport phenomena and a wide range of material behaviours. Despite its success and apparent completeness, the macroscopic theory as usually presented is deficient in several

* Now at Bell Telephone Laboratories, Inc., Murray Hill, New Jersey.

respects, especially in its treatment of mixtures. The procedure for obtaining linear relations between fluxes and forces is arbitrary and ambiguous. 'Curie's theorem' is often used inconsistently, and rational account of the number of independent variables is seldom taken. The mechanics of stress and flow are adequately treated for pure fluids but not for mixtures of fluids; in the latter case only the composite fluid is considered. Terms arising from the kinetic energy of diffusion and the so-called inertial and viscous terms are always neglected in practical applications although explicit formulas are available. In short, the usual theory is really successful only for mechanical equilibrium, when pressure and velocity gradients vanish. In recent years Coleman and Noll[2] and Coleman and Mizel[3] have developed a new approach to thermodynamics built on a firm foundation in mechanics—precisely the area slighted in the conventional theory. In view of its mathematical origins it is not surprising that the new approach to thermodynamics is a model of rigour and that it remedies the deficiencies in mechanics of the conventional non-equilibrium theory, at least in the case of pure components (single materials).

Having great respect for both the practical success and microscopic relevance of the conventional theory and the rigour and completeness of the new approach, we present here a formulation of the thermodynamic theory of mixtures which is developed using Coleman's methods[2,3], but which is aimed at reinforcing and supplementing the previous results. The approach here is quite similar to that of Müller[4], and likewise has its origins in the work of Truesdell[5]. Müller, however, did not obtain the bi-linear form for the entropy production which is central to the earlier theory[1]. We obtain the bi-linear form by choosing expressions for the heat flux slightly different from, but no less general than, those of Müller. For brevity only the barest rudiments of the fundamental theory and only an outline of our methods are presented here. Full details may be found elsewhere[6].

Generalized linear phenomenological equations are deduced rationally by application of the entropy inequality to general expressions, called constitutive relations, which relate dependent variables to a complete set of independent variables. In particular, we assume that the basic local mechanical and thermal variables energy, entropy, heat flux, entropy flux, stresses and interaction forces are determined by the basic independent variables temperature, temperature gradient, component densities, component density gradients, component velocities and component velocity gradients. The physical principles upon which the theory is constructed are: (i) *balance* of mass, energy, linear momentum, and angular momentum; (ii) *equipresence*, which means that an independent variable present in one constitutive relation is present in all unless its presence in a particular constitutive relation contradicts some other physical principle or any assumed symmetry of the material in question; (iii) *material objectivity*, which forbids descriptions which depend on the frame of reference of the observer, no matter how he moves; (iv) *positive entropy production*, which is nothing but the second law of thermodynamics, and which we call the *entropy inequality*. The procedure is to formulate the most general set of constitutive relations consistent with principles (i), (ii) and (iii) and with any other particular assumptions about the material in question. We then establish the necessary and sufficient conditions on the constitutive relations so that the entropy inequality holds identically. The resulting, reduced constitutive relations together with the basic dynamic equations comprise our complete set of macroscopic transport equations.

With Truesdell[5], we assume that the mean motion of the fluid is determined solely by the motions of the components, and that the latter motions are determined by the usual stress properties plus interaction (exchange o momentum). For clarity, we restrict the treatment to binary mixtures there is no conceptual difficulty in extending it to any finite number oı components. Since we wish to achieve the theory which corresponds most directly to the successful practical theory[1], we further restrict the treatment to the special class of mixtures whose constitutive relations are linear in the temperature gradient, in the density gradients, and in objective combinations of the velocities and their gradients. A material of this special class will be called an *ordinary* binary fluid mixture. A conviction that this type of material accurately represents the binary fluid mixtures commonly encountered in the chemist's laboratory is the reason for the label 'ordinary'. The appropriateness of this label can be evaluated only by testing the ability of the constitutive relations to represent accurately the observed responses of the material to variations of independent variables. We neglect entirely the question of chemical reactions.

II. THERMODYNAMIC PROCESS

The generalization for mixtures of the definition of Coleman and Noll[2] is: A process is a *thermodynamic process* for a ν-component mixture if it can be described by the set of $(5\nu + 6)$ functions,

(i) the component densities ρ_α; (ii) the component velocities $\mathbf{v}^\alpha$; (iii) the interaction forces $\mathbf{m}^\alpha$ acting on component α due to interaction with other components; (iv) the external body forces $\mathbf{b}^\alpha$ acting on component α; (v) the partial stress tensors σ^α; (vi) the specific internal energy ε_I; (vii) the temperature T; (viii) the radiative heat supply r (radiation from the external world absorbed by the fluid); (ix) the heat flux $\mathbf{q}$; (x) the specific entropy S; and (xi) the entropy flux $\mathbf{f}$,

which satisfy:

(a) the balance of mass for each component*,

$$\dot{\rho}_\alpha + u_k^\alpha(\partial\rho_\alpha/\partial x_k) + \rho_\alpha(\partial v_k^\alpha/\partial x_k) = 0 \tag{2.1}$$

where

$$\rho = \sum_{\alpha=1}^{\nu} \rho_\alpha, \quad \rho\mathbf{v} = \sum_{\alpha=1}^{\nu} \rho_\alpha \mathbf{v}_\alpha, \quad \mathbf{u}_\alpha = \mathbf{v}_\alpha - \mathbf{v}, \quad \sum_{\alpha=1}^{\nu} \rho_\alpha \mathbf{u}_\alpha = 0 \tag{2.2}$$

and where, for any function g,

$$\dot{g} = (\partial g/\partial t) + v_k(\partial g/\partial x_k) \tag{2.3}$$

(b) the balance of each vector component of linear momentum for each species,

$$\rho_\alpha \dot{v}_j^\alpha - (\partial\sigma_{kj}^\alpha/\partial x_k) + \rho_\alpha u_k^\alpha(\partial v_j^\alpha/\partial x_k) = \rho m_j^\alpha + \rho_\alpha b_j^\alpha \tag{2.4}$$

* A repeated Latin miniscule subscript on a vector or tensor indicates summation from 1 to 3.

(c) the balance of angular momentum for the mixture,

$$\sum_{\alpha=1}^{\nu} \tau_{ij}^{\alpha} = 0, \quad \tau_{ij}^{\alpha} = \tfrac{1}{2}(\sigma_{ij}^{\alpha} - \sigma_{ji}^{\alpha}) \tag{2.5}$$

(d) the balance of energy

$$\rho\dot{\varepsilon}_I + (\partial q_k/\partial x_k) = \rho r - \rho \sum_{\alpha=1}^{\nu} m_k^{\alpha} u_k^{\alpha} + \sum_{\alpha=1}^{\nu} \sigma_{jk}^{\alpha}(\partial v_j/\partial x_k) \tag{2.6}$$

In order to specify a thermodynamic process it suffices to prescribe the $(4\nu + 5)$ functions ρ_α, $\mathbf{v}^\alpha$, $\mathbf{m}^\alpha$, σ^α, ε_I, T, $\mathbf{q}$, S and $\mathbf{f}$ since the remaining $(\nu + 1)$ functions $\mathbf{b}^\alpha$ and r are determined by (2.4) and (2.6). The specific internal energy ε_I is the total specific energy ε_T less the specific kinetic energy[5,6],

$$\varepsilon_I = \varepsilon_T - \sum_{\alpha=1}^{\nu} \tfrac{1}{2}(\rho_\alpha/\rho)\, v_\alpha^2 \tag{2.7}$$

The total stress tensor σ is related to the component stress tensors by[5,6]

$$\sigma = \sum_{\alpha=1}^{\nu} (\sigma^\alpha - \rho_\alpha \mathbf{u}^\alpha \mathbf{u}^\alpha) \tag{2.8}$$

The component stress tensor here is identical with that of statistical mechanical theory[7].

III. LINEAR CONSTITUTIVE RELATIONS FOR AN ORDINARY BINARY FLUID MIXTURE

A material is defined by a constitutive assumption, which is a restriction on the processes admissible in a body consisting of the material. A mixture of viscoelastic materials susceptible of diffusion and heat conduction is completely determined by the values of the $(3\nu + 4)$ functions A_I, S, $\mathbf{q}$, $\mathbf{f}$, ξ^α, τ^α and $\mathbf{m}^\alpha$, where A_I is the specific internal Helmholtz free energy, $A_I = \varepsilon_I - TS$, and ξ^α is the symmetric part of σ^α,

$$\xi_{ij}^{\alpha} = \tfrac{1}{2}(\sigma_{ij}^{\alpha} + \sigma_{ji}^{\alpha}) \tag{3.1}$$

Müller[4] showed that the entropy flux need not be specifically related *a priori* to the heat flux, but instead may be given by a constitutive relation whose form is determined by the basic principles in the same way as other constitutive relations.

The subsequent exposition is restricted to the special material called an ordinary binary fluid mixture. For such mixtures the constitutive relations are ordinary functions of ρ_1, ρ_2 and T and are linear functions of $(\partial\rho_1/\partial x_k)$, $(\partial\rho_2/\partial x_k)$, $(\partial T/\partial x_k)$, u_k^1, d_{jk}^1, d_{jk}^2 and ω_{jk}^1, where

$$d_{jk}^{\alpha} = \tfrac{1}{2}\,[(\partial v_j^{\alpha}/\partial x_k) + (\partial v_k^{\alpha}/\partial x_j)] \tag{3.2}$$

and

$$\omega_{jk}^{1} = \tfrac{1}{2}\{[(\partial u_j^1/\partial x_k) - (\partial u_k^1/\partial x_j)] + \rho^{-1} \sum_{\alpha=1}^{2} [u_j^{\alpha}(\partial\rho_\alpha/\partial x_k) - u_k^{\alpha}(\partial\rho_\alpha/\partial x_j)]\} \tag{3.3}$$

The constitutive relations for an ordinary binary fluid mixture are[6]:

$$
\begin{aligned}
A_I &= A_I(\rho_1, \rho_2, T),\ S = S(\rho_1, \rho_2, T) \\
q_j &= -C_{qT}(\partial T/\partial x_j) - C_{q1}(\partial \rho_1/\partial x_j) - C_{q2}(\partial \rho_2/\partial x_j) - C_{qu}u_j^1 \\
m_j^1 &= -C_{mT}(\partial T/\partial x_j) - C_{m1}(\partial \rho_1/\partial x_j) - C_{m2}(\partial \rho_2/\partial x_j) - C_{mu}u_j^1 \\
f_j &= -C_{fT}(\partial T/\partial x_j) - C_{f1}(\partial \rho_1/\partial x_j) - C_{f2}(\partial \rho_2/\partial x_j) + Ku_j + (q_j/T) \\
\tau_{ij}^1 &= -\kappa\omega_{ij}^1 \\
\xi_{ij}^{\alpha} &= -\pi_\alpha \delta_{ij} + \sum_{\beta=1}^{2} \phi_{\alpha\beta} d_{kk}^{\beta} \delta_{ij} + \sum_{\beta=1}^{\nu} 2\eta_{\alpha\beta}(d_{ij}^{\beta} - \tfrac{1}{3} d_{kk}^{\beta} \delta_{ij})
\end{aligned}
\tag{3.4}
$$

where d_{kk} is the trace of the tensor d^{β}.

These constitutive relations have been deduced[6] from a starting set of constitutive relations of the form

$$R = R[\rho_1, \rho_2, T, (\partial \rho_1/\partial x_j), (\partial \rho_2/\partial x_j), (\partial T/\partial x_j), u_j^1, u_j^2, d_{ij}^1, d_{ij}^2, \omega_{ij}^1] \tag{3.5}$$

Compared with general theory of the conventional sort[1], the most important feature of (3.5) is the explicit enumeration of independent variables. Compared with the theory of Müller[4], the most important feature of (3.5) is the particular choice of independent variables. Our choice permits us to obtain many more consequences from the entropy inequality than does Müller's choice. Our choice leads to unambiguous definitions of various types of equilibrium, and to a bi-linear form for entropy production.

The thermodynamic processes possible for a particular material must of course depend on the material. Since it is the constitutive relations which define a material, we define an *admissible thermodynamic process* as a thermodynamic process which is compatible with the constitutive relations for the material in question. In particular, an admissible thermodynamic process for an ordinary binary fluid mixture must be compatible with (3.4). Application, in Section IV, of the entropy inequality provides restrictions on the phenomenological coefficients of (3.4) but affects neither the choice of independent variables nor the forms of the constitutive relations.

IV. THE ENTROPY INEQUALITY AND ITS CONSEQUENCES

Entropy and entropy flux are related by the balance equation

$$\rho\dot{S} = -(\partial f_k/\partial x_k) + \rho\Phi + (\rho r/T) \tag{4.1}$$

where Φ is the rate of specific entropy production not due to external radiation sources. Specification of an admissible thermodynamic process determines Φ through the constitutive relations for S and $\mathbf{f}$. The second law of thermodynamics in this context is the *Postulate*[2]: For every admissible thermodynamic process the inequality

$$\Phi \geqslant 0 \tag{4.2}$$

must hold at all times and at every point in the material. That is, the entropy production is non-negative, identically in the independent variables. In order to conserve space, we omit the details[6] of the exploitation of (4.2). Repeated application of (4.2) and full utilization of the independence of the variables of (3.4) yield a host of consequences, which we now list:

(a) Equilibrium theorem

An ordinary binary fluid mixture is at equilibrium if and only if each of the independent variables $(\partial T/\partial x_k)$, u_k^1, d_{jk}^1, d_{jk}^2 and ω_{jk}^1 is zero. Vanishing of the density gradients or of the barycentric velocity is not required for equilibrium.

(b) Entropy equations

For an ordinary binary fluid mixture, a deductive consequence of the principles of rational mechanics is

$$S = -(\partial A_I/\partial T)\,\rho_1, \rho_2 \tag{4.3}$$

which is a familiar identity in thermostatics. Each of the first three coefficients of $(3.4)_4$ vanishes, and the entropy flux becomes

$$\mathbf{f} = Ku^1 + (\mathbf{q}/T) \tag{4.4}$$

The final bi-linear form for the entropy production is

$$\rho T\Phi = (C_{qT}/T)(\partial T/\partial x_k)(\partial T/\partial x_k) + \Theta_T u_k^1(\partial T/\partial x_k) + (\rho^2/\rho_2)\,C_{mu}u_k^1 u_k^1$$
$$+ \sum_{\alpha=1}^{2}\sum_{\beta=1}^{2} \phi_{\alpha\beta}\mathrm{d}_{ii}^{\alpha}\mathrm{d}_{jj}^{\beta} + (\rho/\rho_1^2\rho_2)\,\kappa\omega_{ij}^1\omega_{ij}^1$$
$$+ \sum_{\alpha=1}^{2}\sum_{\beta=1}^{2} 2\eta_{\alpha\beta}(\mathrm{d}_{ij}^{\alpha} - \tfrac{1}{3}\mathrm{d}_{kk}^{\alpha}\delta_{ij})(\mathrm{d}_{ij}^{\beta} - \tfrac{1}{3}\mathrm{d}_{kk}^{\beta}\delta_{ij}), \tag{4.5}$$

where

$$\Theta_T = T(\partial K/\partial T)_{\rho_1,\rho_2} + (C_{qu}/T) + (\rho^2/\rho_2)\,C_{mT} \tag{4.6}$$

(c) Viscosity coefficients

For an ordinary binary fluid mixture there appear to be nine (not two) viscosity coefficients, restricted only by the conditions:

$$\kappa \geqslant 0,$$
$$\phi_{11} \geqslant 0, \phi_{22} \geqslant 0, [\phi_{11}\phi_{22} - \tfrac{1}{4}(\phi_{12} + \phi_{21})^2] \geqslant 0,$$
$$\eta_{11} \geqslant 0, \eta_{22} \geqslant 0, [\eta_{11}\eta_{22} - \tfrac{1}{4}(\eta_{12} + \eta_{21})^2] \geqslant 0 \tag{4.7}$$

(d) Heat flux equation

For an ordinary binary fluid mixture, C_{q1} and C_{q2} vanish, and $(3.4)_2$ becomes

$$q_j = -C_{qT}(\partial T/\partial x_j) - C_{qu}u_j^1 \tag{4.8}$$

This is precisely the heat of transport form of the heat flux equation in the practical theory[1d], with C_{qT} the thermal conductivity coefficient, $\rho_1\mathbf{u}^1$ the diffusion flux $\mathbf{j}^1$ and $(-C_{qu}/\rho_1)$ the heat of transport $Q_1{\star}$.

(e) Diffusion flux equation

Rearrangement of $(3.4)_3$ yields

$$\rho_1 u_j^1 = -(\rho_1 C_{mT}/C_{mu})(\partial T/\partial x_j) - (\rho_1 C_{m1}/C_{mu})(\partial \rho_1/\partial x_j) - (\rho_1 C_{m2}/C_{mu})(\partial \rho_2/\partial x_j) - (\rho_1/C_{mu})\, m_j^1 \quad (4.9)$$

For an ordinary binary fluid mixture, (4.2) requires:

$$\begin{aligned} &C_{mu} \geqslant 0,\ [(\rho^2/\rho_2)\, C_{qT} C_{mu} - \tfrac{1}{4}\,\Theta_T^2] \geqslant 0 \\ &C_{m1} = (\rho_2/\rho^3)\, TK - (\rho_2/\rho^2)(\partial K/\partial \rho_1)_{T,\rho_2} - (\rho_2/\rho)(\partial A_I/\partial \rho_1)_{T,\rho_2} \\ &C_{m2} = -(\rho_1/\rho^3)\, TK - (\rho_2/\rho^2)\, T\,(\partial K/\partial \rho_2)_{T,\rho_1} + (\rho_1/\rho)(\partial A_I/\partial \rho_2)_{T,\rho_1} \end{aligned} \quad (4.10)$$

Further reduction of these equations must await identification of the coefficient K which occurs explicitly not only in $(3.4)_4$, (4.4), (4.6) and (4.10), but also in two further consequences of (4.2) for an ordinary binary fluid mixture; namely, the π_α of $(3.4)_6$ must be given by:

$$\begin{aligned} \pi_1 &= \rho\rho_1(\partial A_I/\partial \rho_1)_{T,\rho_2} + (\rho_2/\rho)\, TK \\ \pi_2 &= \rho\rho_2(\partial A_I/\partial \rho_2)_{T,\rho_1} - (\rho_2/\rho)\, TK \end{aligned} \quad (4.11)$$

The π_α may be called partial pressures since

$$\pi_1 + \pi_2 = \rho[\rho_1(\partial A_1/\partial \rho_1)_{T,\rho_2} + \rho_2(\partial A_I/\partial \rho_2)_{T,\rho_1}] = P \quad (4.12)$$

ACKNOWLEDGEMENT

Support of this research by the National Science Foundation is gratefully acknowledged. Moreover, Predoctoral Fellowships to JLB from the Public Health Service and the Lubrizol Foundation are gratefully acknowledged.

REFERENCES

1. (a) J. Meixner and H. G. Reik, in *Handbuch der Physik*, S. Flügge, (Ed). Springer: Berlin (1959). Vol. III/2; (b) J. G. Kirkwood and B. Crawford Jr, *J. Phys. Chem.* **56**, 1048 (1952); (c) S. R. de Groot and P. Mazur, *Non-Equilibrium Thermodynamics*. North Holland Publ. Co.: Amsterdam (1962); (d) D. D. Fitts, *Non-equilibrium Thermodynamics*. McGraw-Hill: New York (1962); (e) R. Haase, *Thermodynamics of Irreversible Processes*. Addison-Wesley: Reading, Mass. (1969).
2. B. D. Coleman and W. Noll, *Arch. Ration. Mech. Anal.* **13**, 167 (1963).
3. B. D. Coleman and V. J. Mizel, (a) *Arch. Ration. Mech. Anal.* **13**, 245 (1963); (b) *J. Chem. Phys.* **40**, 1116 (1964).
4. I. Müller, *Arch. Ration. Mech. Anal.* **26**, 118 (1967); (b) *Arch. Ration. Mech. Anal.* **28**, 1 (1968).
5. (a) C. Truesdell, *R.C. Acad. Lincei*, (8), **22**, 33 (1957), reprinted in *The Rational Mechanics of Materials*, pp 293–305. C. Truesdell (Ed). Gordon and Breach: New York (1965); (b) C. Truesdell and R. A. Toupin, in *Handbuch der Physik*, Vol. III/1, S. Flügge (Ed). Springer: Berlin (1960); (c) C. Truesdell and W. Noll, in *Handbuch der Physik*, Vol. III/3. S. Flügge (Ed.). Springer: Berlin (1965).
6. J. L. Bartelt, *Ph.D. Thesis*. Michigan State University (1968).
7. R. J. Bearman and J. G. Kirkwood, *J. Chem. Phys.* **28**, 136 (1958).

THERMODYNAMIC RESTRICTIONS IN NON-LINEAR RATE TYPE MATERIALS

J. EDMUND FITZGERALD

Department of Mathematical Physics, University College, Cork, Ireland

ABSTRACT

Although one-sided rate type (differential type) materials have received attention in the literature, the restrictions which should be applied have not been finally resolved. The author seeks to clarify this situation, examining first thermomechanical, and then, thermodynamic, processes.

1. INTRODUCTION

Much recent and not so recent literature has appeared purporting to treat one-sided rate type materials (materials of the differential type) wherein the stress and other dependent variables are a function of, say, the deformation gradient or strain and their *n*th order higher time derivative.

Other than problems associated with coordinate frame indifference and material indifference[1] (material isotropy groups)[2] which have essentially been solved except for some remaining disagreement on use of improper rotations[3] the major problem lies in the thermodynamic restrictions applicable to such materials. Two parts of this question need answering. First, does the second law of thermodynamics in the form of the Clausius–Duhem inequality (CDI) apply to materials undergoing irreversible non-equilibrium thermomechanical processes? Secondly, what are the physical implications of the various initial constitutive assumptions with respect to the type of material described?

It is intended herein to offer an alternative to the first question and hopefully clarify the second.

In essence, it will be shown that one-sided rate type materials depending on deformation gradients, temperature and temperature gradients, as well as the higher time derivatives of these quantities will always be reducible to nothing more than non-linear Kelvin solids showing no relaxation properties. For materials of this type the inclusion of hidden variables is essential to the relaxation process (creep, however, can occur without interval variables).

1.1 Notations and definitions

The notations and definitions used herein are similar to those of Truesdell and Noll's NLFT[1]. We denote vectors and spatial points by boldface Latin miniscules; **q**, **x** . . . Sets, bodies and regions are denoted by script majuscules, $\mathscr{B}$, $\mathscr{R}$. Tensors (linear transformations) and material points are given in boldface Latin majuscules, **F**, **T**, **X**. Configurations and mappings are given in boldface Greek minuscules, $\boldsymbol{\chi}$, $\boldsymbol{\xi}$. Where repeated indices are used, summation over 1, 2, 3 is implied. Transpose of $\mathbf{F}$ is $\mathbf{F}^T$ and its inverse is $\mathbf{F}^{-1}$. The trace (spur) is written $\mathbf{trA} = \mathbf{A}_{ii}$ and the determinant of the matrix representation of $\mathbf{A}$ is $\mathbf{detA}$. Present time is taken as $\mathbf{t}$ and previous time is $\boldsymbol{\tau}$.

We define a *body* $\mathscr{B}$ as a smooth manifold (continuum) of elements (*particles*) $\mathbf{X}$ whose coordinates in some reference are also $\mathbf{X}$. *The configuration* $\boldsymbol{\chi}$ of $\mathscr{B}$ are the elements of a set of one-to-one (invertible) mappings of $\mathscr{B}$ into a three-dimensional Euclidian point space $\mathscr{E}$. The *spatial point* $\mathbf{x}$ is called the place occupied by the particle $\mathbf{X}$,

$$\mathbf{x} = \boldsymbol{\chi}(\mathbf{X}); \mathbf{X} = \boldsymbol{\chi}^{-1}(\mathbf{x}) \tag{1.1}$$

and $\mathbf{X}$ above is the particle whose place is $\mathbf{x}$.

The region of space, $\boldsymbol{\chi}(\mathscr{B})$, into which the body is mapped is called the region occupied by the body $\mathscr{B}$ in the configuration $\boldsymbol{\chi}$. *The reference configuration*, its volume element and its surface area element are $\mathscr{R}$, $\mathscr{V}$, $\mathscr{S}$ respectively. The *deformed configuration*, its volume element and its surface area element (where $\boldsymbol{\chi}$ is not the identity mapping) are $\mathbf{r}$, $\mathbf{v}$, $\mathbf{s}$ respectively.

A *motion* of body $\mathscr{B}$ is a one-parameter family of configurations with the real parameter $\mathbf{t}$, time. Thus

$$\mathbf{x} = \boldsymbol{\chi}(\mathbf{X}, \mathbf{t}); \mathbf{X} = \boldsymbol{\chi}^{-1}(\mathbf{x}, \mathbf{t}) \tag{1.2}$$

We assume the existence and continuity of any derivatives wherever needed.

Critical to the entire theory is the *concept of localization*. A curve $\mathscr{C}$ in $\mathscr{R}$ is deformed by 1.1 into a curve $\mathbf{c}$ in $\mathbf{r}$. If $\mathbf{dX}$ is an *element of arc* along $\mathscr{C}$, using the rule for change of variables in an integral,

$$\int_{\mathscr{C}} [\ldots]\, \mathbf{dX}^{\mathbf{K}} = \int_{\mathbf{c}} [\ldots]\, \mathbf{X},_{\mathbf{k}}^{\mathbf{K}}\, \mathbf{dx}^{\mathbf{k}} \tag{1.3}$$

with

$$\mathbf{X},_{\mathbf{k}}^{\mathbf{K}} \equiv \partial \mathbf{X}^{\mathbf{K}} / \partial \mathbf{x}_{\mathbf{k}} \tag{1.4}$$

The spatial field $\mathbf{dx}$ is thus defined in terms of the material field $\mathbf{dX}$ by

$$\mathbf{dX}^{\mathbf{K}} = \mathbf{X},_{\mathbf{k}}^{\mathbf{k}}\, \mathbf{dx}^{\mathbf{k}} \tag{1.5}$$

whose solution for $\mathbf{dx}$ is

$$\mathbf{dx}^{\mathbf{k}} = \mathbf{x},_{\mathbf{K}}^{\mathbf{k}}\, \mathbf{dX}^{\mathbf{K}} \tag{1.6}$$

In equation 1.6 $\mathbf{dx}$ is a field deformed with the material whose determination requires knowledge of the deformation in a neighbourhood of $\mathbf{x}$.

The $\mathbf{x},_{\mathbf{K}}^{\mathbf{k}}$ are the components of the *deformation gradient*, $\mathbf{F}$. Uniqueness of 1.6 is assured by the postulate of a positive definite bounded volume element such that

$$0 < (\mathbf{dv}/\mathbf{dV}) = \mathbf{J} = \mathbf{det\, F} < \infty \tag{1.7}$$

where $\boldsymbol{\rho}$ is the density.

Thus, $\mathbf{dx}$ is the vector at $\mathbf{x}$ into which the vector $\mathbf{dX}$ at $\mathbf{X}$ is deformed by the linear transformation, 1.6.

We shall deal herein only with *simple materials*, i.e. those whose response is affected by only $\mathbf{F}$ and not its gradients†.

We have defined, therefore, the deformation gradient $\mathbf{F} = \mathbf{GRAD}\ \chi$ at and with respect to $\mathbf{X}$ relative to the deformed configuration, $\mathbf{r}$.

The symbol **grad** · is used to denote the gradient at $\mathbf{X}$ with respect to $\mathbf{x}$.

The usual material derivative and spatial derivative definitions are used.

For a homogeneous continuum, considered here, explicit dependence upon $\mathbf{X}$ is not required in any of the ensuing constitutive equations.

Polar decomposition of $\mathbf{F}$ results in

$$\mathbf{F} = \mathbf{RU} \text{ or } \mathbf{F} = \mathbf{VR} \tag{1.8}$$

where $\mathbf{R}$ is an orthogonal rotation tensor and $\mathbf{U}$ and $\mathbf{V}$ are right and left positive definite, symmetric stretch tensors respectively.

The *right and left Cauchy–Green strain tensors* are

$$\mathbf{C} = \mathbf{F}^{\mathrm{T}}\mathbf{F} = \mathbf{U}^2, \mathbf{B} = \mathbf{F}\mathbf{F}^{\mathrm{T}} = \mathbf{V}^2 \tag{1.9}$$

With $\boldsymbol{\kappa}$ as a reference configuration at time $\mathbf{t}_0$ and $\boldsymbol{\chi}(\mathbf{t})$ and $\boldsymbol{\chi}(\boldsymbol{\tau})$ the configurations at times $\mathbf{t}$ and $\boldsymbol{\tau}$, the usual composition of mappings defines the *relative deformation gradient* as

$$\mathbf{F}(\boldsymbol{\tau}) = \mathbf{F}_{(\mathbf{t})}(\boldsymbol{\tau})\,\mathbf{F}(\mathbf{t})$$

where

$$\mathbf{F}(\boldsymbol{\tau}) : \boldsymbol{\kappa} \to \boldsymbol{\chi}(\boldsymbol{\tau})\,;\ \mathbf{F}(\mathbf{t}) : \boldsymbol{\kappa} \to \boldsymbol{\chi}(\mathbf{t})\,;\ \mathbf{F}_{(\mathbf{t})}(\boldsymbol{\tau}) : \boldsymbol{\chi}(\mathbf{t}) \to \boldsymbol{\chi}(\boldsymbol{\tau}) \tag{1.10}$$

so that

$$\mathbf{F}_{(\mathbf{t})}(\boldsymbol{\tau}) = \mathbf{F}(\boldsymbol{\tau})\,\mathbf{F}(\mathbf{t})^{-1} \tag{1.11}$$

We may thus express the *relative deformation gradient rate* evaluated at present time $\mathbf{t}$ as

$$\mathbf{L}(\mathbf{t}) \equiv \dot{\mathbf{F}}_{(\mathbf{t})}(\mathbf{t}) = \frac{\partial}{\partial \boldsymbol{\tau}} \dot{\mathbf{F}}_{(\mathbf{t})}(\boldsymbol{\tau})\big|_{\boldsymbol{\tau}=\mathbf{t}} = \dot{\mathbf{F}}(\mathbf{t})\,\mathbf{F}(\mathbf{t})^{-1} \tag{1.12}$$

which is simply the spatial velocity gradient

$$\mathbf{grad}\ \dot{\mathbf{x}} \tag{1.13}$$

Similarly, the spatial gradient of the nth acceleration is then

$$\mathbf{L}_n(\mathbf{t}) = \frac{\partial \overset{(n)}{\mathbf{x}}}{\partial \mathbf{x}} = \overset{(n)}{\mathbf{F}}_{(\mathbf{t})}(\mathbf{t}) = \mathbf{grad}\ \overset{(n)}{\mathbf{x}} \quad n = 1, 2, \ldots \tag{1.14}$$

with

$$\mathbf{L}_0 = \mathbf{I}\,;\ \mathbf{L}_1 = \mathbf{L}$$

† It may be noted that non-simple materials subjected to pure homogeneous deformations are also covered by our restriction.

Employing polar decomposition on the relative deformation gradient and differentiating, we can then define the *stretching*, **D**, and the *spin*, **W**, as

$$\mathbf{D} = \dot{\mathbf{U}}_{(t)}(\mathbf{t}) = \tfrac{1}{2}(\mathbf{L} + \mathbf{L}^T)$$
$$\mathbf{W} = \dot{\mathbf{R}}_{(t)}(\mathbf{t}) = \tfrac{1}{2}(\mathbf{L} - \mathbf{L}^T) \qquad (1.15)$$

D is obviously the symmetric part of **L** and is sometimes called the rate of deformation tensor.

A useful relationship is easily derived

$$\mathrm{tr}\mathbf{D} = \mathrm{tr}\mathbf{L} = \mathrm{div}\,\dot{\mathbf{x}} = \frac{1}{\mathbf{V}}\frac{\mathbf{dV}}{\mathbf{dt}} \qquad (1.16)$$

where **V** is here the present volume.

An inner product $\mathbf{A}\cdot\mathbf{B}$ can be represented as $\mathrm{tr}\mathbf{A}^T\mathbf{B}$. Also

$$\mathrm{tr}\mathbf{A}^T\mathbf{B} = \mathrm{tr}\mathbf{B}^T\mathbf{A},\ \mathrm{tr}\mathbf{ABC} = \mathrm{tr}\mathbf{BCA}$$

The *Cauchy stress tensor*, $\mathbf{T} = \mathbf{T}(\mathbf{X}, \mathbf{t})$ is the stress per unit area in the deformed configuration. For non-polar materials herein considered, the usual balance of moment of momenta yields $\mathbf{T} = \mathbf{T}^T$, hence **T** is symmetric. We also define the *specific body force* $\mathbf{b} = \mathbf{b}(\mathbf{X}, \mathbf{t})$ per unit mass as a field force extended on the body $\mathscr{B}$ at **X** by causes outside of $\mathscr{B}$.

The thermodynamic variables to be employed are not in accord with the 1948 I.U.P.A.C. and I.U.P.A.P. recommendations because of the obvious conflict with the mechanical variables. We define here:

(1) The *total internal energy* in the body $\mathscr{B}$ as **E** which is an additive set function of the portions of the body with units (mass) (length)^2 (time)$^{-2}$. The localization of **E** leads to the *specific internal energy*, ε, at a point **X** per unit mass with dimension (length)^2 (time)$^{-1}$.

(2) The *total entropy*, $\mathbf{N}(\mathbf{X}, \mathbf{t})$ as above with its localization to *specific entropy*, $\eta = \eta(\mathbf{X}, \mathbf{t})$ per unit mass.

(3) The *absolute temperature*, $\theta = \theta(\mathbf{X}, \mathbf{t})$ which, unless otherwise specified, is the translational temperature.

(4) The *heat flux vector*, $\mathbf{q} = \mathbf{q}(\mathbf{X}, \mathbf{t})$ whose units are energy per unit area per unit time = (mass) (time)$^{-3}$ where an outwardly directed unit normal vector $\tilde{\mathbf{n}}$ is used, the total flux of heat over an element of body surface area **s** is

$$\oint_s \mathbf{q}\cdot\tilde{\mathbf{n}}\,\mathbf{ds} \qquad (1.17)$$

which by the Green–Gauss theorem yields

$$\int_v \mathrm{div}\,\mathbf{q}\,\mathbf{dv} \qquad (1.18)$$

with dimension energy/unit time and represents the *rate at which heat is leaving the* body surface, **s**. The integrated heat flux is **Q**.

(5) The *specific heat supply* $\mathbf{r} = \mathbf{r}(\mathbf{X}, \mathbf{t})$ per unit mass per unit time absorbed by the body $\mathscr{B}$ at **X** from external non-conductive sources or internal point sources (whose existence is not pertinent at this point).

(6) The *internal state vector*, $\boldsymbol{\varphi}$ or $\boldsymbol{\alpha}$, whose scalar components in $\boldsymbol{\varphi} = (\varphi_1, \varphi_2, \ldots \varphi_n)$ are the internal state variables. This vector may be considered as a hidden variable wherein it is capable of energy transfer.

2. THERMOMECHANICAL PROCESSES

The notation used above is due to Coleman and Gurtin[4]. The impetus for the specific approach used resulted from conversations with R. S. Rivlin and J. Meixner.

2.1 Energy inequality restrictions

The set of three mechanical functions, motion $\mathbf{x}$, stress $\mathbf{T}$ and body force $\mathbf{b}$ with the stated localization of motion plus the six thermal functions listed above and defined for all particles $\mathbf{X}$ in $\mathscr{B}$ over all time $\mathbf{t}$ is called a *thermomechanical process* if and only if the set satisfies the usual local laws of balance of momentum

$$\mathbf{div\,T} + \rho\mathbf{b} = \rho\ddot{\mathbf{x}} \tag{2.1}$$

and the *balance of energy*

$$\rho\dot{\varepsilon} = \mathbf{T}\cdot\mathbf{L} - \mathbf{div\,q} + \rho\mathbf{r} \tag{2.2}$$

Thus, a sufficient condition for a thermomechanical process is one wherein only $\mathbf{x}$, $\mathbf{T}$, $\boldsymbol{\varepsilon}$, $\mathbf{q}$, $\boldsymbol{\eta}$, $\boldsymbol{\theta}$ and $\boldsymbol{\varphi}$ are prescribed and $\mathbf{b}$ and $\mathbf{r}$ are found from equations 2.1 and 2.2.

Consider now the integral of 2.2 over some finite interval of time which for convenience will be taken as $\mathbf{t}\varepsilon[0, \mathbf{t}]$. We shall take the state of the body at all times $\mathbf{t}\varepsilon(-\infty, 0]$ as a thermomechanical *fiducial state* wherein we postulate at all points $\mathbf{X}$ and times $\mathbf{t}^\star \leqslant 0$

$$\begin{aligned}
&\boldsymbol{\theta}(\mathbf{X}, \mathbf{t}^\star) = \boldsymbol{\theta}_0(\mathbf{X}, \mathbf{t}^\star) \quad \mathbf{Q}(\mathbf{X}, \mathbf{t}^\star) = \mathbf{Q}_0(\mathbf{X}, \mathbf{t}^\star)\\
&\mathbf{T}(\mathbf{X}, \mathbf{t}^\star) = 0 \qquad\qquad\quad \mathbf{F}(\mathbf{X}, \mathbf{t}^\star) = \mathbf{1}\\
&\mathbf{L} = \dot{\mathbf{L}} = \ldots = \mathbf{L}^{(n)}(\mathbf{X}, \mathbf{t}^\star) = 0\\
&\boldsymbol{\varepsilon}(\mathbf{X}, \mathbf{t}^\star) = \boldsymbol{\varepsilon}_0(\mathbf{X}, \mathbf{t}^\star)
\end{aligned} \tag{2.3}$$

Equation 2.3 implies a homothermal field. $\boldsymbol{\theta}_0$ could be taken equal to zero without loss of generality. Conditions 2.3_3 and 2.3_4 imply no initial stress nor strain, and 2.3_5 implies no motion at times $t \leqslant 0$. Equation 2.3_2, wherein $\mathbf{Q}$ is the energy per unit volume, simply implies a fiducial state for the heat content which, without loss of generality, may be taken as $\mathbf{Q}_0 = 0$.

The relation 2.3_6 is based on the fact that energy may be defined only to within an arbitrary constant[10]. In this case, if we either take $\boldsymbol{\theta} = 0$ and $\mathbf{Q}_0 = 0$ implying $\boldsymbol{\varepsilon}_0 = 0$ or take $\boldsymbol{\varepsilon}_0$ as a *fiducial energy level below which we cannot go by any thermomechanical process*, integration of 2.2 then gives

$$\rho\boldsymbol{\varepsilon}(\mathbf{X}, \mathbf{t}) = \int_{t0}^{t} \mathbf{T}\cdot\mathbf{L}\,d\tau - \int_{t0}^{t} \mathbf{div\,q}\,d\tau + \int_{t0}^{t} \rho\mathbf{r}\,d\tau, \mathbf{t}_0 \leqslant \mathbf{t}^\star \tag{2.4}$$

We now *postulate* with the definitions given above

$$\boldsymbol{\varepsilon}(\mathbf{X}, \mathbf{t}) \geqslant 0 \tag{2.5}$$

which we call the first inequality of thermomechanics or the *energy inequality*. Note that expression 2.5 holds in non-equilibrium conditions.

The thermal implications of 2.5 are rather obvious and perhaps trivial. Letting no mechanical motion ensue we have from 2.5 and 2.4

$$-\int_{t_0}^{t} \operatorname{div} \mathbf{q}\, d\tau + \int_{t_0}^{t} \rho r\, d\tau \geqslant 0 \tag{2.6}$$

Letting

$$-\int_{-\infty}^{t_0} \operatorname{div} \mathbf{q}\, d\tau = -\Delta \mathbf{Q}^{OUT} = \Delta \mathbf{Q}^{IN} \tag{2.7}$$

where $\Delta \mathbf{Q}^{IN}$ is the amount of energy conducted as heat into the body at time $\mathbf{t}_0$. Also we have

$$\int_{-\infty}^{t_0} \rho r\, d\tau = \Delta \hat{\mathbf{Q}}^{IN} \tag{2.8}$$

which is the amount of energy brought into the body by external radiation. Then we write for the energy (in the form of heat) in the body at time $\mathbf{t}_0$

$$\mathbf{Q}_0 = \Delta \mathbf{Q}^{IN} + \Delta \hat{\mathbf{Q}}^{IN} \tag{2.9}$$

During the non-mechanical process from $\mathbf{t}_0$ to the present time $\mathbf{t}$ we have from 2.4 and 2.9 with $\mathbf{t}_0 = 0$

$$\rho \varepsilon(\mathbf{X}, \mathbf{t}) = -\int_{t_0}^{t} \operatorname{div} \mathbf{q}\, d\tau + \int_{t_0}^{t} \rho r\, d\tau \geqslant 0 \tag{2.10}$$

With the interpretation

$$\begin{aligned} -\int_{t_0}^{t} \operatorname{div} \mathbf{q}\, d\tau &= -\Delta \mathbf{Q}^{OUT}(\mathbf{t}_0, \mathbf{t}) \\ \int_{t_0}^{t} \rho r\, d\tau &= \Delta \hat{\mathbf{Q}}^{IN}(\mathbf{t}_0, \mathbf{t}) \end{aligned} \tag{2.11}$$

equation 2.10 becomes

$$-\Delta \mathbf{Q}^{OUT}(\mathbf{t}_0, \mathbf{t}) + \Delta \hat{\mathbf{Q}}^{IN}(\mathbf{t}_0, \mathbf{t}) + \mathbf{Q}_0 \geqslant 0 \tag{2.12}$$

which can be written with $\Delta \hat{\mathbf{Q}}^{IN} = -\Delta \hat{\mathbf{Q}}^{OUT}$ as

$$\Delta \mathbf{Q}^{OUT}(\mathbf{t}_0, \mathbf{t}) + \Delta \hat{\mathbf{Q}}^{OUT}(\mathbf{t}_0, \mathbf{t}) \leqslant \mathbf{Q}_0 \tag{2.13}$$

Thus, no more net heat may be lost through conduction and radiation over a time interval than was originally in the body at the beginning of the interval under zero mechanical processes.

The mechanical implications are given for the condition of no radiation and no heat transfer over the interval $[0, \mathbf{t}]$. Since these heat processes are assumed independent in that we can in principle ensure that

$$\int_{t_0}^{t} \operatorname{div} \mathbf{q} = 0; \int_{t_0}^{t} \rho r\, d\tau = 0 \tag{2.14}$$

this above restriction does not reduce the generality of the results.

Assume for illustrative purposes that the constitutive equation for the stress tensor, $\mathbf{T}$, is that of a one-sided linear rate type material wherein

$$\mathbf{T} = \mathbf{a}_0\mathbf{B} + \mathbf{a}_1\mathbf{L} + \mathbf{a}_2\mathbf{L}_{(2)} + \ldots + \mathbf{a}_n\mathbf{L}_{(n)} \tag{2.15}$$

where $\mathbf{B} = \mathbf{B}^T$ is the left Cauchy–Green symmetric strain tensor and the

$\mathbf{L}_{(n)}$ are the spatial gradients of the nth time derivative of the velocity with the $\mathbf{a}_n$ constants. Equations 2.4, 2.5 and 2.14 then are

$$\varepsilon(\mathbf{X}, \mathbf{t}) = \int_{\mathrm{to}}^{\mathbf{t}} \mathbf{T} \cdot \mathbf{L}\, \mathbf{d}\tau \geqslant 0 \tag{2.16}$$

The expression $\mathbf{T} \cdot \mathbf{L}$ with 2.15 therefore becomes

$$\mathbf{T} \cdot \mathbf{L} = \mathbf{a}_0 \mathbf{B} \cdot \mathbf{L} + \mathbf{a}_1 \mathbf{L} \cdot \mathbf{L} + \mathbf{a}_2 \mathbf{L}_2 \cdot \mathbf{L} + \mathbf{a}_3 \mathbf{L}_3 \cdot \mathbf{L} + \ldots \tag{2.16a}$$

The first term on the right is then $\mathbf{B} \cdot \mathbf{L} = \mathbf{trBL}$ and from elementary operations

$$\mathbf{trBL} = \mathbf{trL}^{\mathrm{T}}\mathbf{B} = \mathbf{tr}(\mathbf{F}^{\mathrm{T}})^{-1}\dot{\mathbf{F}}^{\mathrm{T}}\mathbf{F}\mathbf{F}^{\mathrm{T}} = \mathbf{tr}\dot{\mathbf{F}}^{\mathrm{T}}\mathbf{F} \tag{2.17}$$

The integral of $\mathbf{tr}\dot{\mathbf{F}}^{\mathrm{T}}\mathbf{F} = \dot{\mathbf{F}} \cdot \mathbf{F}$ may be evaluated by considering first

$$\frac{\partial}{\partial \tau} \mathbf{trF}^{\mathrm{T}}\mathbf{F} = \mathbf{tr}\frac{\partial}{\partial \tau} \mathbf{F}^{\mathrm{T}}\mathbf{F} = 2\mathbf{tr}\dot{\mathbf{F}}^{\mathrm{T}}\mathbf{F} \tag{2.18}$$

Thus we may write, using the definition of an integral and $\mathbf{B} = \mathbf{V}^2 = \mathbf{F}\mathbf{F}^{\mathrm{T}}$,

$$\int_{\mathrm{t0}}^{\mathrm{t}} \mathbf{a}_0 \mathbf{B} \cdot \mathbf{L}\, \mathbf{d}\tau = \int_{\mathrm{t0}}^{\mathrm{t}} \frac{\mathbf{a}}{2} \cdot \frac{\partial}{\partial \tau} \mathbf{trV}^2\, \mathbf{d}\tau = \frac{\mathbf{a}_0}{2} \mathbf{trV}^2 = \frac{\mathbf{a}_0}{2}(\mathbf{I}_1 - 3) \tag{2.19}$$

where $\mathbf{trV}^2 = \mathbf{trB}$ is the left Cauchy–Green stretch tensor which is strictly positive definite, hence $\mathbf{a}_0 \geqslant 0$.

The second term, $\mathbf{a}_1 \mathbf{L} \cdot \mathbf{L}$, in the integral is $\int_{\mathrm{to}}^{\mathbf{t}} \mathbf{L} \cdot \mathbf{L}\, \mathbf{d}\tau$, $\mathbf{L}(\mathbf{t}_0) = 0$ (2.20)

which, by the basic definition of an inner product is positive definite in the argument and, hence, the integral is positive definite. Thus $\mathbf{a}_1 \geqslant 0$.

The third term $\mathbf{a}_2 \mathbf{L}_{(2)} \cdot \mathbf{L}$ may be similarly determined using the relation

$$\int \frac{\partial}{\partial \tau} \mathbf{trL}^{\mathrm{T}}\mathbf{L}\, \mathbf{d}\tau = \int \mathbf{tr}\frac{\partial}{\partial \tau} \mathbf{L}^{\mathrm{T}}\mathbf{L}\, \mathbf{d}\tau = \int \mathbf{L} \cdot \dot{\mathbf{L}}\, \mathbf{d}\tau = \mathbf{L} \cdot \mathbf{L}\Big|_{\mathrm{to}}^{\mathbf{t}} \tag{2.21}$$

Therefore $\mathbf{a}_2 \geqslant 0$. All remaining terms of the form $\mathbf{L}_{(n)} \cdot \mathbf{L}$ with $\mathbf{n} > 2$ cannot be shown to be positive definite for all arbitrary $\mathbf{L}_{(\mathbf{n})}$ and $\mathbf{L}$, hence

$$\mathbf{a}_3 = \mathbf{a}_4 = \ldots\ldots = \mathbf{a}_n = 0 \tag{2.22}$$

A comment is in order here. Assuming, for the linear rate expressions of $\mathbf{T}$, other forms of the strain such as $\mathbf{C}$, the right Cauchy–Green strain tensor and $\mathbf{F}$, the deformation gradient, one cannot show that $\mathbf{a}_0 \geqslant 0$. Similarly, in the operations for determining the existence of $\mathbf{a}_2$, the velocity gradient $\mathbf{L}$ only enters through its symmetric part $\mathbf{L}^{(\mathrm{s})} = \frac{1}{2}(\mathbf{L} + \mathbf{L}^{\mathrm{T}}) \equiv \mathbf{D}$ the stretching tensor. Thus, the energy inequality restricts the linear rate Cauchy stress tensor to

$$\mathbf{T} = \mathbf{a}_0 \mathbf{B} + \mathbf{a}_1 \mathbf{D} + \mathbf{a}_2 \mathbf{L}_{(2)}^{(\mathrm{s})} \tag{2.23}$$

where $\mathbf{L}_{(2)}^{(\mathrm{s})}$ is the symmetric part of $\dot{\mathbf{L}}$.

This result has been obtained without recourse to the second law of thermodynamics, the CDI, and does therefore hold rigorously in non-equilibrium conditions. The details were suggested by J. Meixner[5].

One may, with considerably more algebraic effort, generalize the linear rate constitutive equation used previously as an illustrative example. Using a theorem of Rivlin[6], one may show that the symmetric Cauchy stress tensor **T** is expressible as a polynomial isotropic tensor function of, for example, the left Cauchy–Green strain tensor **B** and the symmetric part of the relative deformation gradient rate, **D**. Then we may express this general non-linear relationship as

$$\begin{aligned}\mathbf{T} = {} & \psi_0 1 + \psi_1 \mathbf{B} + \psi_2 \mathbf{L} + \psi_3 \mathbf{B}^2 + \psi_4 \mathbf{D}^2 \\ & + \psi_5(\mathbf{BD} + \mathbf{DB}) + \psi_6(\mathbf{B}^2\mathbf{D} + \mathbf{DB}^2) \\ & + \psi_7(\mathbf{BD}^2 + \mathbf{D}^2\mathbf{B}) + \psi_8(\mathbf{B}^2\mathbf{D}^2 + \mathbf{D}^2\mathbf{B}^2)\end{aligned} \tag{2.24}$$

where the coefficients $\psi_0 \ldots \psi_8$ are polynomials of the set of irreducible invariants of the two symmetric tensors, **B** and **D**. The specific form for the irreducible invariants in this case is

$$\begin{aligned}& \mathrm{tr}\mathbf{B}, \mathrm{tr}\mathbf{B}^2, \mathrm{tr}\mathbf{B}^3, \mathrm{tr}\mathbf{D}, \mathrm{tr}\mathbf{D}^2, \mathrm{tr}\mathbf{D}^3 \\ & \mathrm{tr}(\mathbf{BD}), \mathrm{tr}(\mathbf{BD}^2), \mathrm{tr}(\mathbf{B}^2\mathbf{D}), \mathrm{tr}(\mathbf{B}^2\mathbf{D}^2)\end{aligned} \tag{2.25}$$

Referring to the energy inequality condition 2.16 and, for simplicity, assuming on the basis of physical experience that the coefficients ψ_α are so slowly varying with time as to be approximated by constants (an admittedly totally unwarranted mathematical assumption) it can then be shown by arguments similar to those previously used herein that

$$\mathbf{T}(\mathbf{B}, \mathbf{D}) = \psi_1 \mathbf{B} + \psi_2 \mathbf{D} + \psi_3 \mathbf{B}^2 + \psi_5(\mathbf{BD} + \mathbf{DB}) + \psi_6(\mathbf{B}^2\mathbf{D} + \mathbf{DB}^2) \tag{2.26}$$

A more general consideration produces the result, with constant coefficients,

$$\mathbf{T}(\mathbf{B}, \mathbf{D}) = \sum_{k=1}^{n} \mathbf{a}_k \mathbf{B}^k + \sum_{k=1}^{n} \mathbf{c}_k (\mathbf{D})^{2k-1} + \mathbf{L}_{(2)} \sum_{k=0}^{n} \mathbf{b}_k (\mathbf{D})^{2k} \tag{2.27}$$

In both of the above equations the existence of a rest configuration requires $\psi_0 = 0$. One can remove the $\mathbf{L}_{(2)}$ dependence in the stress, **T**, by use of such schemes as assuming the body force derivable from a scalar potential. However, this assumption, while removing explicit dependence on $\mathbf{L}_{(2)}$, introduces an implicit dependence on **grad grad T** and the gradient of the scalar potential. This device is not to be recommended. Thus, the minimal reduction in dependent variables obtained by the use of the energy inequality yields for the Cauchy stress

$$\mathbf{T} = \mathbf{f}[\mathbf{B}, \mathbf{D}, \mathbf{L}_{(2)}] \tag{2.28}$$

An equivalent form is

$$\mathbf{T} = \mathbf{f}[\mathbf{B}, \mathbf{A}_{(1)}, \mathbf{L}_{(2)}] \tag{2.29}$$

where $\mathbf{A}_{(1)}$ is the Rivlin–Ericksen[2] deformation rate tensor related to **D** and $\mathbf{D}_{(2)}$ by

$$\mathbf{D} = \tfrac{1}{2}\mathbf{A}_{(1)} \quad \mathbf{D}_{(2)} = \tfrac{1}{2}(\mathbf{A}_{(2)} - \tfrac{1}{2}\mathbf{A}_{(1)}^2) \tag{2.30}$$

and to $\mathbf{L}$ and $\mathbf{L}_2$ by

$$\mathbf{A}_{(1)} = \mathbf{L} + \mathbf{L}^T \quad \mathbf{A}_{(2)} = \mathbf{L}_{(2)} + \mathbf{L}_{(2)}^T + 2\mathbf{L}^T\mathbf{L} \tag{2.31}$$

It is this dependence of $\mathbf{A}_{(2)}$ on $\mathbf{L}^T\mathbf{L}$ which precludes $\mathbf{A}_{(2)}$ being compatible with the energy inequality for the cases considered.

2.2 Material isotropy restrictions

The assumed isotropy of the material and the principle of coordinate frame indifference requires that any properly formulated physical law be invariant in form under the pertinent subgroup of unitary transformations. For our situation, requiring invariance of form under the proper orthogonal subgroup of transformations $\mathbf{G}$ (all real rotations) where $\det \mathbf{G} = +1$ precludes stress dependence on $\mathbf{L}$ or $\mathbf{L}_{(2)}$. The forms $\mathbf{D}_{(2)}$ and $\mathbf{A}_{(2)}$ of 2.30 and 2.31 are, however, acceptable. However, the energy inequality restrictions do not allow the $\mathbf{L}^T\mathbf{L}$ term in the stress equation. Thus, expression 2.29 can be reduced to a non-linear function

$$\mathbf{T} = \mathbf{f}[\mathbf{B}, \mathbf{A}_{(1)}] \tag{2.32}$$

We call a thermomechanical process which satisfies the constitutive equations an *admissible thermomechanical process.*

3. THERMODYNAMIC PROCESSES

In order to develop the discussion relative to relaxation in one-sided rate type materials, we shall refer to results of Coleman and Gurtin[4] without presenting the details of development.

3.1 Equipresence

We shall employ herein Meixner's *strong principle of equipresence* wherein all constitutive equations shall initially contain the same terms[1] and each term shall depend on the same order of the derivatives[7]. Thus, since $\mathbf{B}$ is a function of $\mathbf{F}$ which is $\mathbf{GRAD}\ \mathbf{x}$ and $\mathbf{A}_{(1)}$ is, so to speak, $\dot{\mathbf{F}}$ we require the constitutive functionals for stress, $\mathbf{T}$, heat flux $\mathbf{q}$ and any other dependent variables, $\mathbf{P}$, to be of the form using 2.32

$$\mathbf{P} = \tilde{\mathbf{P}}(\mathbf{x}, \dot{\mathbf{x}}, \mathbf{F}, \dot{\mathbf{F}}, \boldsymbol{\theta}, \dot{\boldsymbol{\theta}}, \mathbf{grad}\ \boldsymbol{\theta}, \mathbf{grad}\ \dot{\boldsymbol{\theta}}) \tag{3.1}$$

where we have not included the internal state variable $\boldsymbol{\varphi}$ as yet.

The material isotropy and the localization postulate immediately eliminate specific dependence on $\mathbf{x}$, $\dot{\mathbf{x}}$ so that with 2.32 and $\mathbf{g} \equiv \mathbf{grad}$ relation 3.1 becomes

$$\mathbf{P} = \tilde{\mathbf{P}}(\mathbf{B}, \mathbf{A}_{(1)}, \boldsymbol{\theta}, \dot{\boldsymbol{\theta}}, \mathbf{g}, \dot{\mathbf{g}}) \tag{3.2}$$

Note that $\dot{\boldsymbol{\theta}}$ need not necessarily appear since $\dot{\mathbf{x}}$ did not appear. Rather the gradients and time derivatives of the gradients appear.

3.2 Clausius–Duhem restrictions

While agreeing fully with Meixner's[7] and Rivlin's[3] comments regarding the absence of thermodynamic validity for the Clausius–Duhem entropy rate inequality in the differential form

$$\mathbf{d}\eta \geqslant \sum \mathbf{d}\varepsilon/\theta \qquad (3.3)$$

where we use $\mathbf{d}\varepsilon$ as the increment of energy addition of any sort, Coleman's thermodynamics of non-equilibrium processes rests essentially on the axiom that 3.3 holds at all times. Coleman[4] following a standard procedure asserts that the rate of entropy growth γ of a system equals the rate of specific entropy growth minus the rate of heat addition divided by absolute temperature. The key axiom is then that the rate of entropy growth, γ, is non-negative[10] at every instant of time. Thus

$$\rho\gamma \equiv \rho\dot{\eta} - \left[\frac{\rho \mathbf{r}}{\theta} - \mathbf{div}\left(\frac{\mathbf{q}}{\theta}\right)\right] \geqslant 0 \qquad (3.4)$$

Defining the Piola–Kirchhoff non-symmetric stress tensor as $\mathbf{S} = \rho^{-1}\mathbf{T}(\mathbf{F}^{\mathrm{T}})^{-1}$ in order to remove conveniently the term ρ from the final equation gives the energy balance as

$$\rho\dot{\varepsilon} = \rho \mathbf{r} - \mathbf{div}\,\mathbf{q} + \rho\mathbf{S}\cdot\dot{\mathbf{F}} \qquad (3.5)$$

Eliminating $\rho\mathbf{r}$ from 3.4 and 3.5 after carrying out the operation indicated in $\mathbf{div}\,\mathbf{q}/\theta$ results in

$$\dot{\eta} \geqslant \frac{\dot{\varepsilon}}{\theta} - \frac{\mathbf{S}\cdot\dot{\mathbf{F}}}{\theta} - \frac{1}{\rho\theta^2}\mathbf{q}\cdot\mathbf{g} \qquad (3.6)$$

From Coleman[4] we note that:

$$\begin{aligned} \dot{\eta} &\geqslant 0 \text{ when } \dot{\mathbf{F}} = \dot{\varepsilon} = \mathbf{g} = 0 \\ \dot{\varepsilon} &\leqslant 0 \text{ when } \dot{\mathbf{F}} = \dot{\eta} = \mathbf{g} = 0 \end{aligned} \qquad (3.7)$$

Thus we see that in a typical stress relaxation test with homothermal conditions, the specific internal energy can decrease and the entropy can increase.

Introducing a *Helmholtz free energy* function ψ where

$$\psi = \varepsilon - \theta\eta \qquad (3.8)$$

and taking the partial derivative of ψ where $\dot{\psi} = \sum_{i=1}^{n} \partial_{\varphi i}\psi \cdot \dot{\varphi}_i$ (with φ_i as any selected independent variable) and $\partial_\varphi\psi \equiv \partial\psi/\partial\varphi$ we then equate the coefficients above to those obtained by writing $\dot{\psi}$ in terms of 3.8 and 3.6. The result is then with a constitutive equation of form 3.2

$$\theta\gamma = (\mathbf{S} - \partial_{\mathbf{F}}\psi)\cdot\dot{\mathbf{F}} - (\partial_\theta\psi + \eta)\,\dot{\theta} - \frac{1}{\rho\theta}\mathbf{q}\cdot\mathbf{g} - \sum_{i=1}^{n} \partial_{\varphi i}\psi\cdot\dot{\varphi}_i \geqslant 0 \quad (3.9)$$

Specifically in a form 3.2 we have

$$\psi = \psi(\mathbf{F}, \dot{\mathbf{F}}, \theta, \dot{\theta}, \mathbf{g}, \dot{\mathbf{g}}, \sum_1^n \varphi_i, \sum_1^n \dot{\varphi}_i) \qquad (3.10)$$

with η, $\mathbf{S}$, $\mathbf{q}$ of the same form.

Not only then is the usual result

$$\mathbf{S} = -\partial_{\mathrm{F}}\psi\,;\ \boldsymbol{\eta} = -\partial_{\theta}\psi \tag{3.11}$$

obtained, based on independence of the dependent variables, but in general

$$\partial_{\varphi i}\psi = 0 \text{ for } 1 \leqslant i \leqslant n \tag{3.12}$$

which in our case removes any dependence on $\mathbf{g}$ or any $\boldsymbol{\varphi}_i$ in ψ. Similarly

$$\partial_{\dot{\varphi} i}\psi = 0 \quad 1 \leqslant i \leqslant n \tag{3.13}$$

removing any dependence on $\dot{\boldsymbol{\varphi}}_i$ whatever $\dot{\boldsymbol{\varphi}}_i$ may be including it being $\dot{\mathbf{F}}, \dot{\boldsymbol{\theta}}, \dot{\mathbf{g}}$.

Some authors[8,9] have interpreted the case where derivatives are used to imply that the entropy inequality simply drops the dependence on $\boldsymbol{\varphi}^{(\mathrm{k})}$ to the level $\boldsymbol{\varphi}^{(\mathrm{k}-1)}$. This is not only untrue for independent variables, but logical deduction, i.e. recycling the equation with $\boldsymbol{\varphi}^{(\mathrm{k}-1)}$ dependence, would lead to $\boldsymbol{\varphi}^{(\mathrm{k}-2)}$ dependence until 3.12 or 3.13 conditions prevailed. In general

$$\partial_{\varphi i}^{(\mathrm{k})}\psi = 0 \quad \mathbf{k} = 1, 2, \ldots, \quad i = 1, 2, \ldots$$

It is this above fact that has led[8] to the often quoted statement that one-sided rate materials cannot exhibit stress relaxation. The solution to this paradox is twofold. First, where all the variables in 3.10 are independent then relation 3.10 can be reduced to

$$\psi = \psi(\mathbf{F}, \boldsymbol{\theta}) \tag{3.15}$$

a so-called caloric equation of state. Defining, as usual, the elastic, ${}_{\mathrm{E}}\mathbf{S}$, and dissipative ${}_{\mathrm{D}}\mathbf{S}$ stress components of $\mathbf{S} = {}_{\mathrm{E}}\mathbf{S} + {}_{\mathrm{D}}\mathbf{S}$ yields by equation 3.9

$${}_{\mathrm{E}}\mathbf{S} = \partial_{\mathrm{F}}\psi \text{ and } {}_{\mathrm{D}}\mathbf{S}\cdot\dot{\mathbf{F}} \geqslant 0 \tag{3.16}$$

Implying, since $\mathbf{S}$ and $\dot{\mathbf{F}}$ are jointly continuous, that ${}_{\mathrm{D}}\mathbf{S} \to 0$ when $\dot{\mathbf{F}} \to 0$, i.e. no relaxation under constant strain. Again this is strictly correct and points out the fact that any one-sided formulation in any number of independent variables and their higher derivatives degenerates to nothing more than an equivalence to a two-element non-linear Kelvin–Voigt solid which can exhibit creep but no stress relaxation excepting a jump relaxation when $\dot{\mathbf{F}} \to 0$. Secondly, the situation may be resolved by simply removing the independence from one or more of the variables entering into, say 3.2. This may be done by specifying one of the derivatives of $\boldsymbol{\varphi}_i$ as dependent, i.e.

$$\dot{\boldsymbol{\varphi}}_i = \mathbf{f}(\mathbf{B}, \mathbf{A}_{(1)}, \ldots, \varphi_i, \ldots) \tag{3.17}$$

Then a dissipation, $\boldsymbol{\sigma}$, inequality will arise from 3.9 leading to a term

$$\boldsymbol{\sigma} = -\partial_{\varphi i}\psi\cdot\mathbf{f} \geqslant 0, \quad \therefore \boldsymbol{\sigma} \geqslant 0 \tag{3.18}$$

for homothermal fields, $\mathbf{g} = 0$.

We now see that dropping the artifice of 3.16 and expressing the rate of change of stress, $\dot{\mathbf{S}}$ as

$$\mathbf{S}\dot{\mathbf{F}} = \partial_{\mathrm{F}}\psi\cdot\dot{\mathbf{F}} + \partial_{\varphi i}\psi\cdot\dot{\varphi}_i \tag{3.19}$$

and with $\dot{\mathbf{F}} = 0$ and 3.18 where $\dot{\varphi}_i = \mathbf{f}$, not necessarily integrable

$$\dot{\mathbf{S}}\mathbf{F} = \partial_{\varphi i}\psi \cdot \dot{\varphi}_i \leqslant 0 \qquad (3.20)$$

Thus, the introduction of a dependence in time of even one φ_i leads to 3.20 which states that the stress must decrease in a stress relaxation test. Needless to say, the dependence of $\dot{\varphi}$ requires a proper statement of initial conditions.

ACKNOWLEDGEMENTS

The author wishes to acknowledge the stimulus in the subject given him some years ago when he attended a short course on non-linear field theories under Professor C. Truesdell. The specific impetus for this current paper was occasioned by several stimulating conversations with Professors R. S. Rivlin (relative to the 'burning bush' paradox) and J. Meixner whose fundamental inequality led to the ideas for the energy inequality herein.

I wish to acknowledge Dean M. L. Williams, who granted my sabbatical leave, John Crowley, and Professor P. M. Quinlan, all who made possible my coming to University College, Cork.

This work was carried out in part under Contract F-44620-68-C-0022P001, a Themis contract under the Air Force Office of Scientific Research.

REFERENCES

1 C. Truesdell and W. Noll, 'Non-linear Field Theories of Mechanics' in *Encyclopedia of Physics*, Vol. III, p 13. Springer: Berlin (1965).

2 R. S. Rivlin and J. L. Erickson, *J. Ration. Mech. Anal.* **7** (No. 2), 321–425 (1955).

3 R. S. Rivlin, 'Red herrings and sundry unidentified fish in non-linear continuum mechanics', *CAM Rep. No. 100-a*, Lehigh University, Pennsylvania (September 1969).

4 B. D. Coleman and M. E. Gurtin, *J. Chem. Phys.* **47** (No. 2), 567–613 (1967).

5 J. Meixner, Remarks made at International Congress of Rheology, Kyoto (November 1968).

6 R. S. Rivlin, *J. Ration. Mech. Anal.* **4**, 681–702 (1955).

7 J. Meixner, 'On the foundation of thermodynamics of processes' presented at International Symposium on Foundations of Relativistic and Classical Thermodynamics, Pittsburgh (April 1969).

8 A. C. Eringen, *Non-linear Theory of Continuous Media.* McGraw-Hill: New York (1962).

9 W. Olszack and P. Perzyna, 'On thermodynamics of differential type material', *Proceedings of the I.U.T.A.M. Symposium on Irreversible Aspects of Continuum Mechanics, Vienna, 1966* Springer: Berlin (1968).

10 I. Prigogine, *Thermodynamics of Irreversible Processes*. 3rd ed. Wiley: New York (1967).

D. IRREVERSIBILITY AND QUANTUM MECHANICS

THE CONCEPT OF IRREVERSIBILITY IN STATISTICAL MECHANICS

ROBERT ZWANZIG

Institute for Fluid Dynamics and Applied Mathematics and Institute for Molecular Physics, University of Maryland

ABSTRACT

The interaction between concepts of irreversibility and the development of non-equilibrium statistical mechanics is discussed with particular reference to certain 'paradoxes' that retarded this development. A recently evolved attitude towards the general problem of irreversibility may be considered responsible for a number of practical advances in this field.

My subject is the interaction between concepts of irreversibility and the development of non-equilibrium statistical mechanics. In particular, I will discuss certain 'paradoxes' that seriously retarded this development, and how these paradoxes are viewed in modern work. Then I will describe an attitude towards the general problem of irreversibility, evolved in the last ten or fifteen years, that in my opinion is responsible for a number of important practical advances in non-equilibrium statistical mechanics.

Everyone knows what irreversibility is. On a primitive level, we know that fire burns wood to ashes, that men grow old and die, and that taxes will increase.

On a more advanced level, we know that irreversibility is associated with an inevitable increase of entropy. We know from experience that we are unable to construct devices to decrease the total entropy of our local environment. Whatever we are able to do, the entropy increases. Our *experience* is summarized in the second law of thermodynamics.

On a still more advanced level, the irreversible increase of entropy is given a cosmic generality and a deep philosophical significance. Here, the standard view is summarized in the famous statement by Clausius: '. . . *die Entropie der Welt strebt einem Maximum zu*'. In this form, the second law of thermodynamics is often used to discuss the approach of the entire universe to a state of thermal equilibrium, or to support our intuition that the flow of time has an 'arrow' attached to it.

Many of the difficulties that arise in the statistical mechanical theory of irreversibility can be traced to the sweeping generality of the third view of irreversibility that we have just referred to. At this level, there seems to be a fundamental contradiction between the second law and mechanics. The main point I want to make here is that these difficulties can be avoided by taking a more modest point of view, in which the second law merely summarizes certain human experiences. I do not say that one *must* take the more modest view, but I say that *if* one takes this view, *then* one can develop a practical non-equilibrium statistical mechanics.

My story begins in 1872 with Boltzmann and the kinetic theory of gases. Boltzmann had two concerns. One was a practical one, to be able to predict various transport properties of gases, for example their viscosity coefficients. The other concern was to understand the mechanical basis for the thermodynamic concept of an equilibrium state. He introduced the Boltzmann equation, which is an integro-differential equation for the evolution of the kinetic distribution function $f(v; t)$, the density of gas atoms having velocity v at time t. In the course of his investigations, he discovered that a particular functional of the velocity distribution,

$$H(t) = \int dv \, f \log f$$

had an extremely interesting time dependence. On computing the evolution of $H(t)$ with the Boltzmann equation, he found that it can never increase with time, or

$$(dH/dt) \leqslant 0 \text{ (all } t)$$

It can only decrease or remain constant; and it remains constant only in the state of thermodynamic equilibrium.

Thus $H(t)$ shows the same kind of irreversible behaviour that we expect of the entropy. And, in fact, H is the negative of the entropy for an equilibrium state.

It should be emphasized that Boltzmann's H-theorem is an exact consequence of the Boltzmann equation for $f(v; t)$. Further, we know (at least in retrospect) that the Boltzmann equation provides a correct and useful theory of simple transport processes in dilute gases.

But almost immediately, certain objections were raised to the validity of the H-theorem and the Boltzmann equation. These objections, in various forms, plagued subsequent investigations in non-equilibrium statistical mechanics for many years.

The first objection, usually called the 'reversibility paradox', was raised by Lord Kelvin and by Loschmidt. In modern terms, this paradox may be called a violation of time-reversal symmetry. The fundamental equations of motion of any conservative dynamical system, e.g. a monatomic gas, are Newton's, Lagrange's or Hamilton's equations. These equations are invariant to the substitution of $-t$ for t; or, they are symmetric to time reversal. The H-theorem and the Boltzmann equation violate this symmetry, so they cannot be consistent with any exact dynamical theory. Therefore they cannot be correct.

The second objection to Boltzmann's work was raised by Zermelo and by Poincaré, and is usually called the 'recurrence paradox'. It arises whenever one deals with a finite closed dynamical system.

If a system of interacting particles is confined to a closed region of space, and if their interaction energy has a finite lower bound, then the motion of the system is confined to a finite region of phase space. According to ergodic theory, the trajectory of the system in phase space passes arbitrarily closely to any assigned position on the surface of constant total energy; given sufficient time, it does so arbitrarily often. So any given state of the system will recur to within any assigned accuracy. This indicates that a gas contained in a finite volume cannot approach an equilibrium state and then stay there indefinitely. Any non-equilibrium state that was passed through once will be visited again if one waits long enough.

(In quantum mechanics, the same objection takes the following form. If the system is enclosed in a finite region of space, its energy level spectrum must be discrete. Then the time dependence of any dynamical property is given by an almost-periodic function, and recurrences are guaranteed.)

The objections that we have quoted, and a number of variations on them, led to the general impression that no statistical mechanical theory based on exact dynamics could be consistent with the irreversibility that we observe in nature.

Because of this, workers in the field felt that one must 'do something' to the exact equations of motion before irreversibility would emerge. Very often, lectures on new methods in non-equilibrium statistical mechanics involved heated discussion of the question 'Where did you put the irreversibility into the theory?' Many ingenious answers were given.

One approach that has been popular for a long time is called 'coarse graining'. This has both classical and quantum mechanical forms, but for illustration I use the classical one. It is argued that the exact position of any system in phase space is never observed experimentally and is of no interest. One should divide phase space into cells of finite size, each cell corresponding roughly to some macroscopic description of the state of the system; and one should focus attention not on the detailed distribution within any individual cell, but only on the net content of that cell.

A variant of this view is called 'time-smoothing'. Here it is argued that we are unable to observe experimentally the precise time dependence of any dynamical quantity; because of inherent limitations on our apparatus, only a time average is observed. The time interval used for averaging is supposed to be short compared with characteristic times of macroscopic processes, but long compared with characteristic times of elementary molecular processes. Sometimes time-smoothing is used along with coarse graining.

Objections can be made to these ideas. One objection is that smoothing techniques are based on the assumed limitations of experimental apparatus. We know from experience, however, that measurement techniques are being constantly refined, and are more and more delicate. Any theory based on assumed limitations of this kind is likely to be superseded some day.

Another objection should be mentioned. Generally speaking, smoothing techniques are only productive if also certain *discarding* operations are performed. Somehow, information has to be thrown away. A careful inspection of theories based on these techniques will disclose that smoothing is always accompanied by various approximations that have the effect of discarding information. In my opinion, these approximations (sometimes quite subtle ones) are vital to the success of smoothing techniques. The operation of smoothing is itself only a convenient way of leading up to the necessary approximations.

Another procedure used to 'put irreversibility into the theory' is to suppose that the system is not contained in a closed box. This is justified by the observation that any real system must interact with its surroundings, and they must interact with *their* surroundings, *ad infinitum.*

In some instances, this method for introducing irreversibility is quite efficient. Consider, for example, the spontaneous emission of a photon from an excited atom. If the photon is able to escape from the atom into infinite space, then it will never return to be re-absorbed, and the process is irreversible. If the accessible space is finite, then eventually the photon must be reflected by a wall, so that it can return to the atom and be re-absorbed.

A difficulty with the open system approach is that the only true equilibrium state is that of the entire universe; this is not much help for our own petty concerns.

Another variation is to suppose that the system is contained in a box having walls that are not fixed, but move randomly according to some stochastic process. This procedure will also lead to irreversibility. While it works, however, the stochastically moving walls may often be regarded as irrelevant. A clock that is going to run down and stop will do so whether or not it is located in some perfectly sealed room, and whether or not the walls of the room are jiggling. In a closed system, we expect that the clock will eventually start up again, but this is of no interest to anyone who wants to know what time it is during his own lifetime.

During this long period of investigation into the foundations of non-equilibrium statistical mechanics, significant progress was made with respect to more practical questions. It soon became clear that the Boltzmann equation gave a valid, experimentally verifiable description of transport processes in gases at low enough densities. The theory of Brownian motion was developed. In the early days of quantum mechanics, the theory of transition rates between quantum states (as expressed in the 'golden rule') was worked out. Onsager derived his reciprocal relations and showed how one could make practical use of non-equilibrium thermodynamics.

All these advances in useful technique for handling non-equilibrium systems were made without regard to the fundamental difficulties connected with the 'paradoxes' I just discussed.

As attempts were made to extend the limits of validity of familiar methods, for example to derive a generalization of the Boltzmann equation that would be valid for dense gases or to construct a theory of transport processes in liquids, the impression grew that progress could be made only after the paradoxes were resolved. This is partly the reason why so much attention was given to methods for 'putting irreversibility into the theory'.

But in the 1950s, a striking change in direction and attitude occurred. Van Hove presented a sound derivation of the Pauli master equation for weakly interacting systems, and gave a correct explanation of the validity of the theory of transition rates. Soon after that, he presented the first generalization of the master equation to strongly interacting systems.

At about the same time, diagrammatic techniques were developed for handling perturbation and other expansions to infinite order. Using these techniques, Prigogine and many others made significant progress in non-equilibrium statistical mechanics.

Also at about the same time, Kubo presented his remarkably simple analysis of the response of a many body system to an external field, leading to the time-correlation function expressions for transport coefficients. Kubo's work had a particularly strong impact, perhaps because his approach was so direct and intuitively obvious.

Another important development of the last decade was the introduction and analysis of simple analytically tractable models for non-equilibrium systems. I mention especially work by Rubin, Montroll, Mazur and others on harmonic oscillator models of Brownian motion. The importance of such simple model systems is that they lead to mathematically exact results, and can be used to test methods based on mathematical approximations.

In my opinion, however, the most important development in recent years was a change in point of view. If one looks for some feature common to recent successful theories, one finds that they all have a strongly operational character.

Consider for example Kubo's analysis of linear transport processes. Suppose that we want to find the electrical conductivity of a metal. In Kubo's theory, we construct a canonical ensemble distribution function for a piece of metal at some temperature. To the Hamiltonian describing the metal we add a perturbation term due to the interaction of the metal with an external electric field. We use perturbation theory to find out how the original equilibrium distribution function is modified by a time dependent electric field. Then we use the modified distribution function to compute the average electric current in the presence of the field. It turns out that the current is proportional to the field. Then the coefficient of proportionality must be the electrical conductivity of the metal.

Notice that each step in this procedure corresponds to an operation that one would perform in a laboratory experiment. Selection of a sample piece of metal at some temperature corresponds to starting out with an initial canonical ensemble distribution function. Switching on an external electric field corresponds to adding a perturbation to the Hamiltonian. Measuring the current corresponds to calculating the ensemble average of the current. The measured electrical conductivity is the coefficient of proportionality between the measured current and the applied field.

The success of this procedure is connected with the following statement of belief. *If each step in a statistical mechanical calculation can be put into one-to-one correspondence with a step in some experiment, then the result of the calculation must be the same as the result of the experiment.*

There are, of course, still some serious difficulties to be faced. The expression for electrical conductivity that one obtains this way may be not at all easy to calculate; the time dependence of a fluctuating electric current must be found, and an equilibrium average has to be calculated. But these are well-defined computations, not involving deep questions of principle. This is where simple model systems are useful.

Essentially the same procedure is followed in modern derivations of the quantum mechanical master equation, describing the evolution of diagonal elements of a density matrix. Here we start out with some non-equilibrium system in which the density matrix is initially diagonal; we use operator methods to follow the time dependence of the density matrix at later times; we focus attention on only the diagonal part of the density matrix at later times; and then we do what are essentially just algebraic manipulations to find a generalized master equation. Explicit computation of the coefficients in the master equation is still a hard job, but no questions of principle are involved.

Similar analyses may be carried out for many familiar transport theories, but a detailed description of all of these would be out of place here. However, one can draw a general conclusion. Successful treatments all seem to be based on the same idea—an initial state is defined carefully, dynamical processes are then followed exactly (though sometimes only in a purely formal way), and finally, only certain specific questions are asked about the results.

But we are still left with the nagging question of the paradoxes. How is it that we are able to proceed at all, in view of the asserted contradiction between exact dynamical principles and the second law? Perhaps this question is best answered by considering a simple model for which exact results can be obtained.

This is a model of Brownian motion in one dimension. Let me first remind you of the standard theory of Brownian motion. I will use the Langevin form of the theory. The Brownian particle has a mass M and a velocity $v(t)$ at time t. Its equation of motion is the Langevin equation

$$M(\mathrm{d}v(t)/\mathrm{d}t) = -\zeta v(t) + F(t)$$

where $-\zeta v(t)$ is the frictional force on the particle and ζ is the friction coefficient. The extra force $F(t)$ is a fluctuating force due to interactions of the particle with its environment, and is known only in a statistical way. In particular, it is treated as a gaussian random variable, with zero mean value, and a second moment

$$\langle F(t)\, F(t')\rangle = 2\zeta k_B T \delta(t - t')$$

T is the temperature of the medium and k_B is the Boltzmann constant. The preceding equation is often called a Nyquist formula or a fluctuation–dissipation theorem.

This standard Brownian motion theory leads to the paradoxes. Consider, for example, the time dependence of the average velocity. It is evident that $\langle v(t)\rangle$ decays exponentially from some initial value

$$\langle v(t)\rangle = v(0) \exp(-\zeta t/M)$$

It does not recur, as it should; and time reversal symmetry is violated.

Now we want to compare this standard theory with a modern one, based on a simple model in which dynamical calculations can be performed exactly. The model is due to R. J. Rubin. [His most recent publication on the subject is in *J. Am. Chem. Soc.* **90**, 3061 (1968). This gives references to earlier work.]

Rubin's model is a finite one-dimensional nearest neighbour harmonic crystal, consisting of $2N + 1$ particles with periodic boundary conditions. All particles except the one labelled 0 have a mass m, while particle 0 has a mass M which is much larger than m. The Hamiltonian is

$$H = \frac{1}{2M} p_0^2 + \sum_{\substack{j=-N \\ \neq 0}}^{N} \frac{1}{2m} p_j^2 + \frac{1}{2} K \sum_{j=-N}^{N} (r_j - r_{j+1})^2$$

p_j and r_j denote the momentum and displacement of the jth particle, and K is a force constant.

Because the model contains only coupled oscillators, the equations of motion are all linear and can be solved by matrix methods. In particular, the perturbation due to the heavy particle can be handled exactly.

Now let us construct an experiment. At the initial time $t = 0$, the heavy particle has a given momentum $p_0(0)$. All other momenta, and all displacements, are assumed to have a thermal equilibrium distribution at temperature T. Thus the initial state is well defined in a statistical mechanical sense. What then is the motion of the heavy particle?

At time t, $p_0(t)$ can be expressed as a linear combination of terms, each of which is a product of a known function of time and some initial momentum or displacement. This is a consequence of the linearity of the equations of motion. By working backwards, one can find a generalized Langevin equation for $p_0(t)$

$$(\mathrm{d}/\mathrm{d}t)\, p_0(t) = -\int_0^t \mathrm{d}s\, k(s)\, p_0(t-s) + F(t)$$

In this, $k(s)$ is a time dependent friction coefficient, and $F(t)$ is a fluctuating force. Further, $F(t)$ is a gaussian random variable. (It is a linear combination of initial values of all displacements and all momenta except p_0, and these are supposed to have a Boltzmann distribution.) The mean value of $F(t)$ vanishes, and its second moment is given by a generalized fluctuation–dissipation theorem

$$\langle F(t)\, F(t')\rangle = Mk_BTk(t-t')$$

Notice that the only difference between this theory and the standard Langevin theory is in the time dependence of the function $k(s)$. If this were a delta-function, then the standard theory would be recovered. (The missing factor of two comes from taking one half the delta-function in the convolution over time.)

Rubin succeeded in calculating analytically the quantities needed to obtain $k(s)$; to avoid excessive detail, I will not write the final result here. More important, he was able to solve analytically for the time dependence of the *average* momentum of the heavy particle. This makes it possible to compare the predictions of the exact calculation with those obtained from the standard Langevin theory.

Without going into detail, I will describe in qualitative terms what the results are. First, the average momentum is an even function of time. This is a consequence of the exact character of the calculation. Time reversal symmetry is not violated. Secondly, if the calculations are performed for a finite lattice, then recurrences in the average momentum are found. This again is a consequence of the exact character of the calculation. We must conclude that no paradoxes are evident in this model.

The third and most important property of the average momentum is found when both the mass ratio M/m and the lattice size N are very large. Then the decay of the average momentum is *approximately exponential.* The precise meaning of this statement was worked out by Rubin.

Three significant time scales are observed. The first is a short one, of the order of the reciprocal of the Debye frequency of the uniform lattice,

$$t_1 = \tfrac{1}{2}(m/K)^{\frac{1}{2}}$$

During this interval, the average momentum is not exponential; rather, it is approximately parabolic in time, with a maximum at $t = 0$. But the change in magnitude of the average momentum is only of order m/M during this interval.

The next time scale is much longer

$$t_2 = (M/m)t_1$$

For times of this order of magnitude, approximate exponential decay is found, with the relaxation time t_2. The difference between the exact result and the exponential approximation is small. Rubin gave a numerical estimate of the orders of magnitude involved when the mass ratio is $M/m = 10^4$. He found that '... after 18 relaxation times, the correction to the exponential is less than 10^{-4} of the value of the exponential'. It seems to me that it would be extraordinarily difficult to detect such a small deviation from exponential decay in any real experiment.

The third time scale is defined by

$$t_3 = Nt_1$$

For times of this order of magnitude, recurrences will be seen. If we had done the calculation for an infinite lattice in the first place, this time scale would never appear; but we would still see the same exponential decay. If the lattice is finite and large enough, we would never see recurrences in our own lifetimes.

I emphasize that this theoretical model of Brownian motion does not involve any smoothing process, stochastic element, or any other act of violence on exact dynamics. Irreversibility is not 'put into' the theory anywhere.

So here we have an answer to our question about the apparent contradiction between exact dynamical principles and the second law. As long as we are willing to settle for a theory of irreversibility on a human, experimentally observable time scale, there are no contradictions. If we confine our efforts in non-equilibrium statistical mechanics to problems that are rooted in operationally well-defined experiments, and if we have the courage to do hard calculations, then we are bound to succeed.

THERMODYNAMIC THEORY OF STABILITY, STRUCTURE AND FLUCTUATIONS

G. NICOLIS

Faculté des Sciences, Université Libre de Bruxelles, Belgium

ABSTRACT

The development of the thermodynamics of irreversible processes is outlined with a review of recent work and a discussion of the application of these concepts to physicochemical, biological and hydrodynamic phenomena. The extension of local thermodynamics to include a theory of stability and of fluctuations receives attention. The author concludes with remarks on the stability properties of chemical reactions in open systems and comments on the possible implications of results in the interpretation of fundamental biological phenomena.

---

1. INTRODUCTION

Classical thermodynamics deals with transformations involving equilibrium states. Once the validity of the second law is admitted, one considers exclusively systems which have already reached thermodynamic equilibrium. The behaviour of these systems is then completely described in terms of a set of state functions, the *thermodynamic potentials*, whose extremal properties determine both the equilibrium state itself and its stability properties[1].

Many attempts to enlarge equilibrium thermodynamics in order to include irreversible changes have been made since the second law was formulated in the middle of the previous century. The early considerations were, however, restricted to the treatment of very special irreversible processes, such as thermoelectric effects. In addition, although a number of scientists such as P. Duhem[2] had already conceived the beginnings of a macroscopic physics embracing both equilibrium and non-equilibrium phenomena, it was only recently, and particularly during the last twenty years, that we have witnessed the firm foundations and the rapid growth of the thermodynamics of irreversible processes. The present object will be to present a review of recent developments in irreversible thermodynamics and to discuss the application of these concepts to the study of physicochemical, biological and hydrodynamic phenomena. The first point to explore will therefore be the following. Is it possible to extend the methods of classical thermodynamics to treat all possible phenomena starting from close to equilibrium states and including arbitrary non-linear situations? In fact we will answer this question by defining a set of conditions which will guarantee a first extension of thermodynamics to non-equilibrium situations. It is not claimed that these conditions apply to all irreversible changes and it is quite possible that a consistent thermodynamic theory could be set up under less restrictive conditions.

Let any given thermodynamic system be divided into microscopically large but macroscopically small subsystems each having a given volume V. We also assume that it is meaningful to specify at any given moment in the subsystem the internal energy content, E and the mole fractions, n_i of species i. At equilibrium the thermodynamic quantities such as temperature T, pressure p, chemical potential μ_i of component i, entropy S, are well-defined quantities depending on E, V and n_i. If now equilibrium does not hold, it is necessary to re-define all these quantities. We assume that T, p, μ_i and S for each subsystem of a globally non-equilibrium system depend on E, V and n_i in exactly the same way as in equilibrium. In other terms, one proceeds as if equilibrium prevailed in each subsystem separately. This is known as the assumption of *local equilibrium*. Analytically, this implies first that a local formulation of non-equilibrium thermodynamics is possible. And secondly, that in this formulation the local entropy will be expressed in terms of the same independent variables as if the system were at equilibrium. In other words, if ns is the entropy density, ne the energy density, and v the specific volume the well-known Gibbs relation will hold locally:

$$s = s(ne, v, n_i)$$
$$T\,\mathrm{d}(s) = \mathrm{d}(e) + p\,\mathrm{d}v - \sum_i \mu_i\,\mathrm{d}n_i \qquad (1)$$

The local formulation of irreversible thermodynamics based on equation 1 has been worked out systematically by Prigogine[3]. A few years later the same author established the domain of validity of this local equilibrium assumption by showing that it implies the dominance of dissipative processes over purely mechanical processes. In more specific terms, at a given point, the molecular distribution functions of velocities and relative positions may only deviate slightly from their equilibrium forms[4].

In his original work Prigogine considered the case of weakly coupled systems (behaving as systems of non-interacting degrees of freedom at equilibrium). For such systems entropy may be defined in terms of the local molecular distribution function $f_1(\mathbf{x}, \mathbf{v}, t)$ ($\mathbf{x}$ is the position, $\mathbf{v}$ the velocity of a particle) through the Boltzmann relation

$$ns = -k \int \mathrm{d}\mathbf{v} f_1 \ln f_1 \qquad (2)$$

where k is Boltzmann's constant.

The extension of Prigogine's results to the case of strongly coupled systems is not trivial, due to difficulties in defining entropy for such systems in terms of the molecular distribution functions. Recently, however, Prigogine and co-workers have elaborated a quasi-particle representation of statistical mechanics which permits description of the thermodynamic properties of strongly coupled systems in terms of new entities, the dressed particles. In this representation ns takes the form

$$ns = -k \int \mathrm{d}\mathbf{v} \tilde{f}_1 \ln \tilde{f}_1 \qquad (3)$$

where $\tilde{f}_1$ is now the one quasi-particle distribution function. In general $\tilde{f}_1$ is a complicated functional of f_1 due to the interactions. Using now this entropy definition J. Wallenborn, M. G. Velarde and the author have extended Prigogine's conclusions to strongly coupled systems[5]. Again it was necessary to assume that the (quasi-particle) distribution function may only deviate slightly from its equilibrium form.

Clearly this is a sufficient condition which guarantees the applicability of the thermodynamic methods. It is conceivable that an irreversible thermodynamics may be constructed based on more general conditions. For instance, Coleman[6] has recently worked out a non-local theory adapted to the study of materials with memory. We do not discuss this approach here but only focus attention on the local formulation.

It should be pointed out that the local equilibrium assumption permits treatment of a great variety of problems corresponding to situations far removed from complete thermal equilibrium. For instance very complicated chemical reaction schemes with extremely large affinities may be treated adequately provided that they are not too fast. Similarly all effects described by the Stokes–Navier equations, including hydrodynamic instabilities, are within the domain of the local equilibrium theory.

Here we discuss the applications of the theory to non-equilibrium situations with special emphasis on typically non-linear problems which cannot even be formulated within the framework of classical thermodynamics. We first present in section 2 a brief review of the general theory and of its applications in the linear region. We then discuss, in section 3, the extension of irreversible thermodynamics arbitrarily far from equilibrium and sketch a unified approach to problems as different as non-linear transport phenomena, hydrodynamic instabilities and so on. In section 4 we discuss an extension of local thermodynamics to include a theory of fluctuations. This is necessary, especially for a theory seeking to describe unstable situations. The final sections 5 and 6 are devoted to the study of chemical reactions in open systems and to comments on the possible implications of the results in the understanding of fundamental biological phenomena.

2. GENERAL FORMALISM—THE LINEAR REGION

The starting point is of course the second law of thermodynamics which deals with the entropy change, $\mathrm{d}S$ in a system. Let $\mathrm{d}S$ be split into two parts[3]. We denote by $\mathrm{d}_e S$ the flow of entropy due to interactions with the external world and by $\mathrm{d}_i S$ the change in S due to processes inside the system, or *entropy production*. We have

$$\mathrm{d}S = \mathrm{d}_e S + \mathrm{d}_i S \tag{4}$$

The second law refers to $\mathrm{d}_i S$ only and reads

$$\mathrm{d}_i S \geqslant 0 \tag{5}$$

where the equality applies for reversible changes.

Using the local equilibrium assumption we may now proceed in the calculation of the entropy production per unit time and volume, σ, defined by

$$P = \frac{\mathrm{d}_i S}{\mathrm{d}t} = \int \mathrm{d}V\,\sigma \tag{6}$$

$$\sigma \geqslant 0$$

The calculation consists of developing equation 4 using the Gibbs equation 1 and then substituting de, dv, dn_i from the balance equations of mass, momentum and energy. The final result is[3]

$$\sigma = \sum_{i} J_i X_i \tag{7}$$

σ is therefore a bi-linear form summed over all irreversible processes i, of suitably defined flows, J_i, associated with these irreversible processes, and of generalized forces X_i giving rise to these flows.

It is clear that, as long as the flows are parameters not related to the corresponding forces, the equations of thermodynamics do not permit the explicit study of the evolution of a system subject to well-defined boundary conditions. It is therefore necessary to combine the general balance equations with additional, phenomenological laws relating Js and Xs. Now experiment shows that at thermal equilibrium there is no macroscopic transport of mass, momentum or energy; as a result all currents J_i vanish. On the other hand the conditions of thermal equilibrium imply the absence of constraints such as systematic temperature gradients etc. Therefore the generalized forces X_i vanish at the same time as J_i do. It is thus quite natural to assume that, in the neighbourhood of equilibrium, linear laws between flows and forces will constitute a good first approximation. The phenomenological laws will therefore take the form

$$J_i = \sum_{j} L_{ij} X_j \tag{8}$$

where the sum is over (coupled) irreversible processes and the phenomenological coefficients $\{L_{ij}\}$ are in general functions of the thermodynamic state variables T, p etc. Equations 6 define the linear domain of irreversible processes[7,8].

The coefficients $\{L_{ij}\}$ cannot be arbitrary. It has been known for a long time that the diagonal coefficients are non-negative. On the contrary, it was only in 1931 that Onsager established the first general relations between the non-diagonal coefficients[9]. He showed that it is always possible to choose the flows and forces such that the matrix $[L_{ij}]$ be symmetrical

$$L_{ij} = L_{ji} \tag{9}$$

These are the celebrated Onsager reciprocity relations which were later generalized by Casimir to a wider class of irreversible phenomena[10].

The proof followed by Onsager relates concepts as different at first sight as fluctuations and macroscopic transport phenomena. This was a very important first step towards a justification of irreversible thermodynamics. It is remarkable that more recent work on the foundations of thermodynamics has fully justified the original Onsager relations. At the same time these works, which are based principally on non-equilibrium statistical mechanics, have established the domain of validity of equations 8 and 9 to be the domain of small deviations of the momentum distribution functions from their equilibrium values[11].

The phenomenological laws 8 together with Onsager's relations 9 constitute a convenient framework within which one can study a great number of irreversible phenomena in linear approximation[7,8]. Interesting as it is, this approach is, however, only a particular aspect of irreversible thermodynamics. The local formulation of irreversible thermodynamics has been developed in yet another direction—the search for variational principles. The question one asks is whether there exists a general principle—other than the second law—characterizing non-equilibrium states themselves independently of the details of phenomena occurring in the system. In order to formulate this question quantitatively it is necessary to analyse in some detail the character of a non-equilibrium state in thermodynamics. In an isolated system one has $d_e S = 0$ and the second principle implies that entropy increases until it reaches its maximum value. The system thus tends more or less rapidly to a uniquely determined permanent state which is the state of thermodynamic equilibrium. Consider now instead of an isolated system a closed system which can exchange energy with the external world, or an open system which can exchange both energy and matter. In this case, and provided the external reservoirs are sufficiently large to remain in a time-independent state, the system may tend to a permanent régime other than the equilibrium one. This will be a *steady non-equilibrium state.* Now this régime is no longer characterized by a maximum of entropy (as $d_e S \neq 0$) or by a minimum of free energy. In other terms, the variational principles valid in thermal equilibrium cannot be extended beyond this state. It is therefore necessary to look for new principles which generalize the concept of a thermodynamic potential to steady (or slowly varying in time) non-equilibrium states. To this end we subdivide the domain of non-equilibrium phenomena into two parts: the region close to equilibrium and the region of states arbitrarily far from equilibrium. Here we deal only with the first region, i.e. the linear domain of irreversible processes, and we only consider systems in mechanical equilibrium.

In his classical work on reciprocity relations[9] Onsager had already proposed a variational principle for such non-equilibrium states which he called the *principle of least dissipation of energy.* In this principle it is understood that the thermodynamic forces remain fixed and only the macroscopic currents may vary.

On the other hand Prigogine has shown[3,7] that steady states close to equilibrium are also characterized by an entirely different variational principle according to which, at the steady state, the entropy production per unit time is a minimum

$$d_i S/dt = \text{minimum} \tag{10}$$

The interest of this principle is that it implies the existence of a thermodynamic potential, the entropy production, as a non-equilibrium state function. The physical interpretation of this principle is therefore quite different from Onsager's variational principle. In addition, in the least dissipation principle the flows vary but the forces are fixed, whereas in the minimum entropy production principle the flows vary at the same time with the forces and are only subject to the boundary conditions imposed on the system.

Certainly, both principles determine correctly at the steady state the distribution of currents and forces in the system. However, we mainly deal here with the consequences of the minimum entropy production principle which was the only one to have been extended beyond the linear domain. Before we discuss non-linear problems let us outline some interesting aspects of this principle.

In the first place, it is important to realize that it provides a general *evolution criterion*. Indeed, the validity of the theorem of minimum entropy production together with the second law implies that a physical system will necessarily evolve to the steady non-equilibrium state and that the latter corresponds to a stable situation. On the other hand it can be shown[7] that, under certain conditions, the steady state which, by the theorem, is characterized by a minimum of dissipation or so to say by a maximum of efficiency, is also characterized by a lower value of entropy than at equilibrium: increased efficiency is thus combined with an increasing complexity of structure (as measured by the entropy reduction). In this way the theorem of minimum entropy production provides a link between the concept of '*structure*' and of '*evolution*' towards a dissipative state[12].

The interest of these concepts and the possibility of a connection between them might seem academic in the framework of a linear description of irreversible processes but it acquires a fundamental importance in the non-linear region.

3. NON-LINEAR THERMODYNAMICS

For a long time non-equilibrium thermodynamics was confined to the study of linear problems. There exist, however, a large number of important and even quite frequently occurring phenomena which cannot be described by the methods of linear thermodynamics even in a first approximation. For instance, with chemical reactions it is often necessary to adopt non-linear phenomenological laws. Also whenever a system is not at mechanical equilibrium the coupling between dissipative and convective processes leads to effects of a new type which cannot be treated by the methods of linear theory.

The extension of the local formulation of thermodynamics to the non-linear region has been achieved during the last fifteen years by Glansdorff and Prigogine[7,13,14]. In its present form it comprises three essential aspects: (i) the derivation of general evolution criteria for steady states far from equilibrium, (ii) the search of thermodynamic potentials characterizing these states, (iii) the study of stability of these states. We shall now briefly comment on each of these points separately.

(i) The problem of evolution criteria was solved in two steps, the first involving a discussion of purely dissipative processes[15] the second providing an extension to systems involving mechanical motion[13,16,19]. The final result is as follows. In the whole domain of phenomena which are adequately described by a local theory it is possible to construct a differential expression $\mathrm{d}\Phi$ depending on the state variables such that

$$\mathrm{d}\Phi/\mathrm{d}t \leqslant 0 \tag{11}$$

the equality being applicable at the stationary state. $\mathrm{d}\Phi$ is a combination of dissipative and convective processes. In the absence of convective motion it can be shown that[15]

$$\mathrm{d}\Phi = \int \mathrm{d}V \sum_i J_i \mathrm{d}X_i \equiv \mathrm{d}_x \mathscr{P} \tag{12}$$

i.e. $d\Phi$ is the variation of entropy production per unit time due to a change in the generalized forces. As in the general case the flows J_i are complicated functions of X_is it follows that in principle $d\Phi$ *is a non-total differential*; in other words $d\Phi$ does not represent the variation of a thermodynamic state function. It is only in the limit of linear phenomena and of validity of the Onsager relations that it becomes the differential of a state function, the entropy production. In this case the evolution criterion can be reduced to the theorem of minimum entropy production.

(ii) The fact that $d\Phi$ is not a total differential in the general case gives rise to the problem of the search for a variational principle. This is a very complicated problem which has only recently been properly formulated. The main point is to realize that, except in a number of exceptional cases where suitable integrating factors or suitable classes of transformations make $d\Phi$ a total differential[7,14], it becomes necessary to formulate an *extended variational principle*. This novel point of view gives rise to a function Ψ, the *local potential* according to the terminology introduced by Glansdorff and Prigogine[16], which shares some of the properties of the potentials of classical thermodynamics. However, it is necessary to look at Ψ as a functional of two sets of functions, average ones corresponding to (quasi-steady) solutions of the macroscopic equations and fluctuating quantities. The extended variational principle has therefore to be understood in terms of fluctuation theory. The Euler–Lagrange equations corresponding to Ψ can be reduced then in the average to the equations of macroscopic physics, i.e. provided at the very end the fluctuating functions are set equal to their average values.

Certainly, the property of the first variation of Ψ to generate the equations of evolution is also shared by functionals other than the local potential. For instance for an equation of evolution

$$L(u) = 0 \tag{13a}$$

defined in a domain V of the independent variable x, the functional

$$\Psi_1 = \int_V dx\, u L(\bar{u}) \tag{13b}$$

(where u and $\bar{u}$ are respectively the fluctuating and average values of u) gives rise to equation 13a by means of an ordinary variational procedure. If now the unknown functions u in the local potential and in the functional 13b are approximated, e.g. by a series expansion, the variational equations resulting from the local potential or from 13b are both identical to the equation obtained when 13a is approximately solved by the well-known Galerkin method[13,14].

As a result the Galerkin method gives the same approximations as those based on the local potential. It can be shown, however, that the extended variational procedure is supplemented by *a minimum property* expressing that the excess local potential is positive definite around the non-fluctuating state. This fundamental property which is largely responsible for the physical significance of Ψ has permitted us to establish convergence of the variational procedure[17]. It also made it possible to treat in a unified way many interesting non-linear hydrodynamic and stability problems[18,19].

(iii) The property of $d\Phi$ not to be a total differential and the lack of a true variational principle also imply that steady states far from equilibrium are no longer characterized by the minimum of a thermodynamic potential. As a result *the stability of these states is not always ensured.* This separation between evolution and stability leads thus to the search for independent stability criteria for states far from equilibrium. Recently a complete *infinitesimal* stability theory of non-equilibrium states has been worked out[14,20–22]. The main result is as follows. Within the domain of validity of the local formulation of thermodynamics it is possible to construct a negative definite quadratic form

$$\delta^2 z = \delta^2 s - (\delta \mathbf{v})^2/T < 0 \tag{14}$$

where s is the specific entropy, $\mathbf{v}$ the average hydrodynamic velocity and δ denotes the variation of the corresponding quantity as a result of a fluctuation. It can be shown that whenever the equilibrium stability conditions are satisfied[1], $\delta^2 s$ is itself a negative semi-definite quadratic form even around states far from thermodynamic equilibrium

$$\delta^2 s \leqslant 0 \tag{15}$$

Moreover, in the limit of small fluctuations,

$$(\partial/\partial t)\, \delta^2 z > 0 \tag{16}$$

in all cases the non-equilibrium state is stable. Clearly this formulation is closely related to the ideas underlying Lyapounov's stability theory[23].

The important point is, however, that equations 14 to 16 constitute *a thermodynamic stability criterion.* Let us emphasize once more that in order to derive this criterion it has been necessary to go beyond the fundamental equation 7 and establish excess balance equations for the quadratic fluctuations, $\delta^2 s$ and $(\delta \mathbf{v})^2$, of entropy and kinetic energy.

Equation 16 contains a complicated interplay between purely dissipative and convective processes. In the neighbourhood of equilibrium it can be shown that the stability criterion is trivially fulfilled once $\delta^2 s < 0$. Alternatively, the existence of thermodynamic potentials guarantees the stability of near-equilibrium states except in the neighbourhood of phase transition points. Moreover, it follows from equations 14 to 16 that internal convection processes can never arise in this limit[13]. Far from equilibrium, however, relation 16 does not follow from 14 and 15 and therefore the stability of the system may be compromised even when the equilibrium state is perfectly stable. If an instability occurs the system necessarily tends to a new régime which may correspond to a completely different state of organization of matter. Since equilibrium remains stable we may say that unlike what happens close to equilibrium, this new régime is not a continuous extrapolation of the equilibrium behaviour.

4. FLUCTUATIONS

At this point the problem of the behaviour of fluctuations becomes quite essential. In a system characterized by a large number of degrees of freedom, such as a typical macroscopic body, fluctuations are always present. Therefore at each moment the system is in a kind of 'dynamical equilibrium' which is determined by the response to its own spontaneous fluctuations.

Usually the fluctuations give rise to a response which brings the system back to the reference state. On the contrary, at the point of an unstable transition fluctuations are amplified and give rise to observable effects. Therefore a new structure which may arise beyond an instability always originates in a fluctuation. As a result a purely deterministic description of the system is no longer sufficient and it becomes necessary to extend irreversible thermodynamics in order to include a macroscopic theory of fluctuations.

Let us recall how this problem is solved at equilibrium. In an isolated system Boltzmann's relation defining entropy in terms of number of states available may be inverted to give rise to the classical Einstein formula giving the probability of fluctuations around a *macroscopic* (equilibrium) state[1,23].

$$P \propto \exp\,[\Delta S/k] \tag{17}$$

where ΔS is the entropy change around equilibrium ($\Delta S < 0$ for a fluctuation) and k Boltzmann's constant. For small fluctuations ΔS may be expanded to second order quantities. Since for an isolated system at equilibrium S is a maximum, equation 17 can be reduced to the second variation term

$$P \propto \exp\,[\delta^2 S/k] \tag{18}$$

What is the generalization of expression 18 to non-equilibrium situations? As we say in section 3 the basic property $\delta^2 S < 0$ which is responsible for the validity of expression 18 is also shared by fluctuations around non-equilibrium states provided the local equilibrium assumption is made. It is therefore tempting to build a non-equilibrium fluctuation theory based, in a first approximation, on expression 18, wherein the excess quantities are calculated around *a non-equilibrium reference state*[25–27]. If this conjecture is justified one will be able to predict the behaviour of fluctuations by studying $\delta^2 S$ and the way it evolves in time and obtain therefore a relation between stability properties and fluctuations.

Attempts to justify the local equilibrium fluctuation theory have been made by Lax[28]. More recently Babloyantz and the author have developed a master equation approach to fluctuations which is applicable to systems arbitrarily far from equilibrium and to the study of arbitrarily large fluctuations[29–31]. In all cases treated the extension of expression 18 to non-equilibrium situations was recovered in the limit of stable systems and small fluctuations, provided some well-defined conditions on the relaxation times were satisfied. These conditions refer to a large separation of time scales between the fluctuating system and the external world. They require the former to be much shorter than the latter so that the state of the environment shall not be influenced by the fluctuating system. This condition is in agreement with the idea that in order to maintain a non-equilibrium macroscopic state one should impose a set of given boundary conditions.

The examples studied by Babloyantz and Nicolis refer mostly to chemical reactions in open systems[29,31] and to energy transfer in a system at mechanical equilibrium[30]. At present our group is involved in work on the stochastic theory of fluctuations in the neighbourhood of unstable transitions. It is expected that this work will permit a clarification of the mechanism of establishment of an instability and of the structure of the final state beyond the transition point. Indeed, it is the initial fluctuation which will determine the type of future situation. The macroscopic time evolution assumes therefore an essentially statistical aspect.

5. DISSIPATIVE STRUCTURES

In equilibrium thermodynamics instabilities only occur at phase transition points. The new phase beyond instability has a markedly different structure; in particular it may correspond to a more ordered state. For instance at the paraferromagnetic transition at the Curie point a system exhibiting spherical symmetry is replaced by a new one having a lesser cylindrical symmetry. Consequently the ferromagnet which is being formed has a much higher degree of 'organization'. Such structures, however, are completely independent of the external world. Once they are formed they are self-maintained and do not require an exchange of energy or matter with the environment.

In systems far from equilibrium a new type of instability appears, due to the existence of constraints which are responsible for the maintenance of a steady non-equilibrium state. Can one associate with these instabilities the formation of ordered structures of a new type? Such non-equilibrium structures would differ from equilibrium ones in that their maintenance would necessitate the continuous exchange of energy and matter with the outside world. For this reason Prigogine, who first suggested the existence of these states, has called them *dissipative structures*[32, 33].

Let us formulate the problem in thermodynamic terms in its most general form. Consider a non-isolated system (closed or open) subject to constraints which give rise to a steady non-equilibrium state. In this state the values of different thermodynamic variables such as flows etc. depend parametrically on a number of quantities, $\{X\}$ measuring the deviation of the system from equilibrium. For instance, X may be a gradient of temperature or composition, the overall affinity of a set of coupled chemical reactions, and so on. Let us adopt the convention that the state $\{X = 0\}$ is the state of thermodynamic equilibrium. For $\{X \neq 0\}$ but small the equilibrium régime is continued by the steady states close to equilibrium whose stability is guaranteed once equilibrium is stable. On the contrary, for $\{X \neq 0\}$ and arbitrarily large, although it is always possible to define a continuous extrapolation of the equilibrium régime, the stability of the states belonging to this branch, which will be referred to as *thermodynamic branch*, is no longer ensured automatically. In addition, the uniqueness property of the equilibrium state is not applicable in this case and the system may present more than the stationary state, provided it obeys non-linear laws. One of these stationary states belongs to the thermodynamic branch but is not necessarily stable. It is therefore possible, *a priori*, to have a number of new effects, for instance: the system may not decay monotonically to the steady state belonging to the thermodynamic branch, once it is perturbed from it; in the limit it may even never return to this state but evolve to a time-dependent régime: under similar conditions it may finally deviate and evolve to a new stationary régime corresponding to a branch different from the thermodynamic one. This transition will manifest itself abruptly as an *instability*, i.e. as a fundamentally discontinuous process.

This situation is frequently met in hydrodynamics. In this domain the problem of instabilities is a classical one which has been studied thoroughly since the early years of the present century[13,36]. Recently Glansdorff and Prigogine have shown that the general formulation of non-linear thermodynamics provides a framework which permits a thermodynamic analysis of such instability phenomena. They have also been able to formulate their conclusions in terms very similar to phase transitions. Amongst the different problems treated let us quote: the onset of thermal convection in a horizontal fluid layer heated from below (Bénard problem); stability of waves and the formation of shocks and detonations; stability of parallel flows; and so on[13,14,16,18].

The occurrence of instabilities and the subsequent formation of structures is a much less obvious effect for purely dissipative systems, i.e. systems without mechanical motion. In fact, it is only during the last few years that this possibility has been studied systematically and a theory of dissipative structures has been set up for such systems[32,33,35–37]. The types of problems which were studied most actively refer to open systems undergoing chemical reactions.

This problem presents a special interest because of the possible implications of the results in the understanding of biological phenomena. Indeed, typical biological processes appear to happen under open system conditions (exchange of ions through membrane processes, ADP ⇆ ATP transformations inside the cell and so on). It is therefore tempting to associate biological structures with dissipative structures arising beyond a chemical instability.

Let us first analyse the non-equilibrium phenomena which may arise in open systems in the neighbourhood of instabilities. We may distinguish between three possible types of situation: (i) oscillations around steady states, (ii) symmetry-breaking instabilities and (iii) multiple steady states.

(i) Although oscillations were suggested a long time ago by Lotka and Volterra[38], it is only during the last few years that there has been a great accumulation of data on the occurrence of time oscillations in systems of chemical reactions. Even more interesting, biochemical reactions of fundamental importance have been shown to exhibit oscillatory behaviour in time. Let us quote for example: oscillations in metabolic reactions comprising activation or inhibition (glycolysis[39,40]); oscillations in protein synthesis in the cellular level (β-galactosidase synthesis in *Escherichia coli*[41] and so on[42]). These phenomena have recently been investigated from a thermodynamic standpoint. It has been necessary to distinguish between two fundamentally different types of behaviour: oscillations having the same character as those occurring in conservative systems (Volterra–Lotka type oscillations[42]) and oscillations beyond a chemical instability[44]. At present it appears that oscillation in metabolic reactions can best be explained by the second type of mechanism. In other words, a biochemical oscillating system would be a dissipative structure.

(ii) This conjecture, which suggests the fundamental importance of the concept of dissipative structure, is also substantiated by the study of symmetry-breaking instabilities[32,33,35,36,45–47]. This term refers to the spontaneous appearance of spatial structure in a previously homogeneous system. The important point is that this spontaneous self-organization has interesting implications from the point of view of both the *space order* and the *function* of the system. Systems which are able to transform part of the energy or matter received from the outside world into macroscopically distinguishable internal order may well exhibit this type of phenomenon and therefore be typical examples of dissipative structures. Again biological systems certainly belong to this category. It is therefore very tempting to associate biological structures with chemical instabilities leading to a spontaneous self-organization[47].

(iii) A change in functional behaviour may also arise in systems which keep their macroscopic space order unchanged. As an example a system may flip-flop between two simultaneously stable steady states which differ only in the levels of concentration of different constituents. It is well known that the classical Jacob–Monod model of regulation in protein synthesis gives rise to such types of transitions which may be associated with 'mutations'[48]. Another very interesting application is in the functioning of excitable membranes. Roughly speaking, a biological excitable membrane may exist in two permanent states: one polarized (associated with the maintenance of different ionic concentrations in the two sides) and one depolarized state derived from the former upon passage of a pulse or upon a change in permeability. It has been shown that this depolarization process may be quantitatively interpreted as a transition arising beyond the point of instability of the polarized state. This instability is due to the difference in the ionic concentrations which here plays the role of the constraint keeping the system in a far from equilibrium state[49].

Summarizing, we may say that instabilities in the thermodynamic branch of solutions can lead to time or space organization and to a change in functional behaviour in open systems undergoing chemical reactions. These instabilities can only arise at a finite distance from thermodynamic equilibrium, i.e. their occurrence necessitates a minimum level of dissipation. Structure and dissipation appear therefore to be intimately connected far from equilibrium. At the same time the system becomes more 'flexible' in this region as it can now occur in a multitude of stable states. This is a very important property which may well be the basis for a thermodynamic theory of evolution in biology.

Computer and laboratory experiments have now confirmed the existence of dissipative structures in certain models[45,46] and also in particular biochemical[39,40] and organic[50,51] reactions. Of course, much remains to be done before it becomes possible to evaluate the full impact of the theory in the interpretation of fundamental biological phenomena.

All these complicated new effects can be and have been analysed within the framework of non-linear thermodynamics outlined in sections 3 and 4[14,32,33,43–47,49]. In addition, the stability conditions 14 and 15 provide sufficient criteria for the types of processes which may give rise to dissipative structures. Aside from the condition of finite distance from equilibrium it is shown that non-linear reaction schemes are necessary for the occurrence of instabilities. This certainly covers the most important biochemical reactions where non-linearity appears through cross-catalysis, autocatalysis, activation or inhibition. Additional new information given by the thermodynamic evolution and stability criteria include: fixation of the direction of rotation in cyclic processes around the steady state; hints about the relative increase of dissipation and energy transfer at the unstable transition point; and so on.

6. CONCLUSIONS

We have seen that, even beyond the linear domain, thermodynamics may yield interesting results concerning the evolution of systems to steady non-equilibrium states and the stability properties of the latter. Non-linear thermodynamics constitutes a framework for a unified study of phenomena as different at first sight as hydrodynamic instabilities and the formation of spatial or temporal structures in chemistry.

One of the most important features of dissipative structures analysed in the previous section is that they are separated from the thermodynamic branch by an instability. One can say that there exists a real threshold for organization of matter determined by a minimum value of the thermodynamic constraint keeping the system out of equilibrium, which value depends in an intrinsic way upon the parameters descriptive of the system (kinetic constants, diffusion coefficients etc.).

It would thus be very tempting to think that dissipative instabilities act as a kind of phase transition leading to *a new state of matter*. In this new state we have essentially novel properties of the large-scale system, although the laws referring to the molecular level may remain unchanged.

It is exciting to realize that the analogy between dissipative and biological structures may lead to the idea that life and absence of life are just two states of matter separated by a chemical instability. This point of view implies that life follows the laws of thermodynamics appropriate to far from equilibrium conditions. It may therefore help to reconcile the duality between the living and the inanimate world with the unity of the laws of nature.

REFERENCES

[1] For a review of equilibrium and of linear irreversible thermodynamics see e.g. H. B. Callen, *Thermodynamics*. Wiley: New York (1960).
[2] P. Duhem, *Energétique*. Gauthier-Villars: Paris (1911).
[3] I. Prigogine, *Etude Thermodynamique des Phénomènes Irréversibles*. Desoer: Liège (1947).
[4] I. Prigogine, *Physica*, **14**, 272 (1949).
[5] G. Nicolis, J. Wallenborn and M. G. Velarde, *Physica*, **43**, 263 (1969).
[6] B. D. Coleman, *Arch. Ration. Mech. Anal.* **17**, 1 and 230 (1964).
[7] I. Prigogine, *Introduction to Thermodynamics of Irreversible Processes*, 3rd ed. Interscience, Wiley: New York (1967). A large part of this monograph deals with linear problems. The last two chapters contain an introduction to non-linear thermodynamics and dissipative structures.
[8] S. De Groot and P. Mazur, *Non-equilibrium Thermodynamics*. North Holland: Amsterdam (1961). This book provides a very detailed and clear presentation of irreversible thermodynamics in the linear region.
[9] L. Onsager, *Phys. Rev.* **37**, 405 (1931); **38**, 2265 (1931).
[10] H. B. G. Casimir, *Rev. Mod. Phys.* **17**, 343 (1945).
[11] P. Résibois, *J. Chem. Phys.* **41**, 2979 (1964).
E. G. D. Cohen, J. R. Dorfman and M. H. J. J. Ernst, *Phys. Letters*, **12**, 319 (1964).
[12] I. Prigogine and J. W. Wiame, *Experientia*, **2**, 451 (1946).
[13] *Non Equilibrium Thermodynamics, Variational Techniques and Stability*, ed. by R. Donnelly, R. Herman and I. Prigogine. Chicago University Press: Chicago (1965).
[14] P. Glansdorff and I. Prigogine, monograph in press, mostly devoted to non-linear problems and stability theory.

[15] P. Glansdorff and I. Prigogine, *Physica*, **20**, 773 (1954).
I. Prigogine and R. Balescu, *Bull. Cl. Sci. Acad. Roy. Belg.* **41**, 917 (1955).
[16] P. Glansdorff and I. Prigogine, *Physica*, **30**, 351 (1964).
[17] P. Glansdorff, *Physica*, **32**, 1745 (1966).
[18] R. S. Schechter, *The Variational Method in Engineering*. McGraw-Hill: New York (1967) and references quoted therein.
[19] G. Nicolis, *Advanc. Chem. Phys.* **13**, 299 (1967), where the extended variational principle is applied to kinetic theoretical problems.
[20] P. Glansdorff and I. Prigogine, *Phys. Letters*, **7**, 243 (1963).
[21] P. Glansdorff and I. Prigogine, in *Problems of Hydrodynamics and Continuum Mechanics*. Society for Industrial and Applied Mathematics: Philadelphia (1969).
[22] P. Glansdorff and I. Prigogine, *Physica*, to appear (1970).
[23] L. Cesari, 'Asymptotic expansions and stability problems in ordinary differential equations'. *Ergebn. Mathem.*, N.F., **16**. Springer: Berlin (1962).
[24] L. Tisza and P. M. Quay, *Ann. Phys.* (*N.Y.*), **25**, 148 (1963).
[25] I. Prigogine and G. Mayer, *Bull. Cl. Sci. Acad. Roy. Belg.* **41**, 22 (1955).
[26] I. Prigogine in *Temperature*, **2**, 215 (1955).
[27] I. Prigogine and P. Glansdorff, *Physica*, **31**, 1242 (1965).
[28] M. Lax, *Rev. Mod. Phys.* **32**, 25 (1960).
[29] G. Nicolis and A. Babloyantz, *J. Chem. Phys.* **51**, 2632 (1969).
[30] A. Babloyantz and G. Nicolis, *J. Statist. Phys.* **1** (1969).
[31] G. Nicolis, 'Stochastic analysis of the Volterra–Lotka model', submitted to *J. Math. Phys.*
[32] I. Prigogine, *Bull. Cl. Sci. Acad. Roy. Belg.* **53**, 273 (1967).
[33] I. Prigogine and B. Nicolis, *J. Chem. Phys.* **46**, 3542 (1967).
[34] S. Chandrasekhar, '*Hydrodynamic and Hydromagnetic Stability*. Clarendon Press: Oxford (1961).
[35] I. Prigogine, structure, dissipation and life' in *Theoretical Physics and Biology* edited by M. Marois. North Holland: Amsterdam (1969).
[36] I. Prigogine, 'Dissipative structures in biological systems' *Proceedings of the Second International Conference on Theoretical Physics and Biology*. Versailles (July 1969).
[37] Instabilities in open systems have been considered for the first time by Turing in a different context: A. M. Turing, *Phil. Trans. Roy. Soc. London*, **B237**, 37 (1952).
[38] V. Volterra. *Leçons sur la Théorie Mathématique de la Lutte pour la Vie*. Gauthier-Villars: Paris (1931).
[39] B. Chance, R. W. Estabrook and A. Ghosh, *Proc. Nat. Acad. Sci. Wash.* **51**, 1244 (1964).
[40] E. E. Sel'kov, *Europ. J. Biochem.* **4**, 79 (1968).
[41] W. A. Knorre, *Biochem. Biophys. Res. Commun.* **31**, 812 (1968).
[42] B. C. Goodwin, *Temporal Organization in Cells*. Academic Press: New York (1963).
[43] R. Lefever, G. Nicolis and I. Prigogine, *J. Chem. Phys.* **47**, 1045 (1967).
[44] R. Lefever and G. Nicolis, 'Dissipative instabilities and sustained oscillations' submitted to to *J. Theoret. Biology*.
[45] I. Prigogine and R. Lefever, *J. Chem. Phys.* **48**, 1695 (1968).
[46] R. Lefever, *J. Chem. Phys.* **49**, 4977 (1968).
R. Lefever, *Bull. Cl. Sci. Acad. Roy. Belg.* **54**, 712 (1968).
[47] I. Prigogine, R. Lefever, A. Goldbeter and M. Hershkowitz-Kaufman, *Nature, London*, **223**, 913 (1969).
[48] F. Jacob and J. Monod, *J. Molec. Biol.* **3**, 318 (1961).
[49] R. Blumenthal, J. P. Changeux and R. Lefever, 'Membrane excitability and dissipative instabilities', *C.R. Acad. Sci., Paris*, **270**, 389 (1970).
[50] A. M. Zhabotinsky, *Biofizika*, **9**, 306 (1964).
[51] H. Büsse, *J. Phys. Chem.* **73**, 750 (1969). In this paper an experiment is reported showing the existence of a spatial structure in Zhabotinsky's reaction.

STOCHASTIC PROCESSES AND THE QUANTUM EVOLUTION OF STATES*

R. E. COLLINS AND F. G. HALL

The Department of Physics, The University of Houston, Houston, Texas

ABSTRACT

Beginning with the statistical description of the results of a duplicated experiment as a set of configurations of some observables together with frequencies of occurrence of each configuration it is shown that an evolution of the system by a stochastic process is obtained when the conditioning parameters of the experiment are altered. It is then shown that any such stochastic process has a representation in terms of linear operators in an abstract vector space with a state vector evolving by an isometric operator **S** and commuting Hermitian operators representing observables which evolve by another, unitary, operator **U**. This has the structure of conventional quantum theory as in the interaction, or Dirac picture, but here **S** is *not* unitary as in conventional theory. This is shown to yield a matrix of transition probabilities that is *not* doubly stochastic as in conventional theory and hence the Pauli master equation does not follow. All implications of this completely irreversible quantum theory have not yet been fully explored, but it points to a new viewpoint in irreversible thermodynamics.

INTRODUCTION

A recent series of papers [1–6] has developed the idea that much of the formal mathematical structure of physical theory can be deduced directly from the statistical properties of experimental data. The present paper presents that portion of these studies which bears directly on the problem of irreversible physical processes; specifically we point out what appears to be a major flaw in conventional quantum theory and exhibit the proper connection of the quantum mechanical evolution of states to a stochastic process.

DUPLICATED EXPERIMENTS AND TIME

We consider a duplicated experiment with a physical system in which certain parameters are given fixed values and selected properties of the system are measured. Thus suppose that in order to duplicate exactly the conditions of the experiment we must fix values for $q_1, q_2, \ldots q_K$ and $\eta_1, \eta_2, \ldots \eta_J$; then values of $(x, y, z, \ldots)$ are measured. Distinct results are represented by sets of numbers, say

$$r_n = (x_n, y_n, z_n, \ldots), \; n = 1, 2, 3, \ldots \tag{1}$$

* Supported in part by a Frederick Gardner Cottrell grant-in-aid from the Research Corporation.

such that two results, r_n and r_m, are distinct, $r_n \neq r_m$, if they differ in one or more entries. The variables of r are not all independent, say z may be computed from measured values of x and y. Thus if r contains M independent variables the index n is equivalent to M distinct indices, $n_1, n_2, \ldots n_M$.

If the experiment could be exactly duplicated N times and a particular result r_n were obtained N_n times then we would define the probability for r_n as

$$\Pi_n = \operatorname*{limit}_{N \to \infty} \frac{N_n}{N} \tag{2}$$

Thus a sequence of exact duplications would be summarized as a set of results r_n, $n = 1, 2, \ldots$, and a corresponding set of probabilities, Π_n, $n = 1, 2, \ldots$ In general the spectrum of results is determined by the conditioning parameters, that is $x_n(q, \eta)$, $y_n(q, \eta), \ldots$ are functions of the q_i and η_j, and the probability is conditioned by the q_i and η_j that is, $\Pi_n(q, \eta)$.

This statistical description of an experiment does not preclude an exactly deterministic system for which

$$\Pi_n = \delta_{nn'} = \begin{cases} 1, & n = n'(q, \eta) \\ 0, & n \neq n'(q, \eta) \end{cases} \tag{3}$$

but we maintain a general statistical description in which Π_n is not a Kronecker delta.

In reality not all parameters which may condition the outcome of an experiment can be identified and fixed by the experimenter and it is for this reason that we employ a notation indicating two groups of conditioning parameters, those $q_1, q_2, \ldots q_K$ which are identified and fixed and those $\eta_1, \eta_2, \ldots \eta_J$ which are not fixed. Therefore we introduce a one-parameter labelling $\eta_i(t) = \eta_i$, $i = 1, 2, \ldots J$ such that

$$\mathrm{d}\eta_i = \dot{\eta}_i(t)\,\mathrm{d}t,\ i = 1, 2, \ldots J;\ \mathrm{d}t \neq 0 \tag{4}$$

are never *all* zero. Thus we acknowledge the fact that the external universe is always changing and introduce the notation $x_n(q, t)$, $y_n(q, t), \ldots$ and $\Pi_n\,(a, t)$ for the spectral values of observables and their corresponding probability. Here t is defined as the time[5].

Here we have one description of a duplicated experiment in which the results are explicitly identified as time dependent; we may then investigate further the question of whether duplication then has any real meaning, i.e. when not all conditioning parameters are fixed. But an alternative is to consider a time-ordered sequence of measurements of the observables $r = (x, y, z, \ldots)$ while those $q_1, q_2, \ldots q_K$ accessible to control are fixed. Thus using any one of the n_i, which is never fixed, as the time-ordering reference [solve this $\eta_i = \eta_i(t)$ for t] we may consider the possibility of defining a probability

$$P(r_{n(1)} \to r_{n(2)} \to \ldots \to r_{n(N)} | q)$$

for the time-ordered sequence of results; $r_{n(1)} \to r_{n(2)} \to \ldots \to r_{n(N)}$; i.e. $r_{n(1)}$ is observed at time t_1, $r_{n(2)}$ at $t_2, \ldots$ etc.

In order properly to define such a probability in mathematical terms it is necessary to construct an event space which forms a sigma algebra and a sigma additive measure over this space. In another paper it is shown that such a probability can be properly defined[6] if one introduces a certain equivalence relation between 'paths'. Thus the sequence of results $r_{n(1)} \to r_{n(2)} \to \ldots \to r_{n(N)}$ is a 'simple path'; a compound path is

$$r_{n(1)} \to r_{n(2)} \to \ldots \to (r_{n(k)} \quad \text{or} \quad r_{n'(k)}) \to \ldots \to r_{n(N)}$$

that is, at the kth measurement we are only able to say, either $r_{n(k)}$ or $r_{n'(k)}$ occurs. The equivalence relation that is introduced is

$$\begin{aligned}[][r_{n(1)} \to r_{n(2)} \to \ldots \to (r_{n(k)} \quad \text{or} \quad r_{n'(k)}) \to \ldots \to r_{n(N)}] \\ = [r_{n(1)} \to r_{n(2)} \to \ldots \to r_{n(k)} \to \ldots \to r_{n(N)}] \\ \text{or} \quad [r_{n(1)} \to r_{n(2)} \to \ldots \to r_{n'(k)} \to \ldots \to r_{n(N)}]\end{aligned} \tag{5}$$

That is, the probability for the compound path is required to be equal to the sum of the probabilities for the two simple paths; this defines the equivalence of paths.

With the conditional probability defined by

$$P(r_{n(1)} \to r_{n(2)} \to .. \to C_{(j)} \to .. \to r_{n(N)} | q, C_{(1)} \to C_{(2)} \to .. \to r_{n(j)} \to .. \to C_{n(N)})$$

$$= \frac{P(r_{n(1)} \to r_{n(2)} \to .. \to r_{n(j)} \to .. \to r_{n(N)} | q)}{P(C_{(1)} \to C_{(2)} \to .. \to r_{n(j)} \to .. \to C_{n(N)} | q)} \tag{6}$$

where $C_{(j)}$ stands for $[r_{1(j)}$, or $r_{2(j)}$ or $r_{3(j)}$ or . . .], that is *some* result at t_j, and the above equivalence relation, one obtains, by summing over the spectra of $n(1)$, $n(2)$, . . $n\ (N - 1)$, the form,

$$\Pi_n(q, t_N) = \sum_{m \geq 1} T_{nm} \Pi_m(q, t_1) \tag{7}$$

where $\Pi_n(q, t_N)$ is identified as

$$\Pi_n(q, t_N) = P(C_{(1)} \to C_{(2)} \to \ldots \to C_{(N-1)} \to r_{n(N)} | q) \tag{8}$$

That is, the probability that some result is obtained at each of $t_1, t_2, .. t_{N-1}$ and the specific result $r_{n(N)}$ is obtained at t_N.

Thus a consideration of a time-ordered sequence of measurements leads naturally to a description of the time evolution of the system as a stochastic process. Shortly we will see that this description of time evolution of a system is arrived at by another argument, but first we point out a direct connection of this description to an operator formalism like quantum theory.

PROBABILITY FUNCTIONS IN l^2 AND THE STOCHASTIC OPERATOR

The probability functions introduced above have the properties

$$\Pi_n(q, t) \geqslant 0 \tag{9}$$

$$T_{nm} \geqslant 0 \tag{10}$$

together with

$$\sum_{n\geqslant 1} \Pi_n(q, t) = 1 \tag{11}$$

and

$$\sum_{n\geqslant 1} T_{nm} = 1 \tag{12}$$

both of which must be identities valid for all q and t configurations.

Because of the non-negative properties of Π_n and T_{nm} we can introduce complex functions $c_n(q, t)$ and $K_{nm}(q, t)$ such that

$$\Pi_n(q, t) = c_n^*(q, t)\, c_n(q, t) \tag{13}$$

and

$$T_{nm}(q, t) = K_{nm}^*(q, t)\, K_{nm}(q, t) \tag{14}$$

Then equation 7, with $t_N = t, t_1 = 0$ appears as

$$c_n^*(q, t)\, c_n(q, t) = \sum_{m\geqslant 1} K_{nm}^*(q, t)\, K_{nm}(q, t)\, c_m^*(q, 0)\, c_m(q, 0) \tag{15}$$

Since the phases of the $c_n(q, 0)$ and $K_{nm}(q, t)$ are arbitrary these can be chosen such that

$$c_n(q, t) = \sum_{m\geqslant 1} K_{nm}(q, t)\, c_m(q, 0) \tag{16}$$

as is proved in the appendix theorem of an earlier paper[2].

Because the sequences $c_n(q, t), K_{nm}(q, t), n = 1, 2, \ldots,$ are square summable these have a representation in terms of an abstract vector space and we have here

$$c_n(q, t) = \langle n|q, t\rangle \tag{17}$$

and

$$K_{nm} = \frac{\langle n|\mathbf{K}|m\rangle}{\langle m|\mathbf{K}^\dagger\mathbf{K}|m\rangle^{\frac{1}{2}}} \tag{18}$$

where $\mathbf{K}$ is an abstract operator. These inserted into equation 16 yield

$$\langle n|q, t\rangle = \sum_{m\geqslant 1} \frac{\langle n|\mathbf{K}|m\rangle}{\langle m|\mathbf{K}^\dagger\mathbf{K}|m\rangle^{\frac{1}{2}}} \langle m|q, 0\rangle \tag{19}$$

or

$$|q, t\rangle = \mathbf{S}|q, 0\rangle \tag{20}$$

where

$$\mathbf{S} = \mathbf{KD} \tag{21}$$

with $\mathbf{D}$ being the diagonal Hermitian operator

$$\mathbf{D} = \sum_{m\geqslant 1} \frac{|m\rangle\langle m|}{\langle m|\mathbf{K}^\dagger\mathbf{K}|m\rangle^{\frac{1}{2}}} \tag{22}$$

Thus the stochastic evolution of states leads directly to the evolution of a representative state vector by an operator **S**. In other papers[4,5] we have shown that

$$\mathbf{S}^{\dagger}\mathbf{S} = \mathbf{I} \tag{23}$$

but in general

$$\mathbf{S}\,\mathbf{S}^{\dagger} \neq \mathbf{I} \tag{24}$$

unless **K** is a unitary operator. However, no special properties for **K** need be specified, nor for **D** either. If $\mathbf{D}^{-1}$ exists, then using equation 21 **K** is given by $\mathbf{SD}^{-1}$, but $\mathbf{D}^{-1}$ exists only if $\mathbf{K}^{\dagger}\mathbf{K}$ is a diagonal operator and the determinant, $|D|$, exists; this requires that

$$-2\ln|D| = \sum_{m\geq 1} \ln\langle m|\mathbf{K}^{\dagger}\mathbf{K}|m\rangle \tag{25}$$

be a convergent series. We also show elsewhere[6] that **S** has a semi-group property. Thus **S** is an isometric operator for an arbitrary operator **K** but may be a unitary operator if **K** is itself a unitary operator, i.e. equations 23 and 24 define an isometric operator. One can readily verify that equations 11 and 12 are identically satisfied by virtue of these forms.

If we define another abstract operator $\boldsymbol{\pi}$ acting in this vector space as a Hermitian operator having a complete set of eigenvectors $|n\rangle$, $n = 1, 2, \ldots$, spanning the space and eigenvalues $\Pi_n(q, t)$, $n = 1, 2, \ldots$, that is,

$$\boldsymbol{\pi}|n\rangle = \Pi_n(q,t)|n\rangle,\ n = 1, 2, \ldots \tag{26}$$

then we can show that[5]

$$\boldsymbol{\pi}(q,t) = \mathbf{S}\boldsymbol{\pi}(q,0)\,\mathbf{S}^{\dagger} \tag{27}$$

is isomorphic to equation 15 and hence is the representation of the stochastic process in the abstract vector format.

This analysis, which proceeds from a consideration of a time-ordered sequence of measurements to the representation of the evolution of the system by an isometric operator in an abstract vector space, fails to indicate how observables are to be represented in the formalism; another approach yields this[5].

OBSERVABLES AND THEIR OPERATORS, AN ALTERNATE VIEWPOINT ON TIME EVOLUTION

Since $\Pi_n(q, t)$ can be represented as $c_n^*(q, t)\,c_n(q, t)$ we see that the expectation value of any observable is given by

$$\langle x\rangle = \sum_{n\geq 1} x_n(q,t)\,c_n^*(q,t)\,c_n(q,t) = C^{+}XC \tag{28}$$

where the $x_n(q, t)$ are the measured values of x corresponding to the distinct results r_n. Some $x_m(q, t)$ may thus be equal. Here a matrix notation is introduced with C being the column matrix with elements $c_n(q, t)$ and X the diagonal matrix with elements $x_n(q, t)$. Also the norm condition, equation 11, appears as

$$C^\dagger C = 1 \tag{29}$$

These forms follow directly from nothing more than the non-negative property of $\Pi_n(q, t)$. Since equation 29 can also be written for any other observable, say y, we see that each observable is represented by a diagonal real matrix and these are all of the same order. Furthermore all diagonal matrices commute, that is

$$XY - YX = 0 \tag{30}$$

for example.

Since this, as well as equations 28 and 29, is invariant under unitary transformation, say

$$C' = VC, \quad X' = VXV^\dagger, \quad Y' = VYV^\dagger \tag{31}$$

with V a unitary matrix, we see

$$C'^\dagger C' = 1, \quad \langle x \rangle = C'^\dagger X' C', \quad X'Y' - Y'X' = 0 \tag{32}$$

and hence in an arbitrary basis the collection of all observables is represented by a collection of commuting Hermitian matrices. This result is completely independent of any arguments about the time evolution of the system, it rests solely on the definition of an expectation value and the non-negative property of a probability.

However, these same forms can be written for *any* values of the q_j and t, say with the q_j and t replaced by $q_j + \delta q_j$ and $t + \delta t$. Thus in a matrix format

$$C_\delta^\dagger C_\delta = 1, \quad \langle x \rangle_\delta = C_\delta^\dagger X_\delta C_\delta, \quad X_\delta Y_\delta - Y_\delta X_\delta = 0, \quad \text{etc.} \tag{33}$$

where the δ subscript indicates the incremented arguments.

Then we introduce linear transformation matrices K and U such that

$$C_\delta = KC, \quad X_\delta = UXU^\dagger, \text{etc.} \tag{34}$$

where X_δ as well as X must be diagonal. Furthermore, in order to preserve commutation[5] of spectral matrices, every spectral matrix must be transformed by the *same* U and we must have*

$$U^\dagger U = I \tag{35}$$

Thus U must be an isometric matrix and X and X_δ must in general be of different order.

However, if we demand that as all $\delta q_j \to 0$ and $\delta t \to 0$

$$\text{limit } X_\delta - X \tag{36}$$

then it can be shown[5] that U must be unitary and in particular, for infinitesimal δq_j and δt,

$$U = I + i \sum_{j=1}^{K} P^{(j)} \delta q_j - iH\,\delta t \tag{37}$$

* At the time of this writing we have recognized that $U^+U = \alpha I$ with α a scalar is sufficient so further generalization of the theory is possible.

where the $P^{(j)}$ and H are commuting Hermitian matrices. Then every spectral matrix, as X above, must be of a fixed order and all must commute with U to remain diagonal. Thus the spectral matrices of all observables commute with the $P^{(j)}$ and H, and the $P^{(j)}$ and H themselves correspond to observables.

On the other hand we have no basis for requiring $C_\delta \to C$ as all δq_j and δt go to zero because these quantities contain an arbitrary phase. We point out that the unit norm condition in equations 29 and 33

$$C^\dagger C = 1, \quad C_\delta^\dagger C_\delta = C^\dagger K^\dagger K C = 1 \tag{38}$$

does *not* require K to be unitary or even isometric, because K is uniquely related to C and $K^\dagger K$ inserted into any other matrix product need not leave the product invariant. Equation 38 is satisfied if we impose the less stringent condition on the elements of K,

$$\sum_{n \geqslant 1} K_{nm}^* K_{nm} = 1 \tag{39}$$

and *choose* the arbitrary phases of the $c_n(q, t)$ such that[5]

$$\sum_{l \neq m} \sum_{m \geqslant 1} K_{nm}^* K_{nl}\, c_m^*(q, t)\, c_l(q, t) = 0 \tag{40}$$

Then using equation 34 relating the $c_n(q + \delta q, t + \delta t)$ to the $c_n(q, t)$ we find that equation 15 is obtained simply by setting $t = 0$ and $\delta t = t$.

Thus we find the $K_{nm}^* K_{nm}$ to be the transition probabilities in a stochastic equation and again the state of the system evolves by a stochastic process, but now with translation of the q_j as well as t. This alternate analysis is explored at length in another paper[5].

Most significant is the fact that the state matrix C evolves by one matrix K, for which no special properties are postulated, beyond equation 39, while spectral matrices, like X, evolve by a different matrix, U, which must be unitary or at least isometric.

Carried over to the abstract vector picture we find that in addition to the operators $\mathbf{S}$, $\mathbf{D}$, $\mathbf{K}$ and $\boldsymbol{\pi}$ defined above, we also have a Hermitian operator corresponding to each observable in the experiment, as $\mathbf{X}$ for example, and these all commute. We also have the unitary evolution operator $\mathbf{U}$ expressed in terms of commuting Hermitian operators $\mathbf{P}_j$, $j = 1, 2, \ldots K$ and $\mathbf{H}$, as

$$\mathbf{U} = \exp\left\{ i \sum_{j=1}^{K} \mathbf{P}_j\, \delta q_j - i\mathbf{H}\, \delta t \right\} \tag{41}$$

In general $\boldsymbol{\pi}$ does not commute with the $\mathbf{P}_j$ and $\mathbf{H}$, but operators of all observables, as $\mathbf{X}$ for example, must commute with $\mathbf{U}$ and hence with the $\mathbf{P}_j$ and $\mathbf{H}$ as well. In particular, we have from equation 34

$$\mathbf{X}_\delta = \mathbf{U}\mathbf{X}\mathbf{U}^\dagger \tag{42}$$

in the abstract operator format and this yields

$$i\frac{\partial \mathbf{X}}{\partial t} = \mathbf{H}\mathbf{X} - \mathbf{X}\mathbf{H}, \quad -i\frac{\partial \mathbf{X}}{\partial q_j} = \mathbf{P}_j\mathbf{X} - \mathbf{X}\mathbf{P}_j \tag{43}$$

and we see that no operator containing the q_j and t as parameters may represent a *proper* observable.

IRREVERSIBLE QUANTUM THEORY

What we have here is simply an irreversible quantum theory; the isometric operator **S** yields a unidirectional evolution of the state vector, but operators of observables evolve by the unitary operator **U**. This is essentially the format of the interaction, or Dirac, picture of conventional quantum theory with states evolving by **S** and observables by **U**, but here **S** is not in general unitary.

In this regard we note that from equations 14 and 18 the transition probabilities are given by

$$T_{nm} = \langle n|\mathbf{S}|m\rangle \langle m|\mathbf{S}^{\dagger}|n\rangle = \frac{\langle n|\mathbf{K}|m\rangle \langle m|\mathbf{K}^{\dagger}|n\rangle}{\langle m|\mathbf{K}^{\dagger}\mathbf{K}|m\rangle} \tag{44}$$

and this is *not* a doubly stochastic matrix as in conventional quantum theory unless the operator **K** is unitary. As already noted, **S** is unitary if **K** is unitary, and in fact **S** is then equal to **K** and we then have the expression familiar to us in conventional quantum theory. The stationary states of quantum theory are those for which $\mathbf{S} = \mathbf{K} = \mathbf{I}$, for then $T_{nm} = \delta_{nm}$ is the identity; i.e. there are then no transitions.

CONCLUSION AND DISCUSSION

At the time of this writing we have not yet constructed an isometric operator **S** appropriate to a particular system to illustrate the application of this irreversible quantum theory because the formulation does not yield a general format for the construction of such an operator. Even so we can draw a few specific conclusions. For example, the usual form of the master equation[7], which is based on a unitary evolution operator for states and hence a doubly stochastic matrix of transition probabilities, must be at best only a good approximation to the proper description of the time evolution of the state probability. Furthermore, having shown that there is a direct logical derivation of this formalism from nothing more than the statistical properties of experimental data we open the way to further generalizations of the theory.

REFERENCES

[1] R. E. Collins, 'Embedding concepts in statistical thermodynamics', in the symposium proceedings, *Critical Review of the Foundations of Thermodynamics*, J. A. Brainard, Ed. University of Pittsburgh (in press) (1969).

[2] R. E. Collins, 'Statistical basis for physical laws', *Phys. Rev.* **183**, (No. 5). 1081–1097 (1969)

[3] R. E. Collins, 'A statistical basis for the many-body model of matter', submitted to *Phys. Rev.* (1969).

[4] R. E. Collins, 'Statistical basis for quantum theory; the embedding formulation', submitted to *Phys. Rev.* (1969).

[5] R. E. Collins, 'Generalized quantum theory', submitted to *Phys. Rev.* (25 September 1969). (To be published).

[6] F. G. Hall and R. E. Collins, 'Stochastic processes and their representations in Hilbert space', submitted to *J. Math. Phys.* (1970).

7. W. Pauli, in *Probleme der Moderne Physik, Sommerfeld Festschrift* (1928). [An extensive discussion of the master equation and its relation to quantum theory is given in Chapter 1 of *Kinetic Equations of Gases and Plasmas*. T. Y. Wu, Addison-Wesley: New York (1966).]

IRREVERSIBILITY IN NON-LINEAR OSCILLATOR SYSTEMS†

JOSEPH FORD

School of Physics, Georgia Institute of Technology

ABSTRACT

The statistical behaviour of individual phase-space trajectories for non-linear oscillator systems is demonstrated via computer calculations. These results are interpreted in terms of a mathematical theorem due to A. N. Kolmogorov, V. I. Arnol'd and J. Mojer.

I. INTRODUCTION

In this paper we investigate the classical motion of oscillator systems governed by the Hamiltonian

$$H = \sum_{k=1}^{N} \tfrac{1}{2}\omega_k (P_k^2 + Q_k^2) + \gamma[V_3 + V_4 + \ldots] \tag{1}$$

where N is the number of oscillators, ω_k are the positive frequencies of the harmonic approximation, γ is the non-linear coupling parameter, and V_3, V_4, etc., are cubic, quartic, etc., polynomials in Q_k and P_k. Our intent is to determine those essential properties of Hamiltonian 1 which are crucial for irreversibility. In particular, we focus our attention on the individual trajectories of Hamiltonian 1 and seek to determine when most of these trajectories exhibit stochastic behaviour, by which we mean that a trajectory moves more or less randomly over a sizeable part, perhaps all, of the energy surface. From the viewpoint of thermodynamics, clearly most (Q_k, P_k) sets on a widely stochastic trajectory would correspond to equilibrium. Thus starting the trajectory at a disequilibrium (Q_k, P_k) set would inevitably lead to equilibrium giving the appearance of irreversibility. Indeed, computer calculations for simple examples of Hamiltonian 1 show that, under the proper conditions, the approach to equilibrium is rapid and large deviations from equilibrium are rare.

It is perhaps most convenient to discuss the stochastic properties of Hamiltonian 1 in terms of a theorem due to Arnol'd[1]. Arnol'd rigorously proves that most (in the sense of measure theory) trajectories for Hamiltonian 1 lie on smooth, N-dimensional, integral surfaces (called tori) embedded in the $2N$-dimensional phase space provided, among other things, that:
(1) either γ or, equivalently, the total energy is sufficiently small, and

---

† This work supported in part by the National Science Foundation.

(2) the harmonic frequencies do not satisfy resonant frequency conditions of the form $\sum_{k=1}^{N} n_k\omega_k = 0$ when the integers n_k are such that $\sum_{k=1}^{N} |n_k| \leqslant 4$.

Clearly when Conditions (1) and (2) are satisfied most trajectories are not widely stochastic. The virtue of the Arnol'd theorem for our purposes is that, to a large extent, it actually delimits the conditions for stochastic behaviour. Indeed violation of Condition (1) even for $N = 2$ in general leads to a rather sudden onset of widespread stochasticity[2–4] as the non-linearity becomes strong. Violation of Condition (2) for $N \geqslant 3$ leads to widespread stochasticity even in the limit as the non-linearity tends to zero[5].

Conditions (1) and (2), when satisfied, yield non-stochastic motion because they minimize the effects of resonant interactions. In the langage of quantum mechanics, Condition (2), for $N \geqslant 3$, disallows those resonant, three and four phonon processes so widely invoked in solid state physics[6]. Condition (1) does not allow the higher order phonon processes to affect more than a minority (in the sense of measure theory) of states. On the other hand, violation of Condition (1) and/or Condition (2) gives the non-linear resonances free rein to affect the motion. In order to illustrate these effects, in Section II we demonstrate by example for $N = 2$ that each isolated resonant interaction serves to introduce new stable and unstable periodic orbits into the unperturbed classical motion. When two or more resonances overlap and influence the same trajectory, the system phase space trajectory wanders over the energy surface being scattered, in a sense, by the randomly positioned stable and unstable periodic orbits. In Section III, we demonstrate, again using a simple example, that overlapping, cubic, resonant 'three phonon' interactions can yield stochastic behaviour even in the limit as the non-linearity tends to zero. Section IV then presents our conclusions.

In this paper we actually demonstrate stochasticity only for simple examples and for small N; moreover, we rely heavily on computer calculations. We do this without apology. Indeed we wish to emphasize that irreversibility is not a property requiring infinite N, that irreversibility can be illustrated using simple examples, and that the computer can perform highly informative 'experiments'. Indeed it is the author's belief that the computer, guided by analytical considerations, can contribute heavily to our understanding of statistical mechanics through the detailed study of small systems.

II. STOCHASTICITY FOR $N = 2$

The stochasticity of Hamiltonian 1 for relatively strong non-linearity can be illustrated by studying some simple examples for the case $N = 2$. Let us begin by considering the isolated resonance† described by

$$H = J_1 + J_2 - J_1^2 - 3J_1J_2 + J_2^2 + \beta J_1 J_2^{\frac{3}{2}} \cos(2\varphi_1 - 3\varphi_2) \qquad (2)$$

where β is chosen such that the resonant, angle term is small relative to the

† Hamiltonians 2 and 3 actually violate Condition (2) of Arnol'd's theorem without yielding stochastic behaviour. When Condition (2) is violated using only a single or isolated resonance interaction the system motion is non-stochastic because an isolated resonance always yields motion on smooth, integral surfaces.

pure-J terms. Hamiltonian 2 has for convenience been expressed in action-angle variables and the cosine term represents the resonant 'five phonon' interaction $2\Omega_1 \rightleftarrows 3\Omega_2$. Here Ω_1 and Ω_2 are the initial condition dependent, non-linear frequencies of the motion for $\beta = 0$. When $N = 2$, it is possible to completely survey the motion generated by the Hamiltonian using graphical methods, and we now describe these methods and present results.

Since the Hamiltonian H is a constant of the motion, each system trajectory in the four-dimensional phase space is confined to a three-dimensional subspace. Let us now imagine that this three-dimensional subspace is intersected by a two-dimensional plane, called a level curve plane[2, 4]. If a system trajectory is stochastic then its intersections with this plane will consist of a set of randomly scattered points. If the system trajectory is non-stochastic, i.e. the system trajectory lies on a two-dimensional integral surface, then the intersection points with the plane will form a curve, called a level curve. Plotting the level curve plane intersections for a representative sampling of trajectories will thus reveal the general character of the motion. In *Figure 1*, we plot a typical level curve diagram for Hamiltonian 2. Were

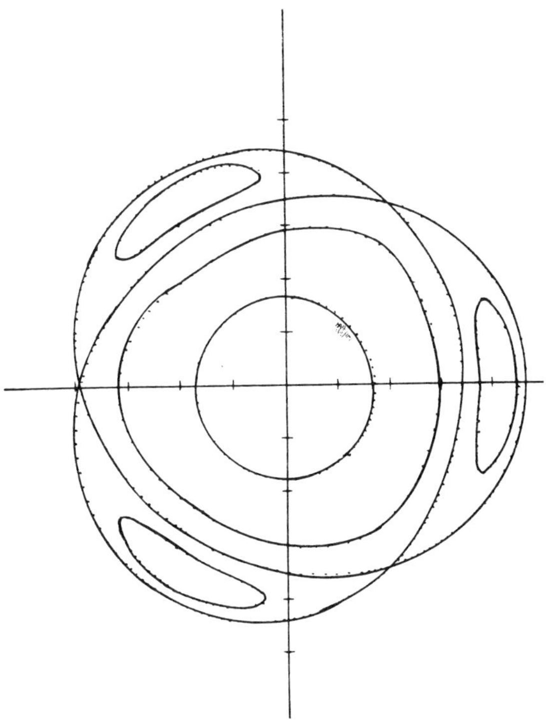

Figure 1. Typical level curves for Hamiltonian 2 showing the distortion in the $\beta = 0$ level curves due to an isolated resonance.

$\beta = 0$, the level curves would be circles centred on the origin; the $2\Omega_1 \rightleftarrows 3\Omega_2$ resonant interaction distorts the $\beta = 0$ level curves in a relatively narrow region of the plane by introducing the new stable (three central invariant points of the crescent regions) and unstable (three self-intersection points) periodic orbits shown. The level curve plane for Hamiltonian 2 exhibits only non-stochastic trajectories because the additional integral $I = 3J_1 + 2J_2$ confines all system trajectories to lie on smooth, two-dimensional integral surfaces.

Next we consider the isolated resonance described by

$$H = J_1 + J_2 - J_1^2 - 3J_1J_2 + J_2^2 + \alpha J_1 J_2 \cos(2\varphi_1 - 2\varphi_2) \tag{3}$$

where the pure-J terms are the same as in Hamiltonian 2 and where α is chosen such that the resonant angle dependent term is small relative to the pure-J terms. This angle dependent term represents the resonant 'four phonon' interaction $2\Omega_1 \rightleftarrows 2\Omega_2$. A typical level curve plane for Hamiltonian 3 appears in *Figure 2*. Here again the resonant interaction has distorted the

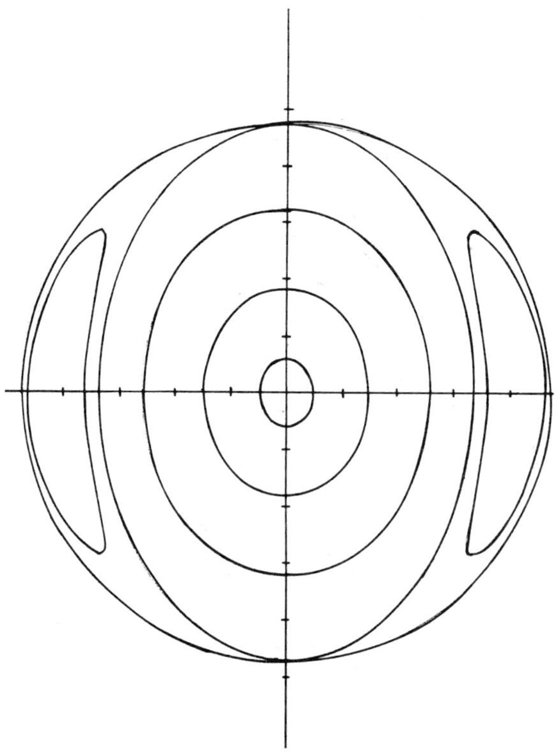

Figure 2. Typical level curves for Hamiltonian 3 showing the distortion of the $\alpha = 0$ level curves for an isolated resonance.

$\alpha = 0$ motion by introducing stable and unstable periodic orbits, but located in a different part of the plane. Also here again the additional integral $I = J_1 + J_2$ ensures smooth level curves everywhere. In Hamiltonians 2 and 3 as α and β increase from zero, the widths of the resonant, crescent regions also increase from zero. Consequently, one anticipates that if both interactions acted simultaneously they might overlap for α and β sufficiently large. We thus now consider the doubly-resonant Hamiltonian

$$H = J_1 + J_2 - J_1^2 - 3J_1J_2 + J_2^2 + \alpha J_1J_2\cos(2\varphi_1 - 2\varphi_2) + \beta J_1J_2^{\frac{3}{2}}\cos(2\varphi_1 - 3\varphi_2) \qquad (4)$$

In *Figure 3* we show the computer obtained level curve diagram typical of Hamiltonian 4 when α and β are small. *Figure 4* shows the computer calculated level curve diagram for α and β sufficiently large that resonance overlap occurs. The isolated dots represent intersection points for a single, stochastic trajectory. As α and β are increased further or as the energy is increased, the stochastic zone increases in size until it almost completely covers the allowed regions of the plane. This stochasticity, due to resonance overlap, here illustrated for the simple Hamiltonian 4 is in general characteristic of Hamiltonian 1 when the non-linearity is large.

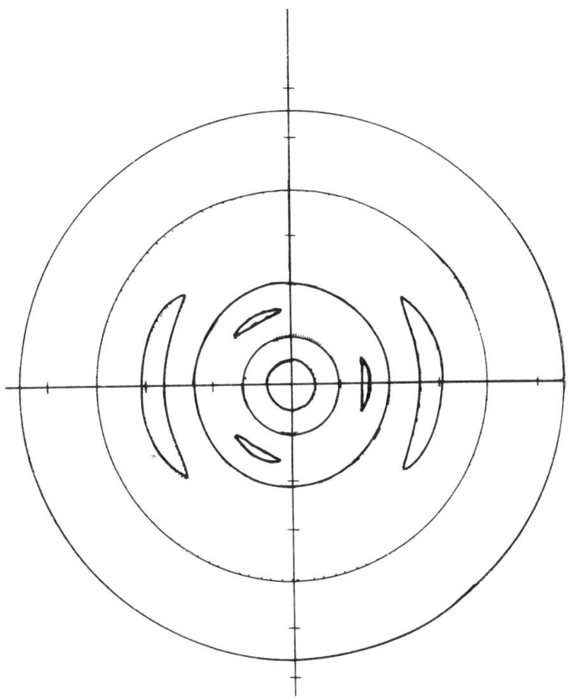

Figure 3. Typical level curves for Hamiltonian 4 for relatively small α and β. Here the two resonance regions do not overlap.

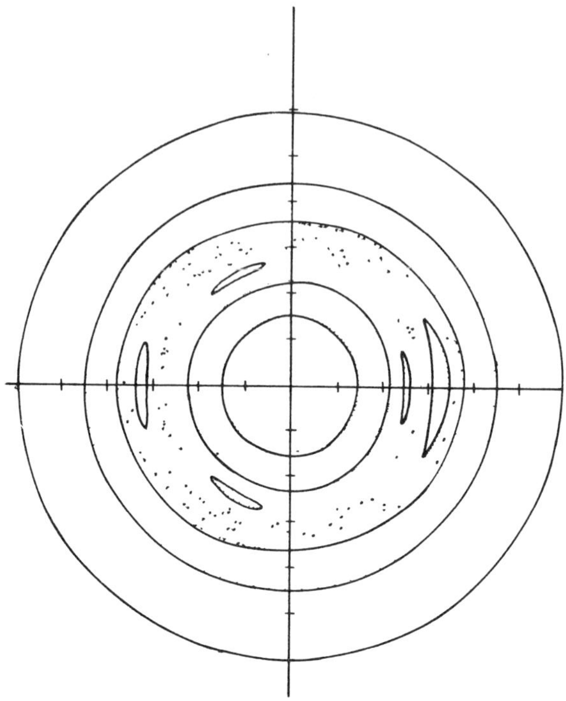

Figure 4. Level curves for Hamiltonian 4 for values of α and β such that overlap occurs. The isolated dots are the intersection points for a single trajectory.

III. STOCHASTICITY FOR $N = 3$

In order to illustrate that stochasticity can occur for Hamiltonian 1 even in the limit as the nonlinearity goes to zero provided $N \geqslant 3$ and Condition (2) is violated, we choose to consider the doubly-resonant, three particle Hamiltonian

$$H = J_1 + 2J_2 + 3J_3 + \gamma[\alpha J_1 J_2^{\frac{1}{2}} \cos(2\varphi_1 - \varphi_2) + \beta(J_1 J_2 J_3)^{\frac{1}{2}} \cos(\varphi_1 + \varphi_2 - \varphi_3)] \quad (5)$$

where the harmonic frequencies $\omega_1 = \frac{1}{2}\omega_2 = \frac{1}{3}\omega_3 = 1$ and these non-linear terms were chosen in order to illustrate the effect of resonant 'three phonon' interactions. In particular the α-term represents the $2\omega_1 \rightleftarrows \omega_2$ process while the β-term represents the $(\omega_1 + \omega_2) \rightleftarrows \omega_3$ process. In order to show that stochasticity can persist even as the non-linearity goes to zero, i.e. as γ goes to zero, we introduce the time dependent canonical transformation

$$J_1 = \bar{J}_1, J_2 = \bar{J}_2, J_3 = \bar{J}_3 \quad (6a)$$

$$\varphi_1 = \bar{\varphi}_1 + t, \varphi_2 = \bar{\varphi}_2 + 2t, \varphi_3 = \bar{\varphi}_3 + 3t \quad (6b)$$

Hamiltonian 5 then becomes

$$\mathscr{H} = \gamma[\alpha \bar{J}_1 \bar{J}_2^{\frac{1}{2}} \cos(2\bar{\varphi}_1 - \bar{\varphi}_2) + \beta(\bar{J}_1\bar{J}_2\bar{J}_3)^{\frac{1}{2}} \cos(\bar{\varphi}_1 + \bar{\varphi}_2 - \bar{\varphi}_3) \quad (7)$$

Since γ is merely a multiplicative factor in Hamiltonian 7, we see that γ affects only the time scale of the motion and not its stochasticity. Moreover, since $I = \bar{J}_1 + 2\bar{J}_2 + 3\bar{J}_3$ is a constant of the motion for this simple Hamiltonian, we may reduce the problem to one having two degrees of freedom and thence determine stochasticity using level curve diagrams. In *Figure 5* we plot typical computer obtained level curves for two different initial

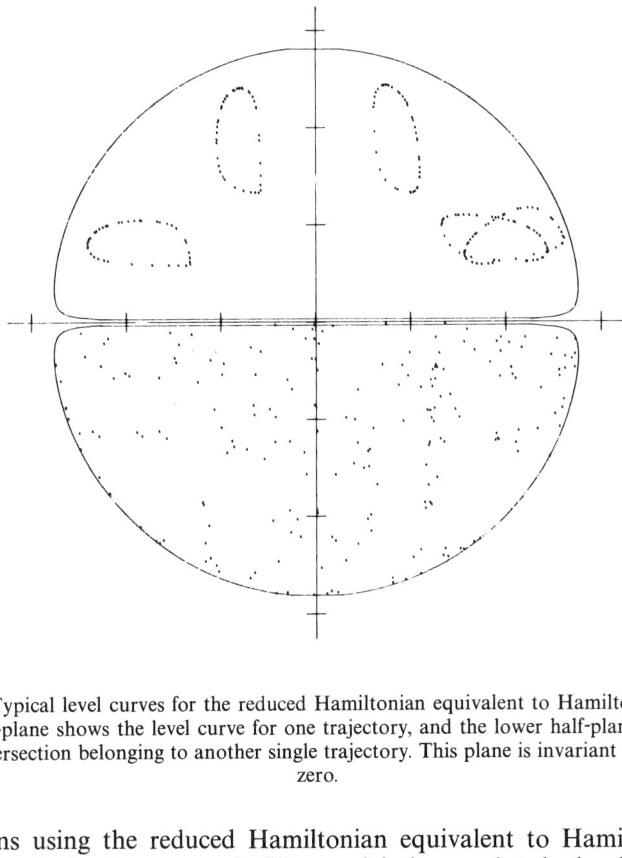

Figure 5. Typical level curves for the reduced Hamiltonian equivalent to Hamiltonian 5. The upper half-plane shows the level curve for one trajectory, and the lower half-plane shows the dots of intersection belonging to another single trajectory. This plane is invariant as γ tends to zero.

conditions using the reduced Hamiltonian equivalent to Hamiltonian 5. Since here the level curves only fill a semicircle, we plot the level curve for one trajectory in the upper half-plane and the level curve for another trajectory in the lower half-plane. Since γ only determines the time scale for these level curves, *Figure 5* is invariant as γ tends to zero, excluding $\gamma = 0$ of course. The bottom half of *Figure 5* exhibits the highly stochastic orbits which can occur. Detailed calculations[5] reveal that, depending on the ratio (α/β), as much as 70 per cent of phase space for Hamiltonian 5 contains stochastic trajectories. Moreover, since the number of overlapping, resonant, cubic interactions increases very rapidly with increasing N, one anticipates that the case $N = 3$ yields the minimum stochasticity.

IV. CONCLUSIONS

In this brief paper, we have attempted to illustrate, using simple examples, that stochastic behaviour characteristic of all non-linear oscillator systems obeying Hamiltonian 1. Stochasticity arises from resonance overlap which can occur for $N \geqslant 2$. Stochasticity occurs for relatively large non-linearity, almost regardless of the values of the frequencies in the harmonic approximation, due to the overlap of individual, non-linear resonances of various order whose widths are large because the non-linearity is large. For extremely small non-linearity, only the cubic 'three phonon' or possibly the quartic four phonon' resonances can overlap provided $N \geqslant 3$ because the width of these resonances, which depends primarily on the suitably commensurate harmonic frequencies, can remain large even when the non-linearity is small. Said another way, if we specify the system 'state' for these non-linear systems by giving the 'energy' of each oscillator, we observe that an initial individual state is resonantly coupled to that density of final states envisioned by the quantum mechanical Golden Rule[7] as leading to irreversible behaviour. Moreover, the stochastic irreversibility discussed here is an inherent property of the mechanical equations of motion. This irreversibility occurs provided that there is widespread resonance overlap, and no additional assumptions of a non-mechanical nature are needed. Thus one has here that beginning understanding of the ultimate source of irreversibility which can contribute significantly to statistical mechanics.

REFERENCES

[1] V. I. Arnol'd, *Russian Math. Surveys*, **18** (No. 6), 85 (1963), Ch. 1, Sec. 9 and 10. Also see V. I. Arnol'd and A. Avez, *Ergodic Problems of Classical Mechanics*, Ch. 4. W. A. Benjamin: New York (1968).

[2] M. Henon and C. Heiles, *Astron. J.* **69**, 73 (1964).

[3] F. M. Izrailev and B. V. Chirikov, *Dokl. Akad. Nauk S.S.S.R.* **166**, 57 (1966) [English transl.: *Soviet Phys.-Doklady* **11**, 30 (1966)]. Also see G. M. Zaslavskii, Statistical irreversibility in non-linear systems', *Preprint 254*, Institute of Nuclear Physics, Novosibirsk, U.S.S.R. (1968); or see N. J. Zabusky and G. S. Deem, *J. Computational Phys.* **2**, 126 (1967).

[4] G. H. Walker and J. Ford, *Phys. Rev.* **188**, 416 (1969).

[5] J. Ford and G. H. Lunsford, *Phys. Rev.* **A1**, 59 (1970).

[6] J. M. Ziman, *Electrons and Phonons*. Oxford University Press: London (1960).

[7] E. Merzbacher, *Quantum Mechanics*. Wiley: New York (1964).

ENTROPY OSCILLATION AND THE *H* THEOREM FOR FINITE SEGMENTS OF INFINITE COUPLED-HARMONIC-OSCILLATOR CHAINS

HARRY S. ROBERTSON AND MANUEL A. HUERTA

Department of Physics, University of Miami, Coral Gables, Florida 33124

ABSTRACT

Following a general discussion of the approach to equilibrium of a finite system in contact with a heat bath, an illustrative calculation is presented in terms of a weakly-coupled, harmonically-bound oscillator chain. A modified Gibbs entropy is defined in terms of ρ_N, the reduced Liouville function of the system, which is obtained from the total Liouville function of the system and heat bath by (in principle) integration over the heat-bath variables. Since the system and heat bath are mutually interacting, some structure is observable in the entropy function as the system evolves from its initial value toward equilibrium, but the entropy ultimately evolves to its correct equilibrium value, despite time-reversible dynamics, because ρ_N spreads from an initially sharp distribution to a final one that is characteristic of the heat bath in equilibrium. The entropy function is presented as an analytically defined, conceptually accurate substitute for Boltzmann's H.

1. INTRODUCTION

Compelling arguments have been offered in support of the point of view that no physical system of thermodynamic interest can properly be regarded as isolated[1,2]. Although the impossibility of shielding a system completely from cosmic rays and fluctuating gravitational fields is seldom questioned, the fundamental importance of these interactions of the system with the outside world is often either ignored or denied. Rather than provoke an unwanted emotional response to any position we might take on the ability of a general isolated system to approach equilibrium as a consequence of its own dynamics, we merely point out that, as in equilibrium statistical mechanics, the approach to equilibrium may be treated with complete validity in terms of open systems, and usually with greater ease than for closed ones. (Canonical theory is surely no less valid than microcanonical theory, and is usually simpler.)

The general system to be studied herein is a collection of interacting particles coupled to an infinite heat bath. The initial positions and momenta of the particles in the system are assumed to be known as well as measuring techniques permit, whereas the initial heat-bath variables are known only statistically. No assertion is made that the heat-bath variables are random, whatever that means; rather the initial choice that the heat-bath variables are canonically distributed is to be regarded only as a statement of our knowledge of them, and not as an assertion that the variables *are* so distributed.

Gibbs[3] used the Liouville function ρ of an N-particle system to define the entropy

$$S = -k_{\mathrm{B}} \int \rho \ln \rho \, \mathrm{d}\Gamma \tag{1}$$

where the integration is over the entire $6N$-dimensional phase space, and at equilibrium the Liouville function is given by

$$\rho = \exp(-\lambda - \beta E) \tag{2}$$

where λ is a normalization constant, $\beta = 1/k_{\mathrm{B}}T$, and E is the energy variable of the N-particle system. Since ρ is a constant (its total time derivative, according to the Liouville theorem, is zero), Gibbs was unable to provide an analytical development of S, as the system evolves from some initial non-equilibrium state, to its final equilibrium value. Although the Gibbs entropy, with ρ given by equation 2, is the correct thermodynamic entropy, there is apparently no provision in Gibbs–Liouville theory for S to get to its equilibrium value, or even to change with time.

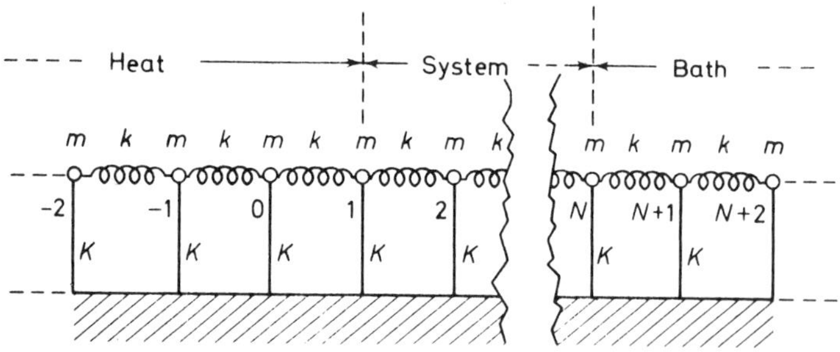

Figure 1. Infinite chain of oscillators, each harmonically bound to its home position by a leaf spring of constant K, and coupled to nearest neighbours by springs of constant k. A finite segment is regarded as the thermodynamic system, and the surroundings as the heat bath.

In the present development, ρ is the Liouville function for the entire system and heat bath; although it formally satisfies the Liouville equation, it is not used explicitly in the subsequent calculations. Instead, a reduced Liouville function ρ_{N} is obtained by integration (in principle) of ρ over the heat-bath variables. This ρ_{N}, no longer a phase-space constant, contains implicitly the interactions of the system with the heat bath, and an entropy defined as in equation 1, except with ρ_{N}, correctly expresses the evolution of the system to equilibrium.

The assumption that the heat-bath variables are canonically distributed can only be made as an *initial* state of knowledge, because the (presumably better known) system variables interact with the heat bath and transiently sharpen our knowledge of the heat-bath variables. (Analogously, a finite cold system in contact with a warmer heat bath transiently cools the heat bath in the neighbourhood of the contact area.) The sharpening of heat-bath variables permits non-thermal transfers of energy across the boundaries of the system in such a way that the entropy evolution is not always a monotonic function of the time. This point is examined in detail in Section 4.

Notable features of the present treatment include the clear separation of dynamics from statistics, freedom from considerations of ergodicity, and easy clarification of the sometimes muddied process by which a system with time-reversible dynamics can evolve to equilibrium. These features are discussed explicitly in terms of a system of weakly-coupled, harmonically-bound oscillators, as shown in *Figure 1*. The dynamics is treated in Section 2, followed by the introduction of statistical features in Section 3. Specific results, particularly for small systems, are given in Section 4, and a short discussion of these results appears in Section 5. More detailed treatments of some aspects of this paper, and more extensive references to related work, appear elsewhere[4–7].

2. DYNAMICS

The Hamiltonian of the system shown in *Figure 1* is given by

$$H \doteq \sum_{n=-\infty}^{\infty} [(p_n^2/2m) + (Kx_n^2/2) + (k/2)(x_n - x_{n+1})^2] \tag{3}$$

The general solution to the equations of motion is of the form

$$x_n(t) = \sum_{r=-\infty}^{\infty} [x_{n+r}(0)\, f_r(t) + p_{n+r}(0)\, g_r(t)/m\Omega] \tag{4a}$$

$$\text{and } p_n(t) = m\dot{x}_n(t), \text{ where} \tag{4b}$$

$$f_r(t) = \pi^{-1} \int_0^{\pi} \mathrm{d}\phi \cos r\phi \cos [\Omega t(1 - 2\gamma \cos \phi)^{\frac{1}{2}}] \tag{5}$$

$$g_r(t) = \Omega \int_0^t f_r(t')\, \mathrm{d}t' \tag{6}$$

$\Omega^2 = (K + 2k)/m$, $\omega^2 = k/m$ and $\gamma = (\omega/\Omega)^2$. An exact treatment of this particular chain is available elsewhere[6], but for present purposes, when $\gamma \ll 1$ and the oscillators are thus weakly coupled, equations 5 and 6 may be quite accurately approximated as

$$f_r(t) = J_r(\gamma\Omega t) \cos (\Omega t - r\pi/2) \tag{7}$$

and

$$g_r(t) = J_r(\gamma\Omega t) \sin (\Omega t - r\pi/2) \tag{8}$$

After a long time, still for $\gamma \ll 1$, equations 7 and 8 differ from the exact expressions of equations 5 and 6 principally in the phase of the trigonometric term, but the essential ideas of the treatment are not affected by use of the approximate expressions.

Equations 7 and 8 exhibit the basic dynamic features that can lead to equilibrium: since $f_0(t)$ is proportional to $J_0(\gamma\Omega t)$, it is evident that the influence of each particle's own initial conditions in determining its future must vanish as $t \to \infty$, so that its ultimate state is determined by the initial conditions of the other members of the chain. Even more strongly, since $J_n(x)$ vanishes as $x^{-\frac{1}{2}}$ for $x \gg n$, it is apparent that any finite segment of the chain must evolve to a state that is determined entirely by the initial conditions of increasingly remote parts of the surrounding heat bath.

The one other feature of note in the dynamics is the time reversibility of equations 4, since the $p_r(0)$s change sign on time reversal. (A film of the motion of the chain would make sense when run in either direction.) Apparently, then, the approach to equilibrium of this system does not lie in the dynamics, though the loss-of-memory feature discussed in the last paragraph is necessary in order that equilibration is not dynamically precluded.

3. STATISTICS

Equations 4 are not statistical; they give definite values of $x_n(t)$ and $p_n(t)$ in terms of the presumably-known initial conditions. It is now appropriate to recognize that these initial conditions are not precisely known, but rather must be described as a distribution $\rho(0)$, the initial Liouville function of the entire system and heat bath.

It is convenient to specify the initial conditions of the system variables in terms of centred Gaussian distributions by writing

$$x_n'(0) = x_n(0) - u_n \tag{9a}$$

and

$$p_n'(0) = p_n(0) - v_n \tag{9b}$$

where u_n and v_n are the initial expectation values of the coordinates and momenta, respectively, of the system variables. The heat-bath initial conditions are distributed essentially canonically, so that $\rho(0)$ may be written

$$\rho(0) = \prod_{n=1}^{N} \frac{\exp\{-[x_n'(0)/\alpha]^2/2\}}{\alpha(2\pi)^{\frac{1}{2}}} \prod_{n=1}^{N} \frac{\exp\{-[p_n'(0)/\delta]^2/2\}}{\delta(2\pi)^{\frac{1}{2}}}$$
$$\times \prod{}' \frac{\exp\{-\beta[p_n^2(0)/2m + m\Omega^2 x_n^2(0)/2]\}}{(2\pi/\beta\Omega)} \tag{10}$$

where α and δ are the initial variances of system variables and $\beta = 1/k_B T_b$, with T_b as the temperature of the heat bath. The symbol Π' denotes a product over all variables outside the system. The heat-bath variables are distributed as if they were uncoupled classical oscillators of mass m and frequency Ω, in canonical equilibrium. The coupling terms $kx_n x_{n+1}$ have been omitted, both because they are small when $k \ll K$ and because they complicate the calculation. (They are included in the exact treatment of this problem[6].)

Any function of the variables $x_n(t)$ and $p_n(t)$ can now be averaged by using equations 4 to express the function in terms of the initial values and integrating with $\rho(0)$ over the initial phase space. Application of this procedure yields

$$\langle x_n(t)\rangle = \sum_{r=1}^{N} [u_r f_{r-n}(t) + (v_r/m\Omega)\, g_{r-n}(t)] \tag{11}$$

and $\langle p_n(t)\rangle = m\langle \dot{x}_n(t)\rangle$. Since equation 11 is a finite sum, it is evident from equations 7 and 8 that $\langle x_n(t)\rangle \to 0$ and $\langle p_n(t)\rangle \to 0$ as $t \to \infty$. Primed variables may now be defined for all t as

$$x'_n(t) = x_n(t) - \langle x_n(t)\rangle \tag{12}$$

and similarly for $p'_n(t)$.

The proper function to represent the phase-space distribution of the system variables is $\rho_N[Y(t)]$, where $Y(t)$ is a column vector of the centred system variables, the transpose of which is $\tilde{Y} = (x'_1 x'_2 \ldots x'_N p'_1 \ldots p'_N)$, and $x'_j = x'_j(t)$. The function $\rho_N[Y]$, which is the reduced Liouville function mentioned in the introduction, is usually not directly obtained from the Liouville function of the entire chain. Instead, the characteristic function for the chain is found as

$$\phi_N(Z) = \int \exp\,[i\tilde{Y}(t)Z]\rho(0) \prod_{n=-\infty}^{\infty} \mathrm{d}x_n(0)\,\mathrm{d}p_n(0) \tag{13}$$

where Z is a $2N$-dimensional Fourier-transform vector. The inverse Fourier transform gives $\rho_N(Y)$ directly as

$$\rho_N(Y) = \int \exp\,[-i\tilde{Z}Y]\phi_N(Z) \prod_{j=1}^{2N} (\mathrm{d}z_j/2\pi) \tag{14}$$

The actual integrations yield the result

$$\rho_N(Y) = (2\pi)^{-N}(\det W)^{-\frac{1}{2}} \exp\,[-\tilde{Y}W^{-1}Y/2] \tag{15}$$

where W is the covariance matrix given by

$$W = (W_{ij}) = (\langle y_i y_j\rangle) \tag{16}$$

and the y_ns are the components of Y.

The reduced Liouville function $\rho_N(Y)$, a function only of the system coordinates and momenta, is the correct distribution to represent the state of a system interacting with a heat bath. A Gibbs entropy can be written in terms of ρ_N as

$$S_N = -k_B \int \rho_N(Y) \ln\,[h^N \rho_N(Y)] \prod_{i=1}^{N} \mathrm{d}x'_i\,\mathrm{d}p'_i \tag{17}$$

where h is, for classical purposes, a constant with units of action, required to make the argument of the logarithm dimensionless. Substitution of ρ_N from equation 15 into equation 17 yields

$$S_N = Nk_B + k_B \ln\,[\hbar^{-N}(\det W)^{\frac{1}{2}}] \tag{18}$$

where $\hbar = h/2\pi$. Thus S_N is seen to be given entirely by the covariance matrix W, the elements of which are time dependent, as in S_N. Specific results for the weakly-coupled, harmonically bound chain are given in Section 4.

4. RESULTS

For simplicity, it is assumed that the initial variances of the system satisfy the relation

$$\delta^2/2m = m\Omega^2\alpha^2/2 = k_B T_0/2 \tag{19}$$

where T_0 is regarded as the initial temperature of the system. The covariance matrix W is now written

$$W = \begin{pmatrix} M & G \\ \tilde{G} & Q \end{pmatrix} \tag{20}$$

where $M = (M_{ij})$, etc., and $M_{ij} = \langle x'_i(t) x'_j(t) \rangle$; $Q_{ij} = \langle p'_i(t) p'_j(t) \rangle$, and $G_{ij} = \langle x'_i(t) p'_j(t) \rangle$. After some modest algebra, these matrix elements are found to be:

$$M_{ij} = (k_B/m\Omega^2)\{T_b\delta_{ij} + (T_0 - T_b)(1, N; n - i, n - j) \cos [(i - j)\pi/2]\} \tag{21}$$

$$G_{ij} = [k_B(T_0 - T_b)/\Omega]\,(1, N; n - i, n - j) \sin [(i - j)\pi/2] = -G_{ji} \tag{22}$$

and

$$Q_{ij} = (m\Omega)^2 M_{ij} \tag{23}$$

where the parenthesized expression denotes

$$(a, b; c, d) = \sum_{n=a}^{b} J_c(\gamma\Omega t)\, J_d(\gamma\Omega t) \tag{24}$$

and the sum is always over the index n, which must appear in c and d. We may now write

$$\det W = \det \begin{pmatrix} M & G \\ \tilde{G} & Q \end{pmatrix} = (m\Omega)^{2N} \det \begin{pmatrix} M & G/m\Omega \\ -G/m\Omega & M \end{pmatrix} \tag{24}$$

The final matrix of equation 24 may be written as a sum of direct matrix products:

$$\begin{pmatrix} M & G/m\Omega \\ -G/m\Omega & M \end{pmatrix} = M \times \begin{pmatrix} 1 & 0 \\ 0 & 1 \end{pmatrix} + (G/m\Omega) \times \begin{pmatrix} 0 & 1 \\ -1 & 0 \end{pmatrix} \tag{25}$$

Rotation in 2×2 space to diagonalize the final matrix of equation 25 permits equation 24 to be written

$$\det W = (m\Omega)^{2N} \det \begin{pmatrix} M + iG/m\Omega & 0 \\ 0 & M - iG/m\Omega \end{pmatrix} \tag{26}$$

But $(M + iG/m\Omega)$ is Hermitean, and $\det (M + iG/m) = \Pi\lambda_i$, where the λ_i are all real. Therefore we obtain

$$|\det W|^{\frac{1}{2}} = (m\Omega)^N \det (M + iG/m\Omega) \tag{27}$$

with the matrix elements given by

$$(M + iG/m\Omega)_{ij} = (k_B/m\Omega^2)\{T_b\,\delta_{ij} + (T_0 - T_b)(1, N; n - i, n - j) \exp [i(r - s)\,\pi/2] \tag{28}$$

Since the coefficient of $T_0 - T_b$ in equation 28 is a finite sum of Bessel-function products that vanish as $t \to \infty$, the matrix $M + iG/m\Omega$ is seen to become scalar, and the equilibrium entropy is given by equation 18, as $|t| \to \infty$, as

$$S_N = Nk_B + Nk_B \ln (k_B T_b/\hbar\Omega) \tag{29}$$

the correct canonical entropy for a system of N independent classical oscillators. At $t = 0$, the matrix is also diagonal, and the initial entropy is easily seen to be

$$S_N(0) = Nk_B[1 + \ln(k_B T_0/\hbar\Omega)] \tag{30}$$

Thus the entropy correctly evolves from its initial value to the final equilibrium value, in accordance with the expectations implicit in Gibbs's formulation of statistical mechanics.

The temporal evolution of S_N is most easily presented in terms of a temperature function $T(N, t)$, such that $T(N, 0) = T_0$ and $T(N, \infty) = T_b$, but at other times the function is to be regarded only as a mathematical convenience. In terms of $T(N, t)$, S_N can be written as

$$S_N/Nk_B = 1 + \ln[k_B T(N, t)/\hbar\Omega] \tag{31}$$

where $T(N, t)$ is calculated from equations 18 and 27. The results for the first few Ns are:

$$T(1, t) = T_b + (T_0 - T_b)\, J_0^2(\gamma\Omega t) \tag{32}$$

$$T(2, t) = T_b + (T_0 - T_b)\,(J_0^2 + J_1^2) \tag{33}$$

$$\begin{aligned} T(3, t) = [\![&\{T_b + (T_0 - T_b)\,[2J_1^2 + (J_0 - J_2)^2]\}\,\{T_b^2 + T_b(T_0 - T_b) \\ &\times [J_0^2 + 2J_1^2 + (J_0 + J_2)^2] + (T_0 - T_b)^2\,[J_0^2(J_0^2 + J_2^2) \\ &+ 2J_1^2(J_0 + 2J_2)]\}]\!]^{\frac{1}{3}} \end{aligned} \tag{34}$$

and

$$\begin{aligned} T(4, t) = [\![&\{[T_b + (T_0 - T_b)\,(J_0^2 + J_1^2 + J_2^2 + J_3^2)] \\ &\times [T_b + (T_0 - T_b)\,(J_0^2 + 2J_1^2 + J_2^2)] - (T_0 - T_b)^2 J_2^2(J_1 + J_3)^2\}^2 \\ &+ \{2(T_0 - T_b)^4 J_2^2(J_1 + J_3)^2\,(2J_0J_2 - J_1^2 + J_1J_3)^2\} \\ &+ \{(T_0 - T_b)^4\,(2J_0J_2 - J_1^2 + J_1J_3)^4\} - \{2[T_b + (T_0 - T_b) \\ &\times (J_0^2 + J_1^2 + J_2^2 + J_3^2)]\,[T_b + (T_0 - T_b)\,(J_0^2 + 2J_1^2 + J_2^2)] \\ &\times (T_b - T_0)^2\,(2J_0J_2 - J_1^2 + J_1J_3)^2\}]\!]^{\frac{1}{4}} \end{aligned} \tag{35}$$

where the arguments of the Bessel functions are all $\gamma\Omega t$. These temperature functions become increasingly complicated as N increases, with no apparent general expression or simplification. The first three are shown in *Figure 2*. The single-particle temperature function $T(1, t)$, which could reasonably be called the temperature of the system, starts at T_0, increases to T_b when $\gamma\Omega t = 2{\cdot}405$, and bounces back to lower temperature, returning to T_b at successive zeros of J_0. These pre-equilibrium swings to $T = T_b$, with subsequent bounces, seem to be at odds with the ideas of the H theorem and with the simplistic 'time's-arrow' concept of entropy. The reason for this behaviour is clear, however, when one realizes that when $J_0 = 0$, the system's initial conditions have (at those instants) no influence on its behaviour, and its motion is determined entirely by the initial conditions of heat-bath variables. At other times, the system's initial influence on the heat bath returns from the heat bath to reduce its temperature, but ever more feebly.

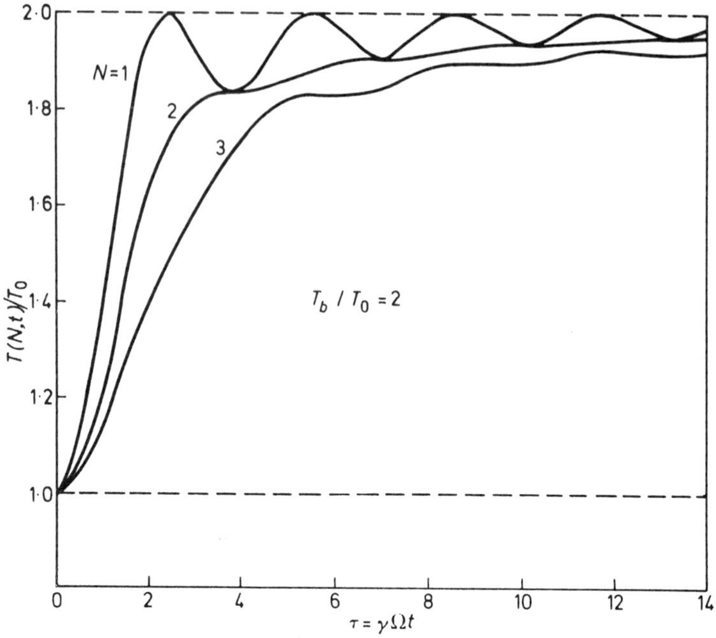

Figure 2. Plots of $T(N, \tau)/T_0$ for $N = 1$, 2 and 3, and $T_b/T_0 = 2$, where $\tau = \gamma\Omega t$.

The two-particle system shows no bounces of the temperature function, since $J_0^2 + J_1^2$ is a monotonic non-increasing function of its argument, but the slope of $T(2, t)$ is zero at values corresponding to the zeros of J_1. At no pre-equilibrium time is the two-particle system completely determined by the heat-bath variables, since internal influence is shared. The three-particle system shows an even smoother temperature function, and it is conjectured that $T(N, t)$ and $S_N(t)$ become increasingly structureless as N increases.

5. DISCUSSION

Boltzmann's H is, except for sign, the conceptual (though not, in general, the theoretically correct) equivalent of the entropy function. The entropy exhibited in equation 17 has the property that it evolves from the initial value, determined entirely by system variables at $t = 0$, to the canonical equilibrium value as $|t| \to \infty$. The evolution, except for $N = 1$ and 2, seems to be sufficiently smooth to satisfy the conceptual content of the H theorem, though the temporal development is somewhat bumpy because of the interaction of system and heat bath. The evolution is time-reversible at $t = 0$, showing that from a given set of initial conditions, *retro*diction is no better than *pre*diction, and that as the system evolves in either direction of time away from the initial state, our *description* of it, ρ_N of equation 14, evolves to a state of knowledge that can only be described as equilibrium.

6. ACKNOWLEDGEMENT

We thank J. M. Blatt for pertinent correspondence. This work was supported in part by the United States Atomic Energy Commission.

REFERENCES

[1] L. Brillouin, *Scientific Uncertainty, and Information*. Academic Press: New York (1964).
[2] J. M. Blatt, *Progr. Theor. Phys.* **22**, 745 (1959).
[3] J. W. Gibbs, *Elementary Principles in Statistical Mechanics*. [republished by Dover: New York (1960)].
[4] H. S. Robertson and M. A. Huerta, *Phys. Rev. Letters*, **23**, 825 (1969).
M. A. Huerta and H. S. Robertson, *J. Statist. Phys.* **1**, 393 (1969).
[5] M. A. Huerta and H. S. Robertson, to be published, preprints as MIAPH-TP-69.21 available upon request.
[6] M. A. Huerta and H. S. Robertson, to be published, preprints as MIAPH-TP-69.22 available upon request.
[7] H. S. Robertson and M. A. Huerta, to be published, preprints as MIAPH-TP-69.20 available upon request.

SECONDARY EFFECTS IN IRREVERSIBLE THERMODYNAMICS

J. HARRIS

Postgraduate School of Chemical Engineering,
The University, Bradford 7, U.K.

ABSTRACT

Linear phenomenological relations are recast to include relaxation effects. The relations are then written in a form suitable for general motion of the system and transformed to a coordinate system which is stationary relative to the observer. Generally, secondary fluxes are then observed which would be important in the fields of heat and mass transfer, for example. The Onsager relations are interpreted as reciprocal relations between the distribution functions of relaxation times. The principles on which these developments are based are that the thermodynamic properties of elements of a material are independent of the properties of neighbouring elements and also of the motion of the element in space, but may depend upon the thermodynamic history of the element.

1. INTRODUCTION

Many processes occur in which a specific physical quantity is transported through a sequence of non-equilibrium states of the system. Such transport processes are of a thermodynamically irreversible nature which is characterized by an irreducible increase in entropy.

The simpler aspects of these irreversible processes are usually treated on the macroscopic scale by *linear phenomenological laws* of which there are many, such as Newton's viscous law relating deformation stress with deformation strain rate in fluids, Fick's law relating flowrate of matter in a mixture with the concentration gradient of that matter, and Fourier's law relating heat energy flowrate with temperature gradient. Where one or more of these phenomena occur simultaneously then coupling occurs and important new phenomena are established such as the coupling between heat conduction and diffusion which gives rise to thermal diffusion.

In the subject of irreversible thermodynamics physical quantities such as temperature gradients and concentration gradients are termed 'forces' and the associated effects such as heat energy and mass flowrate are termed 'fluxes'. The product of 'forces' and 'fluxes' gives the entropy production rate or 'entropy source strength'. The identification of process source strengths and therefore the evolution of a system is the central theme of the subject of irreversible thermodynamics. In a published account de Groot[1] has outlined and interpreted many of the main features of the subject and its applications.

In the following work the theory is formulated in a convected coordinate framework which leads to important new results.

2. THE ONSAGER RELATIONS

In summarizing the previous comments on phenomenological laws it may be stated that the forces are linearly related to the fluxes and allowing for coupling, any force may, generally speaking, stimulate a response in any of the possible fluxes. This statement may be compactly represented by

$$J_\alpha = L_{\alpha\beta} X_\beta \quad (\alpha, \beta = 1, 2, 3, \ldots) \tag{1}$$

where summation over the repeated suffix is implied.

In equation 1 the $L_{\alpha\beta}$ are the phenomenological coefficients and those in which $\alpha = \beta$ are the direct coefficients whilst for $\alpha \neq \beta$ coupled or interference effects occur.

The important Onsager relations state that the *phenomenological coefficients are symmetrical,*

$$L_{\alpha\beta} = L_{\beta\alpha} \tag{2}$$

The proof of these relations is treated by de Groot[1] on the basis of statistical mechanics, microscopic reversibility and regression of fluctuations.

The hypothesis introduced by Onsager into the third part of the proof of the relations 2 is that on the average the decay of a fluctuation of the thermodynamic parameters of a system follows the ordinary linear macroscopic laws. Suppose that the *deviations* of the thermodynamic parameters have the values $a_\gamma(\gamma = 1, 2, 3, \ldots)$ then equation 1 may be written

$$J_\alpha = \bar{\dot{a}}_\alpha = L_{\alpha\beta} X_\beta \tag{3}$$

where the bar over $\dot{a}_\alpha$ denotes time averaged over *microscopic* fluctuations.

The hypothesis then implies that the time scale of the process Γ_p is related to the time scale of fluctuation Γ_f and the molecular time scale Γ_m by the inequalities

$$\Gamma_m \ll \Gamma_p \ll \Gamma_f \tag{4}$$

where

$$\Gamma_{\gamma p} = \mathrm{O}\left(\frac{\bar{a}\gamma}{\bar{\dot{a}}\gamma}\right) \tag{5}$$

and

$$\Gamma_{\gamma f} = \mathrm{O}\left(\frac{\bar{\dot{a}}\gamma}{\bar{\ddot{a}}\gamma}\right) \tag{6}$$

Provided phase-shifting between the forces and fluxes does not impeach any of the fundamentals of irreversible thermodynamics, and this appears to be so, then there is the possibility of relaxing the inequality

$$\Gamma_p \ll \Gamma_f \tag{7}$$

and admitting *linear complex phenomenological* laws of the type

$$J_\alpha^+ = L_{\alpha\beta}^+ X_\beta^+ \tag{8}$$

The above development was implied in an isolated example by de Groot in which the relaxation associated with an internal redistribution of energy was treated.

In considering spatial distributions of the forces and fluxes it is necessary to note that tensor forces can only give rise to tensor fluxes *of the same rank*. This is an important consideration when treating coupled phenomena.

Up to the present, no mention has been made of possible motion of the reference frame in which the forces and fluxes are measured. In the following section the phenomenological equations are written in a reference frame which moves in space and this produces modifications in the phenomenological equations as seen by an observer with different motion.

3. GENERALIZATION OF THE EQUATIONS

(i) Small variable strain rates

In this work spatial distributions of physical quantities will be denoted by Latin subscripts or superscripts whilst classes of physical quantities will continue to be denoted by Greek subscripts as before.

The general tensor form of the linear phenomenological laws of the type 8 for small variable strain rates is

$$J^{+\,ijk}_{\alpha\,rst} = L^{+}_{\alpha\beta} X^{+\,ijk}_{\beta\,rst} \tag{9}$$

because the fluctuation programme of the thermodynamic parameters could often be described by a Fourier series in real cases.

Since the phenomenological laws[9] are written in proper tensor form, which ensures invariance of form under a transformation of coordinates, then they are quite independent of the motion of any reference frame in space. When applied to a continuum in motion, addition of corresponding quantities throughout the whole history of motion is accomplished by writing the phenomenological equations in a reference frame which is convected, rotated and deformed with the continuum. Experimental observations are invariably made in a reference frame which is fixed relative to an observer who does not have the motion of all regions of the continuum. Transformation from the convected to the fixed reference frame will under certain circumstances introduce new terms into the phenomenological laws. But initial isotropy remains in the convected reference frame.

The formal apparatus for transforming rheological equations of state containing time derivatives and integrals has already been treated in some detail by Oldroyd[2]; for tensor quantities of general type and any rank of particular interest in practical cases there are phenomenological laws relating tensors of rank one. The linear differential form of 9 is then

$$\left(\sum_{N=0}^{N=P} \lambda_N \frac{\partial^N}{\partial t^N}\right) J^i_\alpha = L_{\alpha\beta} \left(\sum_{M=0}^{M=Q} \tau_M \frac{\partial^M}{\partial t^M}\right) x^i_\beta \tag{10}$$

where $\lambda_0 = \tau_0 = 1$.

For sufficiently slow fluctuations of the forces and fluxes terms containing higher time derivatives than the first in 10 can be neglected and the truncated linear differential equation is

$$\left(1 + \lambda_1 \frac{\partial}{\partial t}\right) J^i_\alpha = L_{\alpha\beta} \left(1 + \tau_1 \frac{\partial}{\partial t}\right) X^i_\beta \tag{11}$$

The general linear integral equation has the form

$$J^i_\alpha = \int^t_{-\infty} \psi_{\alpha\beta}(t - t') X^i_\beta(x, t') \, dt' \tag{12}$$

where t is current time and t' is non-current time and x^j is the fixed coordinate system. The memory function $\psi_{\alpha\beta}(t - t')$ may take the form corresponding to the same type of function obtained in rheological equations of state, namely

$$\psi_{\alpha\beta}(t - t') = \int_0^\infty \frac{R_{\alpha\beta}(\tau)}{\tau} \exp - \frac{[t - t']}{\tau} \, d\tau \tag{13}$$

In 13 $R_{\alpha\beta}(\tau)$ is the distribution function of relaxation times associated with the α flux and β force.

In generalizing 10, 11 and 12 it is noted that the phenomenological equations are not associated with a fixed point x^j in space but rather with an element of material over all time in the interval $-\infty \leqslant t' \leqslant t$. Consequently the time differentiations and integration must follow the motion of the material and only under the special condition of small material velocities, i.e. creeping flow, can the time derivatives and integrals be interpreted in a simple way.

The differential equation 11 is a special case of the general integral form 12 obtained by substituting into 13 a distribution function of the type

$$R_{\alpha\beta}(\tau) = L_{\alpha\beta}\,(\tau_1/\lambda_1)\,\delta(\tau) + L_{\alpha\beta}\,\{(\lambda_1 - \tau_1)/\lambda_1\}\,\delta(\tau_1 - \lambda_1) \tag{14}$$

For the system characterized by 14 then it may easily be shown that fluctuations of a frequency ω give

$$J^+_\alpha = L_{\alpha\beta}/(1 + \omega^2\lambda_1^2)\,[(1 + \omega^2\tau_1\lambda_1) - i\omega(\lambda_1 - \tau_1)]\,X_\beta \tag{15}$$

(ii) General motions

Generalizations of the linear phenomenological equations 10, 11 and 12 are now considered in which the motion of the continuum in which the processes operate is arbitrary.

In the convected coordinate system introduced by Oldroyd[2] with coordinate surfaces ξ^k = constant embedded in the deforming continuum, a material element which is located at ξ^k at time t occupies the same position at all prior and subsequent times. In this reference frame equation 12 takes on the form

$$\Pi^i_\alpha = \int^t_{-\infty} \psi_{\alpha\beta}(t - t')\,\chi^i_\beta(\xi, t')\,dt' \tag{16}$$

where Π^i_α, χ^i_β are the convected components of the fluxes and forces respectively. The spatial distributions in 16 are written as contravariant tensors, but they might equally well be written as covariant tensors.

Transforming 16 to a coordinate system which is fixed relative to the observer, by the techniques introduced by Oldroyd, then

$$J_\alpha^i = \int_{-\infty}^{t} \psi_{\alpha\beta}(t - t')\, X_\beta^r(x', t') \frac{\partial x^i}{\partial x'^r}\, \mathrm{d}t' \tag{17}$$

The covariant form of the equation is

$$J_{i\alpha} = \int_{-\infty}^{t} \psi_{\alpha\beta}(t - t')\, X_{r\beta}(x', t') \frac{\partial x'^r}{\partial x^i}\, \mathrm{d}t' \tag{18}$$

Oldroyd has also considered generalizations of differential equations[3,4]. Considering equation 11 then the corresponding form in the convected coordinate system is

$$\left(1 + \lambda_1 \frac{\partial}{\partial t}\right) \Pi_\alpha^i = L_{\alpha\beta} \left(1 + \tau_1 \frac{\partial}{\partial t}\right) \chi_\beta^i \tag{19}$$

Transforming 19 back to a fixed coordinate system then the form equivalent to 19 is

$$\left(1 + \lambda_1 \frac{\bar{\delta}}{\bar{\delta} t}\right) J_\alpha^i = L_{\alpha\beta} \left(1 + \tau_1 \frac{\bar{\delta}}{\bar{\delta} t}\right) X_\beta^i \tag{20}$$

where

$$\frac{\bar{\delta} J_\alpha^i}{\bar{\delta} t} = \frac{\partial J_\alpha^i}{\partial t} + u^K J^i_{\alpha,K} - u^i_{,K} J_\alpha^K \tag{21}$$

with an identical form for

$$\frac{\bar{\delta} X_\beta^i}{\bar{\delta} t}$$

It may be noted from 21 that when the velocity field u_k and its spatial gradients are not vanishingly small then the process of transforming time derivatives introduces additional terms into equation 20 and its expanded form becomes

$$J_\alpha^i + \lambda_1 \left[\frac{\partial J_\alpha^i}{\partial t} + u^K J^i_{\alpha,K} - u^i_{,K} J_\alpha^K\right] = L_{\alpha\beta}\left(X_\beta^i + \tau_1 \left[\frac{\partial X_\beta^i}{\partial t} + u^r X^i_{\beta,r} - u^i_{,r} X_\alpha^r\right]\right) \tag{22}$$

Identical results are *not* obtained by taking the covariant equivalent of 20 namely

$$\left(1 + \lambda_1 \frac{\bar{\delta}}{\bar{\delta} t}\right) J_{\alpha i} = L_{\alpha\beta} \left(1 + \tau_1 \frac{\bar{\delta}}{\bar{\delta} t}\right) X_{\beta i} \tag{23}$$

for in this case

$$\frac{\bar{\delta} J_{\alpha i}}{\bar{\delta} t} = \frac{\partial J_{\alpha i}}{\partial t} + u^K J_{\alpha i,K} + u^K_{.i} J_{\alpha K} \tag{24}$$

and this should be compared with 21. There are important implications in these results as will be shown later. Other forms of 21 and 24 may be written which bring out more clearly their fundamental difference. To obtain a true comparison the equations are written in cartesian form in which there is no distinction between covariant and contravariant tensors. It is also convenient to take the velocity gradient in cartesian form.

$$u_{K,i} = \frac{1}{2}\left(\frac{\partial u_K}{\partial x_i} + \frac{\partial u_i}{\partial x_K}\right) + \frac{1}{2}\left(\frac{\partial u_K}{\partial x_i} - \frac{\partial u_i}{\partial x_K}\right) = e_{iK} - \omega_{iK} \tag{25}$$

Then 21 becomes

$$\frac{\bar{\delta} J_{\alpha i}}{\bar{\delta} t} = \frac{\partial J_{\alpha i}}{\partial t} + u_K \frac{\partial J_{\alpha i}}{\partial x_K} - (e_{Ki} + \omega_{Ki}) J_{\alpha K} \tag{26}$$

whilst 24 becomes

$$\frac{\bar{\delta} J_{\alpha i}}{\bar{\delta} t} = \frac{\partial J_{\alpha i}}{\partial t} + u_K \frac{\partial J_{\alpha i}}{\partial x_K} + (e_{iK} + \omega_{iK}) J_{\alpha K} \tag{27}$$

It may be seen from 25 that

$$e_{iK} = e_{Ki} \tag{28}$$

and

$$\omega_{iK} = -\omega_{Ki} \tag{29}$$

and hence 27 differs from 26 by the addition of $2e_{iK}J_{\alpha K}$

The simplest time derivative which takes account of both the translation of the continuum and *also its rotation* is just the common part of 26 and 27 and this is denoted by $\mathscr{D}J_{\alpha i}/\mathscr{D}t$ where [3,4]

$$\frac{\mathscr{D} J_{\alpha i}}{\mathscr{D} t} = \frac{\partial J_{\alpha i}}{\partial t} + u_K \frac{\partial J_{\alpha i}}{\partial x_K} + \omega_{iK} J_{\alpha K} \tag{30}$$

4. SIMPLE SHEARING

It is worthwhile from a practical viewpoint to examine some of the implications of the developments in Section 2 when the continuum is deformed in steady simple shearing motion.

Consider laminar motion of the continuum in which the velocity field has the pattern

$$u_i = (\gamma x_2, 0, 0) \tag{31}$$

where γ is constant. This corresponds to steady simple shearing motion in which the shear planes move parallel to the x_1 axis. Suppose that there is a single thermodynamic force of unity tensor rank (vector), which has the distribution

$$X_K = (0, X_2, 0) \tag{32}$$

where X_2 is a constant. This could be for example a temperature or a concentration gradient in the x_2 direction. For simplicity it is taken that only direct fluxes are generated so that in the array of phenomenological coefficients only L_{11} is non-zero.

Differences occur in the final results according to whether the contravariant differential equation 20 or covariant equation 23 is taken. Treating the contravariant equation first then

1 Direction $$J_1 - \lambda_1 \gamma J_2 = -L_{11}\tau_1\gamma X_2 \tag{33}$$

2 Direction $$J_2 = L_{11}X_2 \tag{34}$$

or $$J_x = L_{11}\gamma X_y(\lambda_1 - \tau_1) \quad (34)$$

$$J_y = L_{11}X_y \tag{35}$$

where to avoid confusion the 1, 2 coordinates are now labelled x, y.

The flux J_y in 35 is the ordinary direct flux produced by the force X_y but the flux J_x in 34 is a *secondary flux* which is only zero if $\gamma(\lambda_1 - \tau_1)$ becomes vanishingly small which it would in a stationary continuum.

The covariant case of the same differential equation gives

$$J_x = L_{11}\gamma X_y(\tau_1 - \lambda_1) \tag{36}$$

$$J_y = L_{11}X_y \tag{37}$$

Rheological equations of state have been formulated in which the partial time derivatives have been translated into the form 30 which allows for convection and rotation of the continuum *but not straining*. In this case it is easy to show that the differential equation 11 gives:

$$J_x = \frac{1}{2}\frac{L_{11}(\lambda_1 - \tau_1)}{(1 + \frac{1}{4}\lambda_1^2\gamma^2)}\gamma X_y \tag{38}$$

$$J_y = L_{11}\frac{(1 + \frac{1}{4}\lambda_1\tau_1\gamma^2)}{(1 + \frac{1}{4}\lambda_1^2\gamma^2)}X_y \tag{39}$$

In this case not only are there both direct and secondary fluxes, but a new feature arises in that they are both non-linear in the shear rate, but *the relation between the force and corresponding flux remains linear.*

Phenomenological relations between scalar forces and fluxes would not exhibit secondary effects because in general motion the time derivatives then are interpreted as the Eulerian time derivative $\mathrm{D}S/\mathrm{D}t$ where

$$\frac{\mathrm{D}S}{\mathrm{D}t} = \frac{\partial S}{\partial t} + u^r\frac{\partial S}{\partial x^r} \tag{40}$$

and S is any scalar quantity.

Tensors of rank two have been treated by Oldroyd in the form of stress/strain rate relations.

5. CONCLUSION

The linear phenomenological relations of irreversible thermodynamics have been broadened to include relaxation effects in the coefficients. The well known Onsager reciprocal relations

$$L_{\alpha\beta} = L_{\beta\alpha} \tag{2}$$

which state that the array of phenomenological coefficients is symmetrical can then be restated as

$$R_{\alpha\beta} = R_{\beta\alpha} \tag{41}$$

The corresponding statement here is that the *distribution of relaxation times is symmetrical.*

The general effect of translation, rotation and deformation of the system is that new terms can occur in the thermodynamic equations of transport processes and these describe secondary fluxes. In rheological equations the secondary fluxes take the form of normal force effects in simple shearing; these have often been reported in the literature.

Scalar thermodynamic forces produce no secondary fluxes and in tensors of rank one the covariant form 23 produces negative secondary fluxes positive secondary fluxes in simple laminar shearing are present in the contravariant form and when time derivatives of the type 30 are used. It has already been noted[2] that in the rheological equations the corresponding covariant form produces results which are *not* in accord with experimental results. The contravariant form produces some of the correct types of effects but time derivatives of type 30 are perhaps the most successful[3,4] in simple equations containing relaxation effects.

The secondary flux does not of course contribute to the evolution of entropy since the scalar product of this flux with the force is zero.

It is clear from equation 20 that 'cross' phenomena can also produce secondary fluxes. That is, β forces can produce secondary α fluxes.

A fundamental principle implicit in this work is that the thermodynamic properties of a material element do not depend upon the properties of neighbouring elements but may depend upon the history of thermodynamic states of the element. The thermodynamic properties of the element are also independent of the motion of the element in space.

6. REFERENCES

1. S. R. de Groot, *Thermodynamics of Irreversible Processes.* North Holland: Amsterdam (1966).
2. J. G. Oldroyd, *Proc. Roy. Soc. A*, **200**, 523 (1950).
3. J. G. Oldroyd, 'Complicated rheological properties' in *Rheology of Disperse Systems.* Pergamon: Oxford (1959).
4. J. G. Oldroyd, *Proceedings* of the International Symposium on Second-order Effects in Elasticity, Plasticity and Fluid Dynamics, Haifa. Pergamon: Oxford (1962).
5. Walters, K. *Quart. J. Mech. Appl. Maths.* **25** (Pt 1), 63 (1962).

E. STATISTICAL THERMODYNAMICS IN ASTROPHYSICS AND RELATIVITY

EXTREME REGIMES OF TEMPERATURE AND PRESSURE IN ASTROPHYSICS

MALVIN A. RUDERMAN*

Department of Physics, Columbia University, New York, N.Y.

ABSTRACT

A survey is given of states of matter and phenomena in extreme régimes of temperature and density found in the astrophysical universe: thermodynamics and stellar evolution, terminal states of stars, temperatures of stars, forms of superdense matter in stars, crystallization of superdense stellar matter, neutron star structure, and low temperature phenomena in neutron stars.

1. INTRODUCTION

We live in a remarkably special part of the universe which can sustain life: the temperature is about 300° K; the density of things around us is in the range 0·1 to 10 g cm^{-3}; we are surrounded by a gas of around 10^{-3} g cm^{-3} and pervaded by a one gauss magnetic field. Even the régimes of these parameters accessible in the laboratory cannot approach those found in parts of the astrophysical universe and even less those found in the speculations of astrophysicists.

The density of the centre of the sun, a typical hydrogen burning (main sequence) star is about 10^2 g cm^{-3}. White dwarf cores cannot exceed 10^9 g cm^{-3}. Neutron star centres are calculated to have 10^{14} to 10^{15} g cm^{-3} and may even be much denser. Greatest of all are the possible singular densities in the biography of a piece of matter. In orthodox relativity theory some world lines of our expanding universe must have originated in a singularity of infinite proper density[1–4]. For a homogeneous isotropic universe all matter had such a genesis. Finally all stars too heavy to become neutron stars will contract indefinitely; but as seen by an observer living on the inward falling surface of such a star it will rapidly pass through its Schwarzschild singularity and attain an infinite density.

The temperature in the solar centre $\sim 10^{7}$°K. Red giant cores are close to 10^{8}°K, about the maximum achievable in the explosion of a nuclear weapon. A highly evolved heavy star, just before it becomes a supernova, reaches a central temperature near 6×10^{9}°K—about half a million electron volts per particle. At such a temperature black body radiation weighs 10^3 g cm^{-3} and coexists with an almost equal density of electron–positron pairs. Various scenarios which try to describe a star during a supernova explosion suggest $T \sim 10^{12}$°K and more.

A similar régime of extraordinarily high temperatures occurs in the orthodox general relativistic account of the creation. There is strong evidence

* Professor Ruderman was unable to be present, and a lecture based on his paper was given by Dr C. Pethick.

that the present universe contains an isotropic homogeneous sea of 2·7°K black body photon radiation. As the universe expands such radiation remains black body but with a temperature decreasing like the inverse Hubble radius of the universe. Extrapolating backward in time for a homogeneous isotropic universe gives a view like that in *Figure 1*. In the beginning there may have

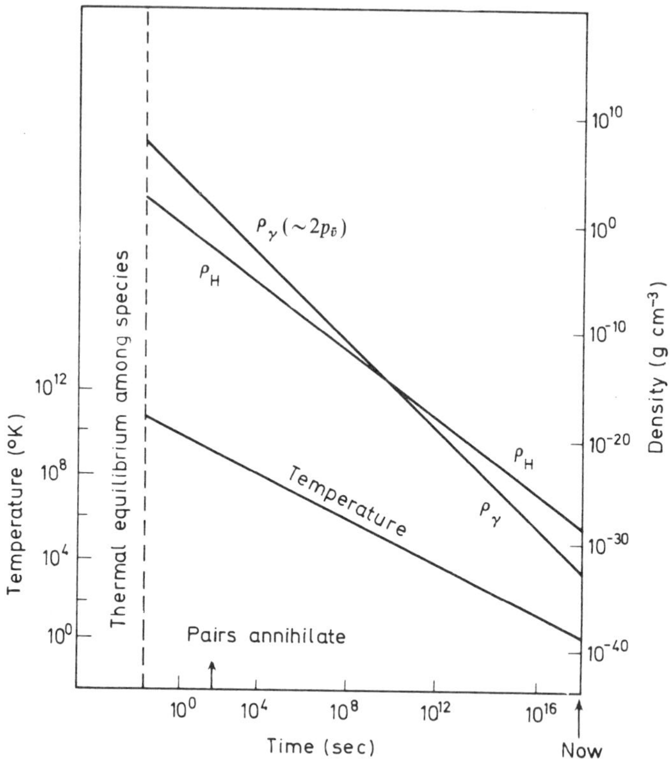

Figure 1. Temperature and density history of a homogeneous isotropic universe. ρ_H is the mass density of protons (hydrogen), $\rho_{\bar{\nu}}$ that of photons, and $\rho_{\nu-}$ that of (anti) neutrinos. Details in ref. 5.

been thermal equilibrium for a fraction of a second. Baryons, antibaryons, mesons, neutrinos, electrons, perhaps quarks or magnetic poles, would have coexisted but after a second or so, when the initial large temperature has fallen to near 10^{10}°K, the weakly interacting neutrinos no longer interact significantly with the remaining matter and are 'frozen out' of the equilibrium. A measurable fraction of quarks and antiquarks ($\sim 10^{-20}$ quarks/nucleon) may remain[5]. Even nuclear reactions cease 10^2 seconds after the big bang when the temperature has dropped to 10^9°K. Electron–positron annihilation then eliminates all positrons and most electrons. The expanding universe then consists of protons, electrons and neutrinos; a considerable fraction of helium ($\sim$20 per cent by mass), and very small fractions of the lighter elements (D, ^{3}He, ^{3}H, ^{7}Be, ^{7}Li).

The very earliest stage, where local thermal equilibrium is presumed to prevail, encourages speculations about possible phenomena that might affect the subsequent evolution of the universe. Some concern mechanisms for establishing the truly remarkable isotropy[6] we see in present cosmic black body radiation[7] from essentially random initial conditions[8]. An extremely vital and entertaining thermodynamic question is the possibility of phase separations. Could a dense universe of nucleons, mesons and antinucleons or perhaps even quarks and antiquarks have spontaneous macroscopically separated phases which partially separate matter from antimatter[9]? The signs and magnitudes of the interactions among nucleons and mesons are indeed such that at certain densities ($\rho \sim 10^{12}$ g cm^{-3}) and temperatures ($T \sim 10^{12}$°K) the free energy might conceivably be lowered by a two phase system one of which is primarily baryons and those mesons attracted to them, the other the charge conjugate, i.e. anti-system. However, there is no adequate quantitative argument that this should be so[10].

Unfortunately the very early stages of the universe are a régime in which not only may the relevant physical laws possibly be incomplete, but the initial conditions almost certainly are. However, the extreme early stages of matter, enormous temperatures, densities, magnetic fields etc. probably are duplicated in the evolution of many stars—only the time sequence being opposite to that of an expanding universe: the star evolves from a hot diffuse gas through ever denser and hotter régimes to one of a number of enormously dense final states. Remarkably the final stages in the death of a star involve at first the highest temperatures known to exist in our present universe; then, usually there follows an extraordinarily rapid cooling which effectively, i.e. in terms of the phenomena which occur, is to a far lower temperature than can be found anywhere else in the natural universe.

2. THERMODYNAMICS AND STELLAR EVOLUTION

In the beginning a star is a mass of gas held in quasi-static equilibrium with the pull of gravity balanced by classical kinetic pressure. Even if it did not radiate it would never be in true thermodynamic equilibrium. A system of classical particles interacting through coulomb and gravitational forces has no lowest energy state. It continually evaporates particles, sending some to infinity and lowering the energy of the remainder. The time scale for such motions is usually sufficiently slow that they are neglected in analyses of stellar structure, at least in earlier stages of stellar evolution. The thermodynamicist might describe such a quasi-static star as a system of negative specific heat. Thirring[11] has recently emphasized that *any* condensed (i.e. bound) system of classical particles which interact among themselves through electrical and gravitational forces should behave as if its specific heat were negative. The total energy

$$E = K + V \tag{1}$$

where K is the total kinetic and V the potential energy of the particle system. For inverse square law forces the virial theorem gives

$$K = -\tfrac{1}{2}V \tag{2}$$

so that

$$E = -K \equiv -\tfrac{3}{2}Nk\langle T\rangle \tag{3}$$

Here N is the particle number and $\langle T \rangle$ the appropriate average temperature. Then the heat capacity $\langle c \rangle$ is defined by

$$\langle c \rangle \equiv \partial E/\partial \langle T \rangle = -\tfrac{3}{2}Nk < 0 \tag{4}$$

A more detailed analysis[11] which includes the virial effects of boundaries shows that both the core of a star and the outer parts separately behave as if they had negative heat capacity. Two systems in thermal contact have no equilibrium when each of them has negative heat capacity. The change in entropy $S(E)$ when a small amount of energy ε is transferred from system 1 at temperature T_1 to system 2 at temperature T_2 is

$$\delta S = \varepsilon \left\{ \frac{1}{T_1} - \frac{1}{T_2} \right\} - \frac{\varepsilon^2}{2} \left\{ \frac{1}{c_1 T_1^2} + \frac{1}{c_2 T_2^2} \right\} \tag{5}$$

where

$$S = S_1(E_1) + S_2(E - E_1) \tag{6}$$

and

$$\frac{1}{T_i} = \frac{\partial S_i}{\partial E_i} \quad \frac{1}{c_i} = -T_i^2 \frac{\partial^2 S_i}{\partial E_i^2} \tag{7}$$

When $T_1 = T_2$, stability, i.e. $\delta S < 0$, follows only when $(c_1)^{-1} + (c_2)^{-1} > 0$. When both heat capacities are negative the hotter system 1 (the stellar core) transfers energy to the cooler system 2 (the stellar exterior). The core grows continually hotter, the exterior continually cooler [cf. *Figure 2*]. The virial

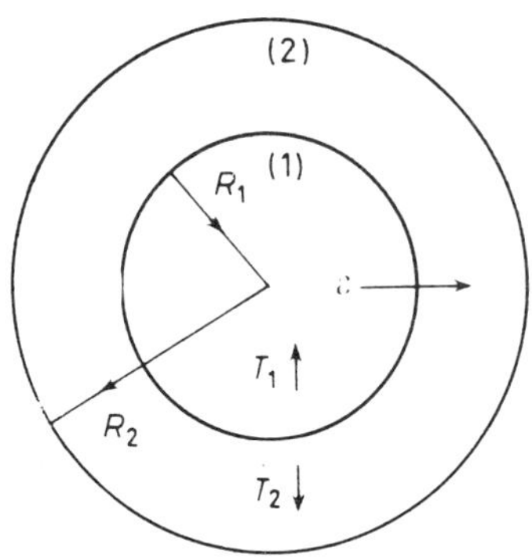

Figure 2. Schematic representation of two negative heat capacity regions of a star.

theorem relates the core temperature T_c to its average density ρ_c. The evolution of a stellar core, as long as it can be described as a system of classical particles, is then approximately given by

$$\rho_c \sim T_c^3 \tag{8}$$

Figure 3 gives the computed evolutionary track for a 16 solar mass ($M_\odot$) star's core as computed by Hayashi, Sogi and Sugimoto[12]. It is well approximated by equation 8. While the core contracts and heats, the outer stellar region cools (since its heat capacity is also negative) and, according to the virial theorem, expands. Thus the evolutionary trend of stellar envelopes is toward cooler, larger surfaces. This is confirmed in the familiar Hertzsprung–Russell diagram for the stellar observables: surface temperature T_e and luminosity L.

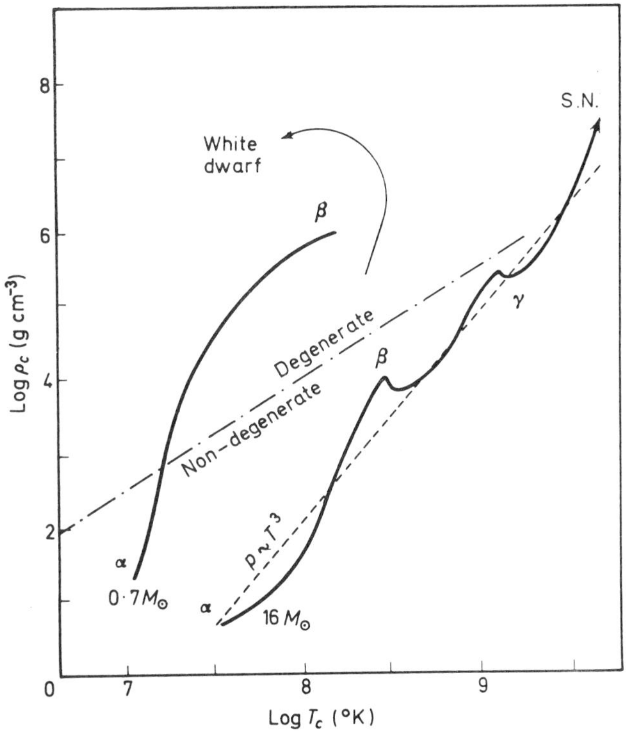

Figure 3. Evolution of stellar cores. α—hydrogen burning; β—helium burning (ref. 12); γ—C, O, Ne burning points. The lighter star ($M = 0{\cdot}7\ M_{\odot}$) quickly becomes degenerate after it leaves the main sequence and evolves very differently.

In *Figure 4* we see that the evolutionary trend of a 16 $M_{\odot}$ star is toward lower surface temperature. Since the stellar radius R_e depends upon L and T_e approximately according to the black body prescription

$$L \sim R_e^2 T_e^4 \tag{9}$$

the star expands greatly as it moves to the right off the main sequence, beginning as a blue star and ending as a red supergiant. Its outer radius expands by a factor of 100 as the atmosphere cools; its core radius decreases by a factor of 30 as the core temperature rises toward 10^{9}°K. A similar evolution—that of two systems of negative heat capacity in thermal contact—may ultimately obtain for entire galaxies[13]. The galactic core of stars shrinks as the stars within it acquire more kinetic energy while the rest of the galaxy expands as the container stars slow up. However, the stellar mean free path is so long that such a qualitative separation into two stellar systems is not maintained. A possible relationship between the expected thermodynamic behaviour of the central parts of classical stellar systems bound by inverse square law forces and observations of very dense very active galactic cores is still obscure.

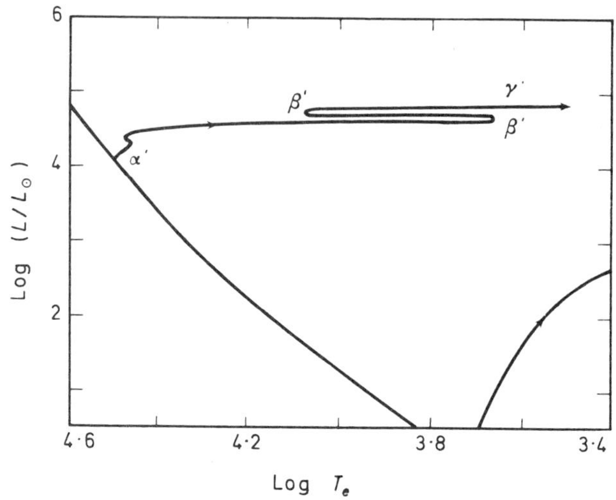

Figure 4. Evolution of stellar surfaces (ref. 12). The points α′, β′, γ′ correspond to the interior points α, β, γ of *Figure 3.* The diagonal line is the main sequence for H-burning stars of all masses. The lower right hand segment corresponds to a 0·7 $M_{\odot}$ star.

3. TERMINAL STATES OF STARS

The evolution of a stellar core through stages of ever increasing densities and temperatures is stopped temporarily by the ignition and consumption of nuclear fuels. During such stages, which occupy most of the stellar lifetime, there is a quite stable balance between energy radiated, energy transferred from the core to the envelope, and energy generated within the core (or its boundary) by exothermal nuclear processes. But the ultimate asymptotic state of a star is achieved only by mechanisms which can permanently stop its contraction. There are three known terminal stellar states: white dwarfs, neutron stars, and gravitational collapse to black holes.

As a stellar core contracts its temperature rises like $\rho^{\frac{1}{3}}$. But the degeneracy energy of non-relativistic electrons rises like $\rho^{\frac{2}{3}}$ (the kinetic motion of electrons contributes most of the central pressure) so that these electrons can ultimately become degenerate. When this happens the kinetic energy of the electrons and therefore the core pressure no longer depends sensitively upon temperature. Such degenerate stars behave like normal condensed laboratory matter where quantum mechanical electron degeneracy is basically responsible for stability. The virial theorem no longer implies a negative heat capacity, and the star core cools as energy is radiated.

Quite generally stable stars which can be the end states of stellar evolution are composed of such cool matter in which thermal pressure makes a negligible contribution. The source of pressure may be electron degeneracy (white dwarfs) or nucleon repulsion when the central density is sufficiently large (neutron stars).

The equation of state of cool matter over all régimes relevant for stellar matter is sketched in *Figure 5*. Only a very small part of it, corresponding to

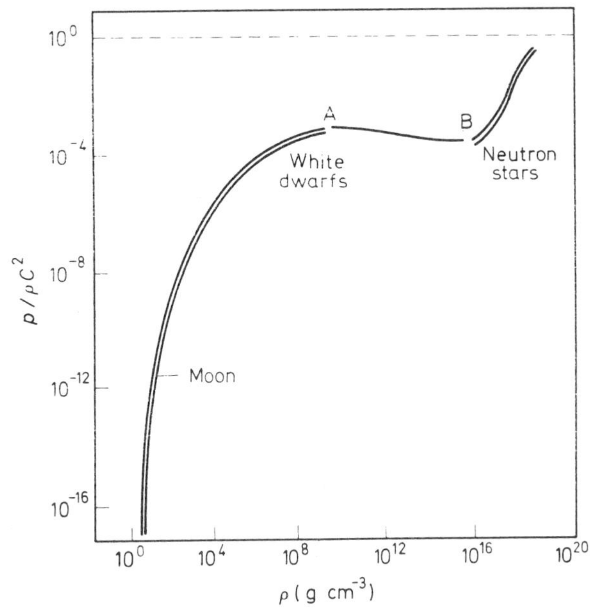

Figure 5. The ratio of pressure (p) to density (ρ) times c^2 as a function of density for cool matter.

matter at the density of the moon's centre is accessible in the laboratory. Above 10^5 g cm^{-3} the electrons which are the main contributor to the pressure are well approximated by a non-relativistic degenerate electron gas. At $\rho \sim 10^7$ g cm^{-3} the electron Fermi energy approaches 10^6 eV and the electrons become relativistic. The matter is then less stiff since the pressure of relativistic degenerate electrons does not rise as rapidly with decreasing volume as does that of non-relativistic ones. As the electron Fermi energy rises above 1 MeV ($\rho \gtrsim 10^7$ g cm^{-3}) the embedded nuclei become unstable against the capture of energetic electrons which convert nuclear protons to neutrons plus neutrinos which readily escape from the star without further interaction. In such a régime the matter is quite compressible since the electron Fermi energy (and hence pressure) ceases to rise substantially with increasing density. Rather the electrons continue to be absorbed by protons until at $\rho \sim 5 \times 10^{13}$ g cm^{-3} all but a few per cent of the nucleons are neutrons. Subsequent increases in density give a rapidly rising pressure in part because of non-relativistic neutron degeneracy and more because of strong short range neutron repulsion. For $\rho \gtrsim 10^{15}$ g cm^{-3} the interaction energy among particles is no longer small next to the mass differences among elementary particles or even next to the total rest masses. Here nothing is really known about the appropriate equation of state for matter. The restriction

$$p < \rho c^2 \tag{10}$$

is necessary if the sound speed at long wavelengths

$$c_s^2 = \mathrm{d}p/\mathrm{d}\rho \tag{11}$$

is to be less than the speed of light. Zel'dovich[14] has suggested a model that attains the high density limit $p = \rho c^2$. However, this limit may not have a sacred role in theoretical physics. Violations do not contradict Lorentz invariance and positive definiteness of energy[15]. But various proposed theories which do permit $p > \rho c^2$ have all involved either an unacceptable instability or a violation of conventional notions of causality[16, 17]. There are of course no experimental data to test either the high density limit of the equation of state or the validity of causality in such a régime.

That part of the equation of state curve of *Figure 5* which is doubled corresponds to stiff matter which satisfies the criterion

$$\mathrm{d}p/\mathrm{d}\rho > \tfrac{4}{3}p/\rho \tag{12}$$

It consists of two disjoint sections. Such matter is stiff enough to constitute the core of cool stable stars which are able to resist being crushed by their own gravitational contractions. The associated stars are represented in *Figure 6*. Dying stars associated with the lower segment of the equation of state curve are white dwarfs and planets. Stars corresponding to the upper segment are called neutron stars even if the main constituents of the core (for $\rho \gtrsim 10^{15}$ g cm^{-3}) are no longer neutrons.

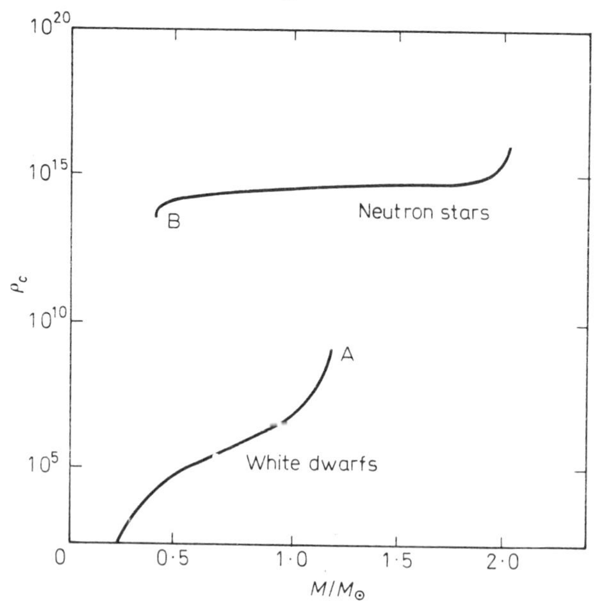

Figure 6. Masses of cool stars, in units of the solar mass $M_\odot$, as a function of central density ρ_c. The points A and B correspond to those of *Figure 5*.

There is an upper bound of about two solar masses ($M_\odot$) above which no cool stable star can exist. Even if matter were to become infinitely stiff ($dp/d\rho \to \infty$), this limit would not substantially change since increased pressure also increases the gravitational attraction between various parts of the star sufficiently to crush the star no matter how stiff it is. Thus if $M \gtrsim 2M_\odot$ the star cannot die as a stable star. Nothing can permanently stop its continual contraction toward infinite density. After the termination of its nuclear burning a heavy star of modest angular momentum will collapse through its Schwarzschild singularity radius ($R \sim 2\,GM\,c^{-2} \sim 10^6$ cm) until at least parts of it reach infinite density[1]. (Quantum corrections to orthodox general relativity might conceivably change this conclusion but not before $\rho \sim 10^{90}$ g cm^{-3}). As seen by an observer on the contracting star the collapse through the Schwarzschild radius and to a singularity takes only a few seconds. But a distant observer sees a star asymptotically approaching the Schwarzschild radius which it never reaches. The light from the stellar surface grows redder and dimmer. The singular densities which it can attain inside the Schwarzschild radius can, in orthodox general relativity, have no observable consequences outside. We shall concern ourselves only with the thermodynamic nature of the states associated with the super-high densities and unique temperatures within the white dwarfs and neutron stars.

4. HOW HOT DO STARS GET?

A star which does not end its life as a white dwarf will probably reach a stage where all the core nuclear constituents have ignited leaving behind only iron peak elements which have the highest binding energy of all nuclear species. This will occur at core temperatures $T_c \sim 5 \times 10^{9}$°K and correspondingly high core densities of perhaps 10^8 g cm^{-3}. The chief mechanism for energy transfer out of such a core is neutrino pair emission from electron–positron recombination. The neutrino emission of one such star $\sim 10^{46}$ ergs/sec greatly exceeds the light emission from the entire galaxy. Such stellar cores become unstable a few minutes after the high central temperature is reached: the core contraction rate, limited by the free fall rate of core matter, is too slow to supply the enormous amounts of energy which must be supplied to the core to maintain quasi-equilibrium. Two effects contribute to this negative energy budget. As the temperature rises the minimum free energy $U - TS$ is achieved not by minimizing U which resulted in burning all nuclei to iron but rather in maximizing the entropy S which is attained by breaking up the iron into its constituent α particles and neutrons. This endothermic undoing of the previous nuclear fusion occurs very rapidly and acts as a refrigerant in the core[18]. Simultaneously the neutrino pair emission rate removes energy more rapidly than the core contraction supplies it[19]. The core then is imploding in almost free fall. The neutrino pair emission processes keep it sufficiently cool[20] that the decreasing gravitational potential energy is converted into inward radial velocity rather than thermal energy. If nothing were to stop the collapse such a star would probably not heat greatly as it approached its Schwarzschild radius. But at least for the lighter stars ($M \lesssim 2M_{\odot}$) nuclear repulsion can stop the collapse and the resulting neutron star is formed with internal kinetic energies of about 0·1 Mc^2. For the classical heat capacity of free neutrons this implies a temperature $T \sim 10^{12}$°K. Aside from the very problematical initial moments of the universe this is the hottest conjectured temperature in any object in the universe.

It has been argued that 10^{12}°K is the highest temperature that can ever be achieved anywhere[21,22]. This thesis is based upon some considerable successes of statistical models in elementary particle physics: the centre of mass energy in a very high energy particle collision is confined for about 10^{-23} sec in a very small interaction volume. This is conjectured to be long enough for thermal equilibrium to obtain. Various features of such collisions, for example the exp $(-p_{\perp}c/kT)$ distribution of transverse momenta $p_{\perp}$ for emitted particles, support such a picture. Non-relativistically $T \sim E_{CM}$; for dominant particle creation, as in black body radiation, $T \sim E_{CM}^{1/4}$; experimentally $T \sim T_0 - KE_{CM}^{-1}$ with $T_0 \sim 1{\cdot}2 \times 10^{12}$°K. Hagedorn[22] has proposed that such a limit, $T < T_0$, follows very naturally from the enormous (exponential) proliferation of new particles and resonances with increasing mass. The number of degrees of freedom in his model increases so rapidly that for thermal equilibrium the energy density ε approaches $\varepsilon = AT_0(T_0 - T)^{-1}$ so that $T_0 = 1{\cdot}2 \times 10^{12}$°K is never exceeded.

No matter what the initial high temperature of a neutron star it will cool by neutrino emission extremely rapidly—to well below 10^{10}°K in less than a second and to near 10^{8}°K within 10^3 years. The very peculiar effects of the huge neutron star magnetic fields upon stellar atmosphere opacity and radiation processes will probably greatly accelerate the stellar cooling, but no quantitative calculations of the cooling rates have been presented.

There is abundant circumstantial evidence that pulsars are rotating neutron stars. The youngest pulsar, the central star of the Crab Nebula, is a remnant of a thousand year old supernova explosion; this neutron star—once the hottest object in the universe—has now cooled so that its interior temperature is of order 10^{8}°K. This is about ten times hotter than the solar centre but we shall see that in terms of phenomena it is now probably one of the coldest places in the universe.

5. FORMS OF SUPERDENSE MATTER IN STARS[23]

The régimes for various forms of superdense matter are sketched in *Figure 7*. At $T \sim 10^{8}$°K and below it is generally crystalline or (above $\rho \sim 10^{14}$ g cm^{-3}) mainly neutron superfluid. A walk into a white dwarf from its surface to its centre begins in a gaseous atmosphere and ends, often, in a solid with $T_c \sim 10^{7}$°K and $\rho_c \sim 10^5$ to 10^8 g cm^{-3}. A similar stroll into a neutron star passes through a much more varied environment [*Figure 8*]. The atmosphere is a few centimetres thick. A few metres below the surface the electrons are highly degenerate and very good thermal conductors so that the star has a constant temperature from that point up to the centre 10^6 cm away. The nuclei arrange themselves into a crystal so that the star has a solid crust usually a kilometre or so thick. The crust disappears because the nuclei do at $\rho \sim 5 \times 10^{13}$ g cm^{-3}. Below this the predominant neutrons

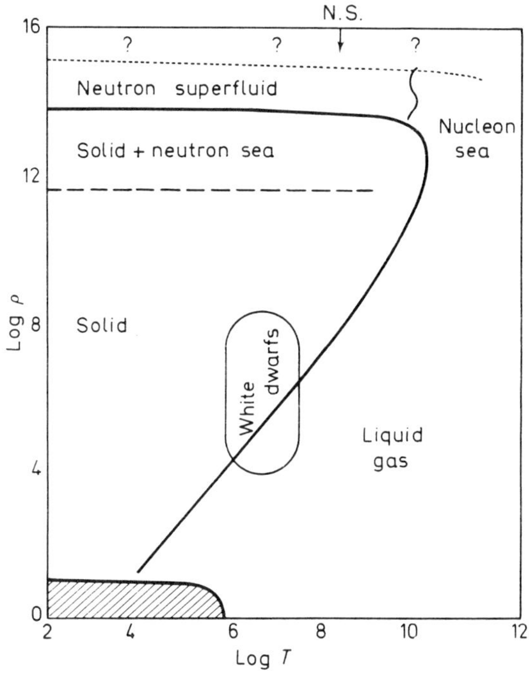

Figure 7. Forms of matter for extreme regions of temperature and pressure. White dwarf core regions are explicitly designated. The crosshatched region is about that attainable in the laboratory.

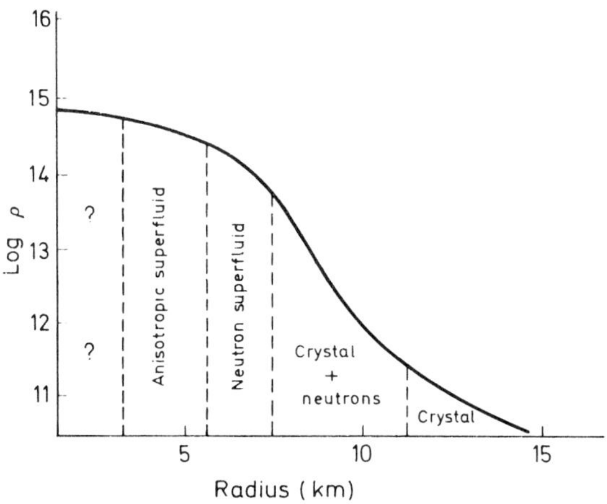

Figure 8. Forms of matter and density in parts of a light neutron star.

are probably a superfluid analogous to the Bardeen–Cooper–Schrieffer electron paired superconductor except that the paired neutrons carry no charge. At about $\rho \sim 3 \times 10^{14}$ g cm^{-3} this superfluid is predicted to take an anisotropic form which has not yet been seen in any laboratory superfluid. As ρ approaches 10^{15} g cm^{-3}, which will happen at the centre of many and perhaps even most neutron stars, the form of the matter, its constituents and the equation of state are not yet known.

6. CRYSTALLIZATION OF SUPERDENSE STELLAR MATTER[24–28]

The core of white dwarfs and the outer regions of neutron stars consist of qualitatively similar material—a highly degenerate electron sea in which are embedded nuclei. A white dwarf's history has probably been such that the matter within it has not yet burned to iron but all helium has been converted to heavier elements; therefore, for the embedded nuclei $2 < Z < 26$. The presumed genesis of neutron stars suggests that all the matter has its lowest free energy; the resulting $Z \geqslant 26$ depends upon the local electron Fermi energy.

When nuclei are embedded in a degenerate electron sea their coulomb fields are screened by the surrounding electrons. In normal laboratory matter almost all of the nuclear charge is completely screened so that the correlation energy between nearest neighbour nuclei is only a few electron volts, characteristic of that for single net charges. The relevant interaction energy

$$\frac{(Ze)^2}{r_z} \rightarrow \frac{(Ze)^2}{r_z} \exp\left(-\frac{r_z}{r_{sc}}\right) + \text{Osc.} \tag{13}$$

where r_z is the separation between neighbouring nuclei, r_{sc} the screening radius (Debye radius), and Osc. a small oscillating term. In laboratory matter $r_{sc} \ll r_z$, but as matter is squeezed to much higher densities the Fermi energy increases so much that the electron kinetic energy exceeds not only the electron–electron energy but also the larger interaction between the electrons and the coulomb fields of the nuclei. (Non-relativistic Fermi energies grow like $\rho^{\frac{2}{3}}$ while interaction energies increase less rapidly, like $\rho^{\frac{1}{3}}$.) Therefore with increasing density the nuclear coulomb fields exert an ever smaller perturbation upon the high momentum electrons around them. When the electron Fermi energy exceeds 10^6 eV ($\rho \lesssim 10^7$ g cm^{-3}), the screening radius $r_{sc} \approx 6\, r_z\, Z^{-\frac{1}{3}}$. Even for iron ($Z = 26$) nearest neighbours and even next nearest neighbours see essentially the unscreened coulomb fields of surrounding nuclei. Over short distances the electron sea behaves like an unpolarizable uniform negative background in which the bare nuclei are embedded. The correlation energies among nuclei become quite enormous. Thus for iron with $\rho \sim 10^8$ g cm^{-3} (the centre of a rather heavy white dwarf) the repulsive coulomb energy is about 10^6 eV, about a million times greater than that for iron at normal densities. It is this enormous correlation energy which causes crystallization of superdense matter even at a temperature of many hundreds of millions of degrees.

If the electron screening is neglected completely, the pure coulomb interaction among nuclei gives a system which has been worked on extensively for thirty five years. At zero temperature the nuclei form a body-centred cubic lattice whose thermal properties have been extensively tabulated. Because the theories of melting are so far from being definitive, mostly because the solid/liquid transition does not greatly change the properties of matter, the melting temperature of such a lattice is not well known. Dimensionally

$$kT_m = (1/\Gamma)(Ze)^2/r_z \tag{14}$$

where Γ is a pure number, independent of the charge Ze and of the nuclear separation r_z. Thus if Γ is known for any coulomb lattice it is known for all. Most substances melt when their interaction energy is about one per cent of the particle interaction energy so that $\Gamma \sim 10^2$. A rough estimate[28] applying Lindemann's rule to the lattice excitations gives $\Gamma \sim 60$. Van Horn[26] estimates $\Gamma \sim 52$ for a theoretical model and 150 from a more precise application of Lindemann's rule[29]. A computer experiment[30] on a classical coulomb gas of 32 particles indicated an instability at $\Gamma = 126$ which resembled the finite particle analogue of a phase transition. Then for white dwarf matter

$$T_m \sim 10^7 (\rho/10^6)^{\frac{1}{3}} (Z/8)^{\frac{5}{3}} \,^\circ\mathrm{K} \tag{15}$$

Typical white dwarf centres have $T_c \sim 10^7\,^\circ$K, close to their melting temperatures. An unanswered question which may have some observational consequences for white dwarfs is whether in this high density régime the melting transition remains a first order one and the size of the heat of transition if it is. A significant heat of fusion ($\sim kT_m$ per nucleus) might observably retard white dwarf cooling rates in the transition region.

7. NEUTRON STAR STRUCTURE

The composition of matter in some density régimes relevant to neutron star interiors is given[31] in *Figure 9*. (It is assumed that the matter is in its lowest energy state.) The numerical density of nuclei, n_z, varies only between 10^{33} and 10^{34} cm^{-3} as ρ varies from 10^{11} to 5×10^{13} g cm^{-3}. The nuclei arrange themselves in a lattice whose melting temperature is given[32] in *Figure 10*. Also plotted is the crystal Debye temperature. Since the expected temperature inside the neutron star is at most a few times $10^8\,^\circ$K, the outer neutron star layer is very far below its melting temperature; it is also a quantum solid with very little heat content. The neutron star 'crust' is much more solid than that of the earth: its terrestrial analogue would be a thick shell of iron at a temperature of at most a few tens of degrees. Inhomogeneities from various frozen-in nuclear species and possible phase separations may give a rich 'geology' to such cold crusts.

At $\rho \sim 5 \times 10^{11}$ g cm^{-3} free neutrons coexist with the nuclei. With increasing density the fraction of neutrons which are free increases until $\rho \sim 5 \times 10^{13}$ g cm^{-3}. There all nuclei disappear and the neutrons constitute a degenerate neutron sea whose Fermi energy $\sim$20 MeV. Pairs of neutrons at the top of the Fermi sea attract each other. Relevant neutron–neutron

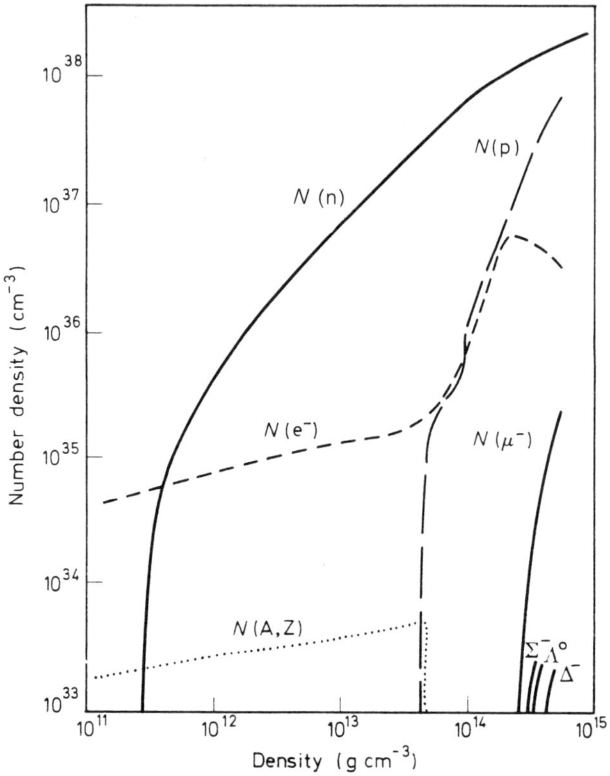

Figure 9. Constituents of 'cold' matter as a function of density (ref. 31).

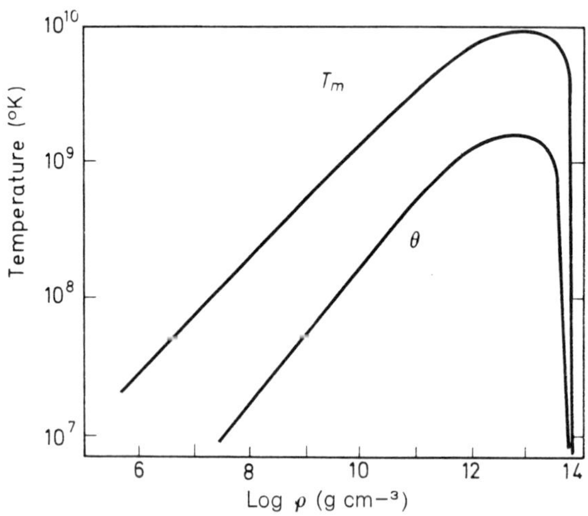

Figure 10. Melting temperature T_m and Debye temperature θ for matter at various densities. Both temperatures drop discontinuously where $N(A, Z)$ does.

scattering phase shifts are represented in *Figure 11* as a function of the wave number of either neutron in the centre of mass system. The $^1S_0$ phase shift is attractive for $k_{CM} < 1{\cdot}4 \times 10^{13}$ cm^{-1} corresponding to $\rho \lesssim 1{\cdot}5 \times 10^{14}$ g cm^{-3}. At higher densities this phase shift becomes repulsive and only the $^3P_2$ interaction remains attractive. According to the Bardeen–Cooper–

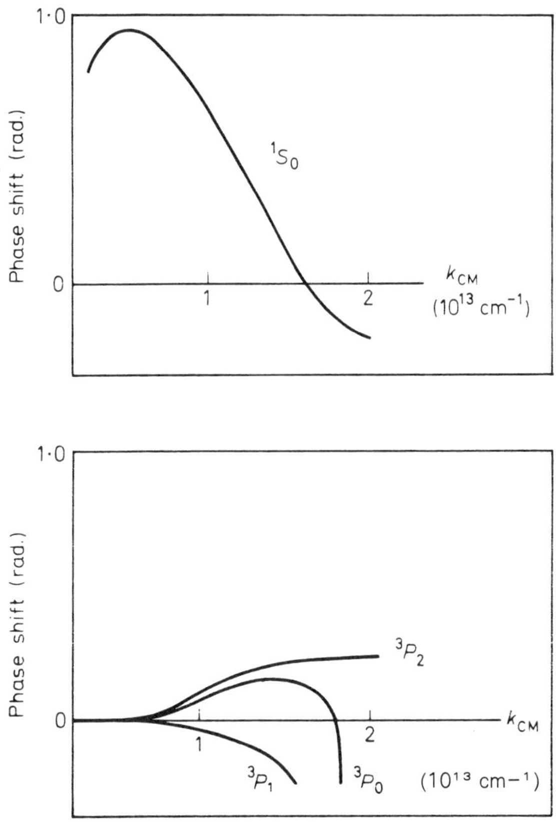

Figure 11. Phase shifts for neutron–neutron scattering as obtained from laboratory proton–proton scattering experiments. k_{CM} is the wave number of either neutron in the centre of mass frame.

Schrieffer theory of electron superconductivity any attractive phase shift at the top of the Fermi sea is sufficient to give a gap in the single particle excitation spectrum. The gap gives superfluid (but not superconducting) properties to the neutrons[33–37]. An estimate of the transition temperature into this state[35–37], based upon the phase shifts of *Figure 11*, is given in *Figure 12*. The computed superfluid transition temperatures are much more than an order of magnitude greater than the expected temperature within the neutron star. Therefore the stellar interior should be filled with a very 'cold' neutron superfluid. Below $\rho \sim 1{\cdot}5 \times 10^{14}$ g cm^{-3} such a superfluid has conventional properties like that of superfluid ^4He. At higher densitie where the neutron pairing is attractive only in a $J = 2$ state the superflui gap is anisotropic; $\Delta \sim (\frac{1}{3} + \cos^2 \theta)$. There will be an anisotropic compressibility associated with this gap. The direction $\theta = 0$ is determined by internal stresses within the star to be in the radial direction.

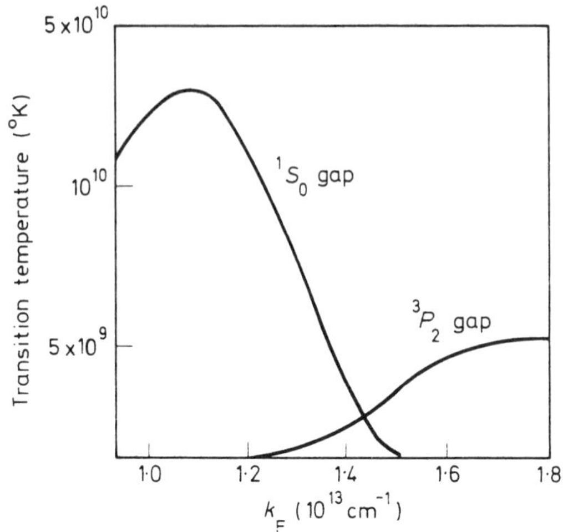

Figure 12. Estimated transition temperatures into the superfluid state for a degenerate neutron sea as a function of k_f, the wave number at the top of the sea (refs. 35–37).

Coexisting with the superfluid neutrons are degenerate seas of electrons and protons, numerically a few per cent as dense as that of the neutrons. The Fermi energy of the degenerate electrons must be just sufficient to prevent both the decay $n \rightarrow p + e + \bar{v}$ and its inverse $e + p \rightarrow n + v$, otherwise the neutron sea would be unstable. The needed electron Fermi energy, E_F, is about 10^2 MeV; because the electrons are light and relativistic, their numerical density, and that of the protons, is much less than that of the neutrons by a factor $(E_F/2m_nc^2)^{\frac{3}{2}}$. These highly relativistic electrons are an extremely good electrical conductor (neutron star internal magnetic fields have a decay time $\sim 10^{13}$ years)[38]. The protons are likely to be superconducting. Although the few protons can be ignored in describing the interaction between neutron pairs at the top of their Fermi sea the converse is not true. If, however, the neutron sea polarizability is neglected and the free proton–proton interaction is used, predicted proton superconducting transition temperatures are in excess of 10^9°K. The superconducting protons do not expel the large ($\sim 10^{12}$ gauss?) neutron star magnetic fields but rather form a type II superconductor which channels the magnetic field within it[39].

In the lighter neutron stars the superfluid-superconducting matter extends to the centre. In the heavier ones the central density reaches and exceeds 10^{15} g cm^{-3}. If interactions are ignored the lowest energy state compatible with the Pauli principle consists of electrons, protons and neutrons together with μ-mesons, hyperons, resonances etc. The interaction energie between particles are comparable to the mass differences and even the enti rest masses. The properties of such a relativistic conglomerate are unknown and may be quite peculiar.

8. LOW TEMPERATURE PHENOMENA IN NEUTRON STARS

The terrestrial laboratory analogue of a neutron star is a very cold thick spherical iron shell filled with superfluid helium and a possible unknown central core. The liquid helium would be at less than a few $\times 10^{-2}$ °K, fantastically cold, in terms of its transition temperature. The neutron fluid is almost incompressible; its sound speed is $10^{-1}c$. Therefore at $T \sim 10^{8}$ °K there are almost no phonons, the fraction of the fluid which in the canonical nomenclature is 'normal' (ρ_n/ρ), being only 10^{-14}. The analogous ratio for laboratory superfluid helium is attained at well below 10^{-2} °K. The neutron star interior, if it is indeed such a 'cold' superfluid, has a very small heat capacity which resides mainly in the degenerate electrons, about 10^{-6} the heat capacity of a classical neutron gas of equal density. At 10^{8} °K the neutron star interior is phenomenologically the coldest known place in the universe.

The association of rotating neutron stars with pulsars offers the possibility of actually observing in neutron stars phenomena unique to very low temperature rotating superfluids. The precision with which pulsar periods can be measured, especially in the short period Crab and Vela pulsars, permits the detection of angular velocity variations of less than 1 in 10^{10} (corresponding to, say, a change in moment of inertia caused by a variation in shape or radius of 10^{-4} cm). Discontinuous increases have been observed in the angular frequency (Ω) of the Vela pulsar ($\Delta\Omega/\Omega = 2 \times 10^{-6}$)[40–41] and the Crab pulsar ($\Delta\Omega/\Omega = 4 \times 10^{-9}$)[42] which are very likely the result of some sudden event which slightly speeds up the spinning crust (starquake?). As the increased crust angular velocity is shared with the much more massive superfluid neutron interior Ω tends to return toward what it would have been without the discontinuity. The time scale for this 'healing' of the crust–interior angular frequency mismatch is of order a few years in the Vela pulsar. These long spin up times are characteristic of those calculated for cold superfluid interiors[43]. The highly conducting crust is coupled to the interior electron–proton sea by the very strong magnetic field which is presumed to pervade the entire star; therefore the electrons and protons co-rotate with the crust. If the protons and neutrons were not superfluid the interaction time between them would be only of order 10^{-15} sec because of proton–neutron collisions. But such transfers of momentum to individual neutrons is essentially forbidden in the superfluid state. However, the rotating neutron superfluid must flow irrotationally: it can mimic rigid body rotation through a paraxial array of moving quantized vortex lines. The fluid velocity satisfies $\nabla \times \mathbf{v} = 0$ except at the centre of each vortex. For rotating neutron star interiors the core radii $\sim 10^{-12}$ cm and the separation between quantized vortex lines is $\sim 10^{-2}$ cm. Only when averaged over many vortex lines can the average fluid velocity $\langle \mathbf{v} \rangle$ satisfy the usual rigid rotation relation $\nabla \times \langle \mathbf{v} \rangle = 2\boldsymbol{\Omega}$. Those neutrons within the vortex cores can still absorb momenta from p–n or e–n collisions, but they are only 10^{-18} of all the neutrons present. When both protons and neutrons are cold superfluids the main mechanism for coupling between the rotating electron–proton sea and the rotating neutrons is collisions of the relativistic electrons with the neutron magnetic moments. The characteristic time for interaction depends sensitively upon the temperature and the gap energy but is characteristically of order a year for $T \sim 10^8\,^\circ$K. If the protons are not superfluid the interaction time is reduced by about 10^6. The above model for the apparent 'healing time' after the pulsar frequency jumps is tenable only with superfluid neutron interiors.

A further possibility for observing effects of the cold rotating neutron superfluid is through the excitation of normal modes of the vortex lattice array. In the co-rotating reference frame the superfluid free energy is minimized (for cylindrical geometry) by a regular triangular lattice array at rest with respect to the container walls. There exists one very slow excitation mode of the array in which each vortex line remains parallel to the (cylinder) axis but there is a redistribution of density of vortices and superfluid angular momentum. Tkachenko[44] has shown that the displacements, **s**, of the vortex lines from their equilibrium positions satisfy a wave equation

$$\ddot{\mathbf{s}} - C_v^2 \nabla^2 \mathbf{s} = 0 \tag{16}$$

with wave velocity

$$c_v = (\hbar\Omega/8m_n)^{\frac{1}{2}} \sim 1 \text{ cm sec}^{-1} \tag{17}$$

(m_n is the neutron mass). Dyson[45] has shown that a necessary subsidiary condition is

$$\nabla \times \dot{\mathbf{s}} = -2\Omega\nabla\cdot\mathbf{s} \tag{18}$$

The Tkachenko–Dyson equations give a fundamental mode[46] whose period τ is proportional to the neutron star radius ($R \sim 10^6$ cm)

$$\tau \sim 140R/\Omega^{\frac{1}{2}} \tag{19}$$

This period is close to one of a few months reported[47] for a very small wobble in the Crab pulsar ($\Omega \sim 200 \text{ sec}^{-1}$). No other normal mode seems to have such a long period, but free nutations from small deformations supported by crust rigidity, or planet induced motions, or even an artifact of the necessarily marginal data analysis may account for the observations. It would be amusing if a new low temperature 'sound' phenomenon not yet seen in the laboratory were first discovered in the coldest natural place in the universe, the $10^8\,^\circ$K interior of a neutron star.

REFERENCES

1 R. Penrose, *Phys. Rev. Letters*, **14**, 57 (1965).
2 S. Hawking, *Phys. Rev. Letters*, **15**, 689 (1965).
3 S. Hawking and G. Ellis, *Phys. Rev. Letters*, **17**, 246 (1965).
S. Hawking and G. Ellis, *Astrophys. J.* **152**, 25 (1968).
4 S. Hawking and D. Sciama, *Comments on Astrophysics*, **1**, 1 (1969).
5 R. Wagoner, W. Fowler and F. Hoyle, *Astrophys. J.* **148**, 3 (1967).
6 Ya. Zel'dovich, L. Okun' and S. Pikel'ner, *Soviet Phys. Uspekhi*, **8**, 702 (1966).
7 R. Partridge and D. Wilkinson, *Phys. Rev. Letters*, **18**, 557 (1967).
8 C. Misner, *Nature, London*, **214**, 40 (1967).
9 R. Omnès, *Journal de Physique*, **30**, Suppl. C3 (1969).

[10] C. Bouchiat, ORSAY preprint.
[11] W. Thirring, CERN preprint.
[12] C. Hayashi, R. Hoshi and D. Sugimoto, *Progr. Theor. Phys.—Suppl.* **22** (1962).
[13] D. Lynden-Bell and R. Wood, *Mon. Not. R. Astron. Soc.* **138**, 495 (1968).
[14] Ya. Zel'dovich, *Soviet Phys. JETP*, **14**, 1143 (1962).
[15] S. Bludman and M. Ruderman, *Phys. Rev.* **170**, 1176 (1968).
[16] M. Ruderman, *Phys. Rev.* **172**, 1286 (1968).
[17] S. Coleman, Preprint of Lectures of 1969 International School of Physics, "Ettore Majorana" (1969).
[18] F. Hoyle and W. Fowler, *Ap. J.* **132**, 565 (1960).
[19] H-Y. Chiu, *Ann. Phys. (N.Y.)*, **26**, 364 (1964).
[20] F. Hoyle, W. Fowler, E. Burbidge and G. Burbidge, *Ap. J.* **139**, 909 (1964).
[21] R. Hagedorn, *Suppl. Nuovo Cimento*, **6**, 311 (1968) and *loc. cit.*
[22] R. Hagedorn, *Journal de Physique*, **30**, Suppl. C3, 79 (1969).
[23] M. Ruderman, *Journal de Physique*, **30**, Suppl. C3, 152 (1969).
[24] D. Kirzhnits, *Soviet Phys. JETP*, **11**, 365 (1960).
[25] E. Salpeter, *Ap. J.* **134**, 669 (1961).
[26] H. Van Horn, *Ap. J.* **151**, 227 (1968).
[27] M. Ruderman, *Nature, London*, **218**, 1128 (1968).
[28] L. Mestel and M. Ruderman, *Mon. Not. R. Astron. Soc.*, **136**, 27 (1967).
[29] H. Van Horn, *Phys. Letters*, **28A**, 706 (1969).
[30] S. Brush, H. Sahlin and E. Teller, *J. Chem. Phys.* **45**, 2102 (1966).
[31] W. Langer, L. Rosen, J. Cohen and A. Cameron, *Astrophys. Spa. Sci.* **5**, 259 (1969).
[32] M. Ruderman, *Nature, London*, **218**, 1128 (1968).
[33] A. Migdal, *Soviet Phys. JETP*, **10**, 176 (1960).
[34] V. Ginzburg and D. Kirzhnits, *Soviet Phys. JETP*, **20**, 1346 (1965).
[35] M. Ruderman, *Proceedings* of the Fifth Eastern Theoretical Physics Conference, D. Feldman, Editor. Benjamin: New York (1967).
[36] R. Kennedy, L. Wilets and E. Henley, *Phys. Rev.* **133**, B 1131 (1964).
[37] M. Hoffberg, A. Glassgold, R. Richardson and M. Ruderman, *Phys. Rev. Letters*, **24**, 775 (1970).
[38] G. Baym, C. Pethick and D. Pines, *Nature, London*, **224**, 674 (1969).
[39] G. Baym, C. Pethick and D. Pines, *Nature, London*, **224**, 673 (1969).
[40] P. Reichley and G. Downs, *Nature, London*, **222**, 229 (1969).
[41] V. Radhakrishnan and R. Manchester, *Nature, London*, **222**, 228 (1969).
[42] P. Boynton, E. Groth, R. Partridge and D. Wilkinson, Princeton Pulsar Symposium (6 November 1969) Proceedings.
[43] G. Baym, C. Pethick, D. Pines and M. Ruderman, *Nature, London*, **224**, 872 (1969).
[44] V. Tkachenko, *Soviet Phys. JETP*, **23**, 1049 (1966).
[45] F. Dyson, Private communication.
[46] M. Ruderman, *Nature, London*, **225**, 619 (1970).
[47] D. Richards, G. Pettengill, G. Counselman and J. Rankin, *I.A.U. Circ. No. 2180* (30 October 1969).

TIME ASYMMETRY IN ELECTRODYNAMICS AND COSMOLOGY

J. V. NARLIKAR

Institute of Theoretical Astronomy, University of Cambridge, Cambridge, England

ABSTRACT

The concept of the 'arrow of time' is discussed in relation to thermodynamics, electrodynamics and cosmology. Time symmetric electrodynamics, the absorber theory of radiation and quantum transitions receive attention to develop a working theory applicable to a universe with a perfect future absorber and an imperfect past absorber, such as the steady state cosmological model. It is stated that the approach adopted herein establishes a strong connection between the electrodynamic and cosmological arrows of time, and points a way towards linking these arrows with the thermodynamic one.

1. THE ARROWS OF TIME IN PHYSICS

What do we mean by the arrow of time? In a somewhat vague manner, we may associate it with the sense in which we relate our subjective experiences to the space-time manifold. It would be rash to attempt a precise definition that will satisfy all scientists and philosophers! It is possible, however, to discuss the subject in a rigorous manner in the more restricted field of fundamental physics.

Given a space/time diagram, we can arbitrarily choose a direction along the time axis, by identifying one end with the past and the other with the future*. This enables us to arrange events in a chronological order. Consider a set of events A,B,C,... say, which can occur again and again. If these events always occur with the same chronological order we are able to use these events themselves to fix the time sense. This would not be possible if these events occurred in a random order at different times. Our physical experience reveals events of both types—although it is the events of the former type that are responsible for the arrow of time. What are these events?

These events can be broadly grouped under three categories, thermodynamics, electrodynamics and cosmology. As an example of the events in the first category, consider a hot body and a cold body in contact with each other. If we observe the system 'a little later', we find the hot body a little cooler and the cold body a little hotter than 'before'. In other words, if we measure the temperatures of the two bodies at different instants we can order these instants chronologically. Such events are usually known as

* I am excluding here cosmological models with closed time-like lines.

irreversible events. In electrodynamics, if we observe an accelerated electric charge we notice that it loses energy. Thus the energy flow from the charge can be used to fix a chronological order. In cosmology the expansion of the universe enables us to do the same. If we photograph two nebulae at instants t_1 and t_2, their separation from each other would tell us whether $t_2 > t_1$ or $t_2 < t_1$. Thus events in each category determine an arrow of time.

The three arrows described above appear to arise from quite different sources. We may raise the question: 'Why do these arrows point the way they do?' In other words, is there any connection between these different arrows of time? Before we search for a solution of the more fundamental question: 'Why an arrow of time?' it may well be profitable to investigate whether the arrows of time in thermodynamics, electrodynamics and cosmology are mutually related.

Opinions differ as to how deep this connection is. Some physicists believe that the connection, if any, is very superficial. I want to discuss the opposite point of view, particularly in connection with electrodynamics and cosmology.

2. TIME SYMMETRIC ELECTRODYNAMICS

Returning to the problem of the radiating charge, we may enquire as to the origin of the time asymmetry. Maxwell's equations are time symmetric. Corresponding to any solution of these equations we may obtain another by changing t to $-t$. Now the problem of radiating charge is solved by means of retarded solutions of Maxwell's equations. If instead we use advanced solutions we would describe a charge *receiving* energy. Mathematically both the solutions (and any linear combination of them) describe the electromagnetic fields associated with an accelerated charge. Only the retarded solution is taken to represent the actual situation. So long as there is no other direction of time we cannot distinguish between the pure retarded and pure advanced solutions. Given another arrow of time, however, say from cosmology, the distinction between the retarded and advanced solutions becomes real. Why do we choose retarded solutions? The usual answer is 'causality'. However, to say that we choose retarded solutions in order to conform with causality does not take us any further in our understanding of the electrodynamic arrow of time. So long as the choice of retarded solutions is made on such arbitrary grounds, we will not be able to make progress in this direction. It appears that to achieve this we have to resort to a different theory of electrodynamics which does not permit so much freedom of choice.

Such a theory already exists in literature, and in its crude form even preceded Maxwell's theory. This is the theory of direct interparticle action. In this theory electric charges interact directly with one another and the electromagnetic field has no independent existence. Mathematically we may describe the theory through an action principle due to Fokker[1]. The action is given by

$$S = -\sum_{a} \int m_a \mathrm{d}a - \sum_{a<b}\sum e_a e_b \iint \delta(s_{AB}^2)\, \mathrm{d}a_i\, \mathrm{d}b^2 \tag{1}$$

Here $a, b, c, \ldots$ label the charged particles, e_a, m_a being the charge and mass of particle a. A denotes a typical point on the world line of a, $\mathrm{d}a^i$ denoting the coordinate displacements and da the proper time element at A, along the world line. We have used flat space coordinates with metric $\eta_{ik} = \text{diag.}\,(-1,\ -1,\ -1,\ +1)$. The vector indices are raised and lowered according to η_{ik}. The delta-function in equation 1 has as its argument s_{AB}^2, the square of the interval between A and B, two typical points on the world lines of a and b. Thus A and B interact only if they are connected by a light ray. This is the relativistic generalization of the Newtonian concept of instantaneous action at-a-distance. The double sum in 1 does not include the terms $a = b$; the self-action is therefore absent. We have taken the velocity of light as unity.

The 4-potential generated by charge a at a point X is given by

$$A_i^{(a)}(X) = e_a \int \delta(s_{AX}^2)\, da_i \tag{2}$$

and it satisfies *identically* the gauge condition

$$A^{(a)i}_{;i} = 0 \tag{3}$$

and the wave equation

$$\Box\, A_i^{(a)} = 4\pi j_i^{(a)} \tag{4}$$

where $j_i^{(a)}(X)$ is the current due to charge a at X

$$j_i^{(a)}(X) = e_a \int \delta_4(A, X)\, da_i \tag{5}$$

We may also define the field tensor corresponding to 2

$$F_{ik}^{(a)} = A_{k;i}^{(a)} - A_{i;k}^{(a)} \tag{6}$$

By virtue of relations 3 and 4 this satisfies the Maxwell equations

identically.

This is a fundamental difference between the above theory and the Maxwell field theory. In the latter, Maxwell equations are genuine equations. Thus to a solution of equation 7 we can add *any* solution of

$$F^{(a)iki}{}_{;} = 0 \tag{8}$$

On the other hand in the above theory equation 7 is an identity. The $F^{(a)ik}$ are really given by equations 2 and 6. If we denote the usual retarded solution generated by charge a to be $F_{\text{ret}}^{(a)}$, equation 6 corresponds to

$$F^{(a)} = \tfrac{1}{2}[F_{\text{ret}}^{(a)} + F_{\text{adv}}^{(a)}] \tag{9}$$

Thus there is no freedom of choice; we are *forced* to accept equation 9.

It was this lack of freedom that proved to be a stumbling block to the theory for a long time. Charges are *observed* to produce retarded fields; not the time symmetric fields given in 9. If we consider the equation of motion of charge a, obtained by setting $\delta S = 0$ for a small variation of the world line of a, we find that a force like the Lorentz force acts on a. But it arises from the fields of all other particles in the universe, i.e. from a field

$$\sum_{b \neq a} \tfrac{1}{2}[F_{\text{ret}}^{(b)} + F_{\text{adv}}^{(b)}] \tag{10}$$

The advanced fields in equation 10 appear to present an embarrassment to the theory.

However, an ingenious way out of this difficulty was found by Wheeler and Feynman[2] with their absorber theory of radiation.

3. THE ABSORBER THEORY OF RADIATION

Wheeler and Feynman argued that the interaction is time symmetric only between pairs of particles. The possibility still exists that the collective interaction of a large number of particles could lead to time-asymmetric results. The actual universe is a collection of large numbers of charged particles. An asymmetry in the large scale behaviour of the universe is therefore expected to lead to a local asymmetry in electrodynamics. According to Wheeler and Feynman the requirement to be met by the universe in order to produce the necessary asymmetry is that it shall act as a perfect absorber of all electromagnetic disturbances generated within it. That is, if we set the charge a in motion, all the disturbances related to this should be absorbed by the universe:

$$\sum_b \tfrac{1}{2}[F^{(b)}_{\text{ret}} + F^{(b)}_{\text{adv}}] \to 0 \qquad \text{as} \quad r \to \infty \tag{11}$$

faster than $1/r$, where r is the distance from a.

Since advanced and retarded fields have supports on different branches of the light core, expression 11 implies that

$$\sum_b \tfrac{1}{2}F^{(b)}_{\text{ret}} \to 0, \quad \sum_b \tfrac{1}{2}F^{(b)}_{\text{adv}} \to 0 \qquad \text{as} \quad r \to \infty \tag{12}$$

and hence that

$$\sum_b [F^{(b)}_{\text{ret}} - F^{(b)}_{\text{adv}}] \to 0 \qquad \text{as} \quad r \to \infty \tag{13}$$

However, the quantity in 13 has no sources and hence it satisfies the homogeneous wave equation. Since it vanishes at infinity faster than $1/r$, it must vanish identically. Therefore the field 10 acting on the particle a has the form

$$\sum_{b \neq a} \tfrac{1}{2}[F^{(b)}_{\text{ret}} + F^{(b)}_{\text{adv}}] = \sum_{b \neq a} F^{(b)}_{\text{ret}} + \tfrac{1}{2}[F^{(a)}_{\text{ret}} - F^{(a)}_{\text{adv}}] \tag{14}$$

The first term on the RHS denotes the retarded interaction of all other particles while the second term is the radiative reaction in the form given by Dirac[3]. Thus the theory appears to provide the correct answer if the condition 11 of perfect absorption is satisfied.

There was, however, one snag in the argument developed so far. Wheeler and Feynman applied this to the static Euclidean universe with a uniform distribution of charges. Such a universe satisfies 11 and hence leads to 14. However, this universe is time symmetric and hence cannot distinguish between the past and the future light cones. Hence the time-reversed form of 14, viz.

$$\sum_{b \neq a} \tfrac{1}{2}[F^{(b)}_{\text{ret}} + F^{(b)}_{\text{adv}}] = \sum_{b \neq a} F^{(b)}_{\text{adv}} + \tfrac{1}{2}[F^{(a)}_{\text{adv}} - F^{(a)}_{\text{ret}}] \tag{15}$$

is also valid.

Wheeler and Feynman recognized this and argued that the asymmetry which allows one to distinguish between 14 and 15 comes from thermodynamics. The thermodynamic arrow of time is ascribed to asymmetrical initial conditions. It is these conditions which make 14 highly probable and 15 highly improbable in the sense of statistical mechanics. Thus according to Wheeler and Feynman the electrodynamic and thermodynamic arrows of time are strongly related.

A new feature into this argument was introduced by Hogarth[4] many years later. He pointed out that the above argument ignores the cosmological time arrow. In an expanding universe the past and future light cones do not behave symmetrically where absorption is concerned. Electromagnetic waves going into the future are red-shifted while those going into the past are blue-shifted. Also, in some models of the universe the matter densities in the past and the future are different. Thus one cannot argue that if 14 holds so would 15 or vice versa. Hogarth's calculations show that in certain models of the universe, e.g. the open Friedmann models, 14 does not hold but 15 does. Also in the steady state model 14 holds but not 15. In other words, the observed alignment of electrodynamic and cosmological time arrows could be explained only in certain cosmological models, and not others.

Hogarth's argument had to be improved in two ways, however. First, he had taken over the Fokker action 1 in flat space and used it in the Robertson–Walker cosmological spaces of the expanding cosmological models. Since these spaces are conformally flat, and the electromagnetic equations are conformally invariant, the above step was justified. For completeness it was necessary, however, to rewrite equation 1 in the curved space of general relativity. Secondly, the use of the refractive index, which plays an important part in the calculation, involved thermodynamics which Hogarth was trying to avoid. In a later work Hoyle and Narlikar[5] took account of both these points, although their conclusion did not differ from Hogarth's. I summarize below the main points of the argument.

Suppose we want to arrive at 14. Then the condition to be satisfied by the universe is that it must be a perfect absorber of all disturbances connected with the motion of a along the future light cone of a. That is

$$\sum_b F^{(b)}_{\text{ret}} \to 0 \qquad \text{as} \quad r \to \infty \tag{16}$$

faster than $1/r$. However, expression 16 requires that we have a definite process of absorption. If we stick to pure electrodynamics this is provided by radiative damping. Now the damping term has the correct sign if 14 holds, and the wrong sign (implying growth of the disturbance) if 15 holds. To avoid ambiguity therefore it is necessary to assume that 14 holds to begin with and then to show that the theory is consistent with this assumption. Thus one has to work within a self-consistent cycle of argument.

When the charge a is set in motion, its retarded field sets other charges in motion along its future light cone. Since 16 implies a damping of the retarded field of a as the wave goes away from it, this also implies that the advanced fields arising from the motion of all other charges b provoked by the retarded field of a must also be damped. Hence we have

$$\sum_b F^{(b)}_{\text{adv}} \to 0 \qquad \text{as} \quad r \to \infty \tag{17}$$

faster than $1/r$. Notice that 17 arises from 16 which is the more basic condition. If we were discussing the consistency of 15 we would require 17 to hold along the past light cone of a and then deduce 16. Also, for 17 to follow from 16 it is essential that the motion of charge a be bounded in space-time, a requirement not met by the 'run-away' solutions. Such solutions are therefore not possible in this theory.

From 16 and 17 we then arrive at 14 through steps similar to those used by Wheeler and Feynman. It is instructive, however, to see the precise part played by the universe.

Suppose we are in a universe in which 14 holds. Consider a charge a at time $t = 0$. The future light cone of a has $t > 0$, and the past light cone $t < 0$. Now we rewrite the LHS of 14 in the following form

$$\sum_{b \neq a} \tfrac{1}{2}[F^{(b)}_{\text{ret}} + F^{(b)}_{\text{adv}}] = \tfrac{1}{2} \sum_{\substack{b \neq a \\ t<0}} F^{(b)}_{\text{ret}} + \tfrac{1}{2} \sum_{\substack{b \neq a \\ t>0}} F^{(b)}_{\text{adv}} \tag{18}$$

The second term on the RHS of 18 is the advanced contribution of all the particles b on the future light cone of a, which are excited by the retarded disturbance from a. This contribution arrives instantaneously as the retarded wave leaves a. We shall call it the 'response' of the universe and denote it by $R^{(a)}$. A comparison with the RHS of 14 shows that

$$\begin{aligned} R^{(a)} &= \tfrac{1}{2} \sum_{\substack{b \neq a \\ t<0}} F^{(b)}_{\text{ret}} + \tfrac{1}{2}[F^{(a)}_{\text{ret}} - F^{(a)}_{\text{adv}}] \\ &= \sum_{\substack{b \neq a \\ t<0}} \tfrac{1}{2}[F^{(b)}_{\text{ret}} - F^{(b)}_{\text{adv}}] + \tfrac{1}{2}[F^{(a)}_{\text{ret}} - F^{(a)}_{\text{adv}}] \\ &= \sum_{\substack{b \\ t \leqslant 0}} \tfrac{1}{2}[F^{(b)}_{\text{ret}} - F^{(b)}_{\text{adv}}] \end{aligned} \tag{19}$$

In 19 we have used the fact that $F^{(b)}_{\text{adv}} = 0$ for $t < 0$.

Equations 19 denotes the response required from the universe in order to arrive at 14 which is in conformity with observations. As mentioned before the steady state universe provides the appropriate response, but not the usual big bang models.

4. QUANTUM CONSIDERATIONS

We have so far considered classical electrodynamics. Can these ideas be extended to quantum electrodynamics? Recent work[6] shows that it is possible to have a quantum theory of direct interparticle action. The shortage of space does not permit a detailed discussion of these new ideas. I will be content with a brief mention of salient features.

The quantum analogue of the radiating electron occurs in the spontaneous transition of an atomic electron. While the induced transitions are both upward and downward with symmetrical rates, the spontaneous transitions are always downward, accompanied by emission of radiation. Can this be related to the expansion of the universe?

In the Maxwellian quantum electrodynamics the phenomenon of spontaneous transition is explained by first quantizing the electromagnetic field. The vacuum state in this theory has non-trivial properties as a result of the field quantization rules, and these play an important part in arriving at the spontaneous transition probability. For example, if the electron is making transitions between two energy levels E_m and E_n, $E_n > E_m$, in the presence of a field of n quanta of frequency $\nu = (E_n - E_m)/h$ the probabilities of downward and upward transitions are related by

$$\frac{P(E_n \to E_m)}{P(E_m \to E_n)} = \frac{n+1}{n} \tag{20}$$

In the direct particle theory there are no fields and hence no field quantization or quanta in the above sense. So far as induced transitions are concerned the theory gives the same result as field theory. But what about spontaneous transitions? Returning to the classical expression 14, the first term on the RHS is responsible for induced transitions whereas the second term leads to spontaneous transitions. Since the second term is part of the response $R^{(a)}$ of the universe, it is to the universe that we must look for an explanation of spontaneous transitions.

It turns out that the calculations are most easily performed by using the path integral method of Feynman[7,8]. This provides the quantum analogue of 14 and 19. For example, the response of the universe is non-zero even when there is no net electric field from all other particles. Thus the electron is influenced by the universe and makes a transition (spontaneously).

The physical reason for this may be understood in the following way. A downward spontaneous transition is accompanied by the emission of an electromagnetic wave which causes transitions in the future absorber. An expanding universe acts as a 'cold environment', with all absorber particles in their ground states. The wave emitted by the source particle a is absorbed by the absorber particles which make upward transitions, induced by this wave. Complete absorption is necessary if the whole cycle is to work self-consistently. It is also easy to see why spontaneous upward transitions do not occur. These would require augmentation instead of absorption of the electromagnetic waves emitted by the source. The universe in the future light cone does not permit this—since the particles are mostly in their ground state and cannot jump down farther to enhance the incoming radiation.

Apart from absorption, the universe also plays an important part through dispersion. If we Fourier analyse the outgoing wave from the source, the different components travel at different velocities through the intergalactic medium. The intergalactic density and other parameters are such that large phase differences are built up between neighbouring frequencies, with the result that the different components become randomly phased. This random phasing is usually *assumed* in field quantization. In the above work this extra assumption is not required.

I have so far not mentioned thermodynamics explicitly. But a connection between the thermodynamic and cosmological arrows now begins to appear. An expanding universe acts as a sink—a fact which appears to be connected with the asymmetry of upward and downward transitions considered above. This asymmetry leads to a result like 20 even in the direct particle theory. It is well known that the law of black body radiation follows from 20. As yet these ideas are somewhat tentative, but they serve to indicate that a connection between thermodynamics and cosmology may be built up.

5. CONCLUSION

The ideas described above work in a universe with a perfect future absorber and an imperfect past absorber. Of the well known cosmological models only the steady state model meets this requirement. If we regard the present astronomical evidence as unambiguously against the steady state theory we must abandon this approach altogether. On past record of astronomical observations, and with the present uncertainties attached to the recent data, such an unequivocal conclusion is not justified[9].

On the other hand we may ask whether there are any gains in the above approach. As discussed above this approach establishes a strong connection between the electrodynamic and cosmological arrows of time, and also points a way towards linking these arrows with the thermodynamic one. Such a connection cannot be established with Maxwellian electrodynamics which allows too much freedom of choice. Moreover, as the recent work[10] shows this approach is better than the usual quantum electrodynamics in handling the so-called radiative corrections.

I end with a few remarks on the question 'why an arrow of time'. If we were able to establish a strong connection between the different arrows of time, physical, biological etc., we could then argue in the following way. An absolute direction is immaterial: the important concept is the relative orientation of the different arrows. For example, if we can say 'we grow old because the universe expands', then this is equivalent to saying that we grow young as the universe contracts! The question 'Why do we grow older, not younger?' without reference to other arrows of time has no meaning.

REFERENCES

[1] A. D. Fokker, *Z.Phys.* **58**, 386 (1929).
A. D. Fokker, *Physica*, **9**, 83 (1929); **12**, 145 (1932).
[2] J. A. Wheeler and R. P. Feynman, *Rev. Mod. Phys.* **17**, 157 (1945).
[3] P. A. M. Dirac, *Proc. Roy. Soc. A*, **167**, 148 (1938).
[4] J. E. Hogarth, *Proc. Roy. Soc. A*, **267**, 365 (1962).
[5] F. Hoyle and J. V. Narlikar, *Proc. Roy. Soc. A*, **277**, 1 (1963).
[6] F. Hoyle and J. V. Narlikar, *Ann. Phys. (N.Y.)*, **54**, 207 (1969).
[7] R. P. Feynman, *Rev. Mod. Phys.* **20**, 367 (1948).
[8] R. P. Feynman and A. R. Hibbs, *Quantum Mechanics and Path Integrals*. McGraw-Hill: New York (1965).
[9] F. Hoyle, *Proc. Roy. Soc. A*, **308**, 1 (1968).
[10] F. Hoyle and J. V. Narlikar, *Nature, London*, **222**, 1040 (1969).

COSMIC EVOLUTION AND THERMODYNAMIC IRREVERSIBILITY

DAVID LAYZER

Department of Astronomy, Harvard University, Cambridge, Massachusetts

ABSTRACT

This paper seeks to explain and relate three macroscopic arrows of time: the thermodynamic arrow, defined by entropy-generating processes in closed systems, the historical arrow, defined by information-generating processes in certain open systems, and the cosmological arrow, defined by the cosmic expansion.

INTRODUCTION

SEVERAL speakers at this conference have referred to the zeroth law of thermodynamics. Cosmology also has its zeroth law. It states that cosmologists are fermions: no two of them can be in the same state of mind at the same time. Earlier in this conference Dr Narlikar explained how the expansion of the universe introduces an asymmetry into local descriptions of radiation processes. He argued that the use of retarded rather than advanced solutions of Maxwell's equations to describe radiation processes supports steady-state cosmology but contradicts conventional relativistic cosmology. On the other hand, Dr Narlikar's theory does not relate the cosmic expansion directly to thermodynamic irreversibility. It ties the electromagnetic arrow of time firmly to the cosmological arrow, but leaves the thermodynamic arrow suspended in mid-air.

The views I wish to elaborate are rather different. First, I believe that the electromagnetic arrow has nothing directly to do with cosmology, but is determined by the thermodynamic arrow in the manner elucidated by Einstein in 1909[1]. Einstein pointed out that the retarded and advanced descriptions of radiation processes occurring in any finite region of space-time are completely equivalent, but the auxiliary conditions in the two descriptions differ in kind. In the retarded description all *macroscopic* radiation sources must be specified, while in the advanced description the *microscopic* absorption processes must be specified *in detail*. In practice one uses the retarded description because one does not have microscopic information about the absorbing matter. For the same reason, if one wishes to describe an irreversible process such as diffusion or heat conduction at the macroscopic level one must describe it as occurring in the 'forward' direction of time. In short, Einstein's argument demonstrates that the asymmetry of macroscopic radiation processes results from precisely those properties of matter in bulk that give rise to other macroscopically irreversible phenomena.

On the other hand, I believe that current discussions of thermodynamic irreversibility, though essentially correct, are incomplete, and need to be supplemented by cosmological considerations. To explain this view, let me recall—very schematically—the sequence of steps leading from a reversible microscopic description of an N-body system to an irreversible macroscopic description.

THERMODYNAMIC IRREVERSIBILITY

The first step is to introduce statistics. Instead of saying that an N-body system is in a definite state k, specified by $6N$ coordinates and momenta (or by a state vector ψ_k), we specify a probability distribution $\{p_k\}$ or a density matrix ρ. The information associated with this description is defined by

$$I = S_{\text{max.}} - S \tag{1}$$

where the entropy S is defined by

$$S = -\sum p_k \ln p_k \quad \text{or} \quad S = -\operatorname{Tr}\{\rho \ln \rho\} \tag{2}$$

In equation 1 $S_{\text{max.}}$ is the maximum value of S consistent with the macroscopic constraints on the system. It is well known that, by virtue of Liouville's theorem, S is a constant of the motion for an isolated N-body system; dynamical evolution neither creates nor destroys information. Therefore the passage to a statistical description does not disturb the temporal symmetry of the description.

The next step is coarse-graining. One combines the microstates k into aggregates α within which macroscopic variables such as the energy do not vary appreciably. Let $\{\bar{p}_\alpha\}$ denote the set of coarse-grained probabilities and $\bar{\rho}$ the coarse-grained density matrix. With the coarse-grained probability distribution (or density matrix) we associate the coarse-grained entropy $\bar{S}$ and a corresponding information measure $\bar{I}$, which I shall call the *macroinformation.* The *microinformation* I' and the corresponding entropy S' are defined by

$$S = \bar{S} + S' \tag{3}$$

$$I = \bar{I} + I' \tag{4}$$

It is easy to show that the microscopic entropy S' is just the average entropy of the aggregates α*. Since the total information I is constant, any change in the macroinformation of the system is accompanied by an equal and opposite change of the microinformation. But merely introducing a distinction between macroscopic and microscopic information does not of course disturb the temporal symmetry of the description.

* The entropy S_α of an aggregate is defined in terms of the conditional probabilities $p(k/\alpha) = p_k/\bar{p}_\alpha$ and in the averaging process that defines S' the quantities S_α are weighted by the probabilities $\bar{p}_\alpha$. Thus

$$S' = \sum_\alpha \bar{p}_\alpha\{-\sum p(k|\alpha) \ln p(k|\alpha)\} \tag{5}$$

Temporal asymmetry is introduced by the third and most crucial step†. Fifteen years ago van Hove[2] proved that if the Hamiltonian of a closed system satisfies a certain rather general condition and if the off-diagonal elements of the coarse-grained density matrix $\bar{\rho}$ vanish *at a particular instant*, then the coarse-grained entropy $\bar{S}$ will subsequently increase monotonically until it assumes its greatest possible value, when the system will be in a state of thermodynamic equilibrium. The essential feature of van Hove's theorem is that it relates the irreversible increase of entropy in a closed system to a property of the initial state. Several similar theorems have subsequently been proved[3]. All of them state that if microinformation (suitably defined) is initially absent in a closed system, then the macroinformation will subsequently decrease monotonically. The thermodynamic arrow is thus defined by a unidirectional flow of information from macroscopic into microscopic degrees of freedom.

Modern theories of irreversible processes have yielded valuable insight into the detailed mechanisms responsible for the approach to equilibrium, as well as powerful methods for calculating transport coefficients. The following discussion proceeds from the assumption that these theories are essentially correct. But if they are correct, they cannot be complete.

For example, a theory that deals only with closed systems obviously cannot explain why irreversible processes occurring in *different* closed systems should define the same arrow. Again, the coarse-graining procedure, and hence the dividing line between macroscopic and microscopic information, is largely arbitrary in the theories under consideration. Finally, these theories offer no justification for the assumption that microinformation is initially absent. Indeed, the very meaning of this assumption is unclear. The absence of microinformation in a theoretical description does not imply—according to currently held views—that it is unattainable in some objective sense, but only that it is uninteresting or hard to get, or both. This seems to imply that thermodynamic irreversibility is at least in part a psychological phenomenon—a position that most physicists would probably be unwilling to accept, though it has been advocated by some.

BOREL'S ARGUMENT

The recognition that no finite physical system can be truly isolated from the rest of the universe seems at first sight to offer an attractive solution to these difficulties. The Hamiltonian of every ostensibly closed system contains a finite contribution representing the interaction between the system and its environment. So long as we focus attention on a definite system, omitting from our description all dynamical variables referring to particles and fields outside the system, the interaction Hamiltonian is not fully known and hence has a stochastic character. This interaction can have a profound effect on the microscopic state of a system, even if it has been shielded as carefully as possible from outside influences. Thus E. Borel[4] calculated that moving 1 g of matter through 1 cm at the distance of Sirius would destroy

† For sake of simplicity, there is here no discussion of the thermodynamic limiting process—a matter of considerable technical interest which, however, is not directly relevant to the present discussion.

microinformation about the state of a macroscopic gas in 10^{-6} second. Such calculations afford a basis for an objective physical interpretation of the probabilities that figure in statistical descriptions of microscopic systems. They also imply that microscopic reversibility is not a property of finite 'closed' systems but only of the universe as a whole. But for this very reason they provide at best only a partial explanation of the thermodynamic arrow.

Borel's calculation shows that in a 'closed' gaseous system microinformation flows quickly into a very large number of external degrees of freedom. Such calculations can be used to justify assumptions about the initial absence of microinformation of the kind that figure in modern theories of irreversible processes. But the interaction between a system and its environment does not impose a particular direction on this flow of information, much less a common direction for all 'closed' systems. We are still faced with the paradox of a microscopically reversible* universe whose temporal structure, viewed macroscopically, is radically anisotropic.

THE HISTORICAL ARROW

So far I have discussed one aspect of this anisotropy, the thermodynamic arrow (defined by entropy-generating processes in 'closed' systems), and alluded to a second aspect, the cosmological arrow (defined by the cosmic expansion). But I have not mentioned what is perhaps the most conspicuous class of processes serving to define time's arrow: those that generate *information* in open systems. Such processes are central to all biological systems. They play an indispensable role in growth, in biological evolution, and in the phenomena of memory and consciousness. But they are not confined to living systems. A record of the Moon's past is written in its pitted surface; the internal structure of a star, like that of a tree, records the process of aging; and the complicated forms we observe in spiral galaxies reflect the evolutionary processes that shape them. The complex overlapping array of all these evolutionary records defines a third arrow of time, the *historical arrow*.

The thermodynamic arrow points in the direction of increasing entropy; the historical arrow points in the direction of increasing information. Equivalently, we may define the historical arrow through the statement that the present state of the universe (or of any sufficiently large subsystem of it) contains a partial record of the past but none of the future.

Because the thermodynamic arrow refers ideally to closed systems while the historical arrow manifests itself only in open systems, their coexistence presents no problem. Entropy-generating and information-generating processes normally proceed simultaneously and more or less independently in every complex system. In living systems, of course, harmful entropy-producing processes are usually offset by countervailing information-producing processes.

COSMOLOGY AND MACROSCOPIC PHYSICS

I shall now sketch a theory that seeks to relate the three arrows of time—

* I am here neglecting the quantitatively small departures from microscopic time reversibility suggested by experiments on the decay of the K^0 meson.

the thermodynamic, the historical, and the cosmological—to one another, and to derive them all from a common postulate. This postulate—a slightly strengthened version of Einstein's cosmological principle—concerns the spatial structure of the universe. It states that no statistical property of the universe serves to define a preferred position or direction in space; the spatial structure of the universe is statistically homogeneous and isotropic*.

Before discussing the implications of this postulate, I should say a few words about the relationship between cosmology and macroscopic physics. From one point of view, cosmological theories are not essentially different from other physical theories. The cosmologist, like the astrophysicist, must make certain assumptions that cannot be verified directly; he must develop the consequences of these assumptions using relevant physical theories; and he must ultimately make predictions that are explicit enough so that they can be contradicted by appropriate observations or experiments. A good theory, whether in cosmology or any other branch of physics, enables one to draw more or less rigorous and quantitative inferences, in agreement with experience, from simple and natural assumptions—which need not themselves be capable of direct verification. Thus the inaccessibility of stellar interiors to direct observation has not prevented the development of highly credible theories of stellar structure. Similarly, the impossibility of directly verifying postulates about the universe as a whole does not in itself doom cosmology to speculative status; the postulates themselves matter less than the quality and quantity of the inferences that can be drawn from them.

Yet there is a basic difference between cosmology and astrophysics. It stems from the fact that the universe is not one member of a class of similar objects characterized by a certain range of physical parameters, but is unique and all-embracing. The physical parameters that characterize a particular star, such as the Sun, have no special significance; other stars are more or less massive, have more or less angular momentum, are richer or poorer in metals. But while it is possible to construct mathematical models of the universe characterized by different sets of parameters, there is only one *correct* model. Hence its defining properties have special significance. In fact we may reasonably accord them the status of physical laws rather than auxiliary conditions.

If we adopt this point of view, we may reasonably employ the usual empirical criteria of simplicity and economy as guides in formulating appropriate cosmological postulates. The postulate of statistical homogeneity and isotropy seems to be the simplest assumption that does not conflict with any currently-accepted physical law.

IMPLICATIONS OF THE COSMOLOGICAL PRINCIPLE

Two well-known consequences of the cosmological principle are especially relevant to the present discussion.

First, the cosmological principle implies that the space-time continuum can be uniquely resolved into space plus time. For it is obvious that if statistical homogeneity and isotropy prevail in a given frame of reference,

---

* Einstein's cosmological principle, as it is normally employed, states that the universe can be represented, in a first approximation, by a uniform and isotropic distribution of matter.

they cannot prevail in any frame of reference that is in motion—uniform or accelerated—with respect to this frame. The only symmetry-preserving transformations are spatial rotations and translations.

This conclusion may seem to have a paradoxical quality. Before Einstein, space and time were distinct. Special relativity fused them into a single four-dimensional continuum, replacing the concept of absolute space by that of the inertial frame of reference. Finally, through a further fusion of the concepts of gravitational force and inertial acceleration, general relativity succeeded in encompassing inertial and non-inertial frames of reference in a single mathematical formalism. The cosmological principle does not undo this work. Einstein's field equations still govern the *local* structure of a four-dimensional continuum in which space, time and gravitation are indissolubly blended; but at the cosmological level of description the old distinction between space and time re-emerges in new dress. And indeed it can be shown that the local inertial frame of reference—the frame in which Newton's theory is approximately valid—is one whose motion with respect to the preferred cosmological frame is unaccelerated. Thus the cosmological frame replaces absolute Newtonian space and time.

The second well-known consequence of the cosmological principle is the cosmic expansion. Einstein's equations (in the simplest and most widely accepted form of his theory) admit no static solutions satisfying the cosmological principle. The theory predicts that the universe expands (contracts) uniformly from (toward) a singular state of infinite density in the finite past (future). The rate of expansion depends on the equation of state and on the mean spatial curvature. For simplicity and definiteness, I shall assume in the following discussion that the spatial curvature is negative or zero. This implies that the universe is spatially infinite. A universe with positive spatial curvature, though unbounded has finite volume and is also, in a certain sense, finite in time. A discussion of this case would raise certain subtle and controversial points not essential to the main discussion.

THE STRONG COSMOLOGICAL PRINCIPLE AND INDETERMINACY

So far I have made use only of the weak form of the cosmological principle. The strong form, which postulates complete statistical homogeneity and isotropy, has a remarkable implication that does not seem to have been noted previously. Consider, by way of illustration, a statistically uniform (Poisson) distribution of points on a straight line; this is the simplest example of a statistically homogeneous (and isotropic) distribution. Let the line be divided into cells of equal length h, the analogues of quantum cells in six-dimensional phase space. The Poisson distribution is defined by a single parameter, $\bar{n}$, the mean occupation number of a cell. A particular *realization* of a given Poisson distribution is characterized by an infinite sequence of integer occupation numbers, e.g. . . . 00210111 . . . . Such realizations have a unique pair of properties, not shared by realizations of Poisson distributions on finite or semi-infinite segments. (a) From a single realization one can calculate the value of the statistical parameter $\bar{n}$ with arbitrary precision. This property, an immediate consequence of the law of large numbers, implies that a single realization contains all the statistical information needed to construct it. (b) Two realizations characterized by the same statistical parameter are operationally indistinguishable, since any finite sequence of occupation numbers must occur with precisely the same frequency in each sequence. Thus it is obviously impossible to devise an operational matching procedure that would distinguish between two realizations having the same statistical properties.

These properties of one-dimensional Poisson distributions obviously apply to statistically homogeneous distributions of points in six-dimensional phase space, provided that a suitably defined correlation distance is finite. (This property is needed to ensure the ergodicity of the distribution.) Let us agree to regard the values of quantities that figure in a complete statistical description of a statistically homogeneous and isotropic universe as constituting macroinformation, and all remaining information as microinformation. What has just been shown is that, under our assumptions, *there is no microinformation; the uncertainties implicit in a statistical description of an infinite universe satisfying the strong cosmological principle are irreducible.*

For example, if such a universe is in a state of thermodynamic equilibrium, it is completely characterized by its temperature and density; all other observable quantities can then be calculated. It is true that a hypothetical observer could measure the actual positions and velocities of molecules in a given region, but since all such regions are on exactly the same footing, and their statistical properties are calculable in advance, such measurements would convey no information in a technical sense.

The preceding argument depends essentially on the quantal character of the microscopic description. For in a classical (i.e. non-quantal) universe the distance between any two particles at a given instant serves to define a given statistical realization completely and to distinguish it from all other possible realizations of the same statistical description. Thus microinformation always exists in a classical universe.

The present conclusion goes beyond the implication of Borel's and similar calculations, that certain kinds of microinformation diffuse very rapidly from ostensibly closed systems. The diffusion process does not destroy information, it merely redistributes it. Thus the total quantity of microinformation in the universe remains constant. According to the present picture, however, microinformation is simply absent.

GROWTH OF MACROINFORMATION IN COSMIC EVOLUTION

Consider a universe filled uniformly with non-interacting particles. Let $\mathbf{p}$ denote the momentum of a particle as measured in a frame of reference that is locally at rest at the particle's instantaneous position. (The corresponding velocity $\mathbf{v} = c^2\mathbf{p}/E$ is thus the velocity relative to the uniformly expanding or contracting substratum.) The internal energy density u and the temperature T are defined by

$$u = \tfrac{3}{2}nkT - n\langle E\rangle \qquad (6)$$

where n denotes the mean particle density, and the particle energy $E = \sqrt{(m^2c^4 + c^2p^2)}$. An elementary kinematic calculation shows that the momentum of a free particle varies with time according to the simple law $\mathbf{p} \propto a^{-1}$, where a is the cosmological scale factor defined by the relation ρa^3 = constant. Thus the momentum of a free particle in an expanding universe continually decreases*. It follows that the temperature and the specific internal energy decrease with time in an expanding universe. In this important respect the universe does not behave like a closed system. We should therefore not be surprised to find that its thermodynamic behaviour differs from that of a closed system.

If the particles are ultra-relativistic (for example, if they are photons), $E \propto p$, so that $T \propto a^{-1}$. For non-relativistic particles on the other hand, $E \propto p^2$, so that $T \propto a^{-2}$. Thus, for a given rate of expansion, a relativistic gas cools less rapidly than a non-relativistic one.

Now consider a mixture of non-relativistic gas and radiation. Suppose that at some initial instant the mixture is in thermodynamic equilibrium at the temperature T_0. Suppose further that there is negligible interaction between the gas and the radiation. Then, as the universe expands (or contracts) away from the initial state, a temperature difference will develop between the two components†.

The cosmic expansion (or contraction) preserves the mean entropy per particle of each constituent, so that the specific entropy of the mixture does not change. But the *maximum* specific entropy increases monotonically in both directions of time. For if the thermalization rate were suddenly to become much greater than the expansion rate, the gas and the radiation would assume a common temperature, and in the process the specific entropy would increase. Hence at any given instant the actual specific entropy is less than its maximum possible value. In the general case, when the thermalization rate is neither vanishingly small nor infinitely large compared with the expansion rate, the specific entropy of the mixture will increase, but not as rapidly as maximum specific entropy. Since information is defined as the difference between the actual entropy and the maximum entropy (subject to given macroscopic constraints), this example shows that *expansion or contraction from an initial state of thermodynamic equilibrium generates both specific entropy and specific information.*

This conclusion obviously applies under much more general assumptions about the state and composition of the cosmic medium. The essential elements of the argument are (a) that the 'initial' state is one of maximum specific entropy (zero information), and (b) that the rate of cosmic expansion or contraction—which is of order $\sqrt{(6\pi G\rho)}$, where ρ denotes the mean cosmic density—may be comparable to or greater than the rates of processes that tend to produce the state of local thermodynamic equilibrium. Because the cosmic expansion or contraction is not quasi-static, it generates departures from local thermodynamic equilibrium and hence generates information. At the same time, irreversible processes generate entropy.

* Since the frequency of a photon is proportional to its momentum, it diminishes with time in an expanding universe. This is the basis of the cosmological redshift–distance relation.

† It is easy to show that for extreme relativistic and extreme non-relativistic gases the thermal character of the momentum distribution is preserved under expansion or contraction.

THE ARROW OF TIME

Suppose that at some epoch the universe was in a state of global thermodynamic equilibrium. Let us tentatively identify time's arrow with the direction in which cosmic entropy and information are generated, anticipating that this will turn out to coincide with the direction in which entropy and information are generated locally. Then the hypothesized state of thermodynamic equilibrium would indeed appear to be an initial state from which the universe must either expand or contract: the two possibilities are equally consistent with our considerations up to this point. Thus according to the present considerations, and in contrast with Dr Narlikar's conclusion, there is no *direct* link between the cosmic expansion and the thermodynamic arrow.

But we are not, of course, free to postulate that at some arbitrary epoch the universe was in a state of thermodynamic equilibrium. For such an assumption to be plausible, the cosmic expansion rate must be vanishingly small compared with local relaxation rates. A simple calculation shows that this condition is likely to be satisfied only near the singular state of infinite density. For it can be shown that the expansion rate $H \equiv \dot{a}/a \sim t^{-1}$ as $t \rightarrow 0$ $(\rho \rightarrow \infty)$. On the other hand, two-body reaction rates vary as $n\mathrm{f}(T)$, where $n \propto a^{-3}$—at least for a certain range of values of a—and $\mathrm{f}(T)$ is a function of temperature. In a universe dominated by relativistic particles, $a \sim t^{\frac{1}{2}}$, while in one dominated by non-relativistic particles $a \sim t^{\frac{2}{3}}$. For interactions other than the Coulomb interaction, the two-body reaction rate is a non-decreasing—or at least not a rapidly increasing—function of temperature, which in turn is a decreasing function of time. It follows that in the limit $t \rightarrow 0$ the cosmic expansion rate does become infinitely slow compared with two-body reaction rates. It is therefore plausible to postulate that the cosmic expansion began from a state of near-thermodynamic equilibrium; the initial stages of the expansion are quasi-static, even though the expansion rate varies as t^{-1}.

To summarize the argument up to this point: From the assumption that no statistical property of the universe serves to define a preferred position or direction in space, we deduce that a *complete* description of the universe can be couched in statistical terms, and so contains no microinformation. On physical grounds, we postulate that thermodynamic equilibrium prevails in the limit $\rho \rightarrow \infty$ $(t \rightarrow 0)$, and we define this as the initial state. Then macroscopic and microscopic information are both absent initially. The cosmic expansion generates entropy and information.

In a universe that expands from an initial singularity, every 'closed' system has a finite past and a more or less definite beginning in time. Given a sufficiently complete cosmogonic theory, one could in principle predict the statistical properties of all 'closed' systems and describe the processes by which they came or will come into being. In any case, a given 'closed' system will contain a certain quantity of macroinformation, but no microinformation (since none was present initially and the cosmic expansion generates only macroinformation, by definition). Thus the present theoretical scheme leads to precisely the sort of assumptions that have been introduced in modern theories of irreversible processes. And if this scheme could be developed into a quantitative theory, it would afford an explicit definition of macroinformation and macroscopic variables for every class of 'closed' systems.

In a general way, it is clear that the historical arrow must be related to the growth of information in the universe as a whole. In most open systems, however, the growth of information results directly from a redistribution of information within an effectively isolated parent system rather than from the cosmic expansion. But a detailed discussion of the historical arrow lies outside the scope of this article.

The scheme I have just outlined supports the intuitive ideas that the world is unfolding in time and that the future is never wholly predictable. For, as we have seen, the specific information content of the universe increases steadily in the direction away from the initial state of maximum cosmic density. This implies that the present state of the universe cannot contain enough information to define any future state. The future grows from the past, as a plant grows from a seed, yet it contains more than the past.

COSMIC EVOLUTION

Can the kind of cosmological assumptions I have discussed support a theory of cosmic evolution that accounts for the observed properties of the astronomical universe?

Having postulated an initial state of thermodynamic equilibrium, we are left with just one free parameter at our disposal: the temperature at a given epoch or density. Fixing this parameter involves a choice between two main possibilities, usually referred to as the hot universe and the cold universe.

In the hot universe, energy resides chiefly in electromagnetic radiation during the early stages of the expansion. The cosmic microwave background (which is thought to result from a thermal radiation field with $T = 2{\cdot}7°K$) is interpreted as a remnant of the primordial radiation field; this allows the initial temperature to be calculated. One can then go on to calculate the relative abundances of elements heavier than hydrogen that would be formed during the early stages of the cosmic expansion. The most crucial prediction is that of the helium abundance, which turns out to be about 28 per cent[5]. Observational tests of this prediction are extremely difficult. There is some evidence for the existence of stars whose atmospheres contain substantially less than the predicted primeval abundance, but astronomers do not yet accept it as conclusive[6].

The stability of the hot universe against local density fluctuations has been carefully studied by a number of authors, with consistently negative results. If only thermal fluctuations were present when the cosmic medium had the density of a nuclear fluid, significant local inhomogeneities could never have developed. This conclusion has forced proponents of the hot universe to postulate that substantial density fluctuations either were present initially or were created at supernuclear densities by physical processes whose experimental verification lies outside the scope of current experimental techniques.

The alternative approach is to postulate an initial state cold enough to allow the formation of substantial density fluctuations. One must then find an alternative explanation for the cosmic microwave background. The requirements for such an explanation imposed by the observed quantity and quality of the radiation field are rather stringent, but can perhaps be satisfied by a hypothesis that links the background radiation field to large-scale explosive events occurring within galaxies during their formative stages[7].

If the initial temperature is sufficiently low, the universe may freeze into solid molecular hydrogen—a possibility first suggested by Zeldovich[8]. Continued expansion would cause the solid cosmic medium to shatter into fragments, whose masses, as a simple calculation shows, would be comparable to those of planetary satellites.

I have developed an approximate and somewhat speculative theory of the ensuing evolutionary stages[9]. The 'gas' composed of solid-hydrogen fragments is unstable against a form of turbulence driven by local gravitational forces. The turbulence creates a wide spectrum of density fluctuations. Ultimately self-gravitating systems begin to separate out. The least massive systems separate out first, then clusters of these systems, then clusters of these clusters, and so on. In this way a hierarchy of self-gravitating systems comes into being, and is still in process of formation at the present time.

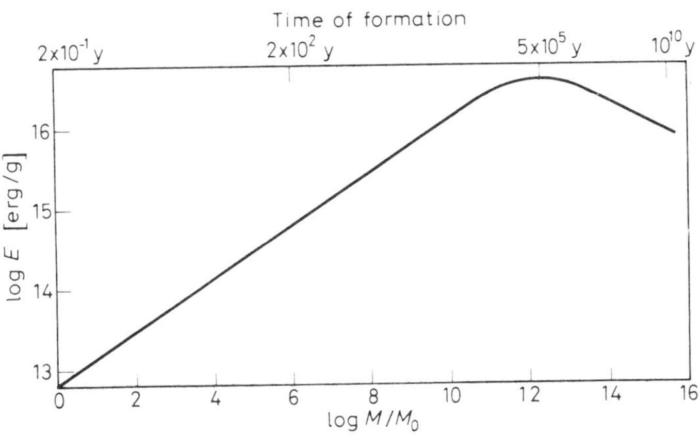

Figure 1. Specific binding energy versus mass for newly formed self-gravitating systems, according to an approximate cosmogonic theory (ref. 9).

Although the theory is speculative and approximate, it predicts a definite relation between the average binding energy per unit mass and the mass of newly formed self-gravitating systems, shown in *Figure 1*. The most tightly bound systems correspond to galaxies and compact galaxy clusters, which accordingly are predicted to have masses around 10^{12} solar masses and binding energies around 10^{16} erg/g. These and other semi-quantitative predictions agree surprisingly well with observational estimates over a mass range of more than fifteen decades. These tentative and approximate results encourage one to believe that further theoretical studies along the lines I have sketched may ultimately lead to a quantitative understanding of cosmic evolution as well as a qualitative understanding of time's arrow.

REFERENCES

[1] A. Einstein, *Phys. Z.* **10**, 185 (1909).

[2] L. van Hove, *Physica*, **21**, 512 (1955).

[3] For example, I. Prigogine, *Non-equilibrium Statistical Mechanics*, Interscience: New York (1962).
W. Kohn and M. Luttinger, *Phys. Rev.* **108**, 590 (1957); **109**, 1892 (1958).
M. Kac, *Proceedings of the Third Berkeley Symposium*, (J. Neyman, ed.), Vol. III, p 171. University of California Press: Berkeley (1956).

[4] E. Borel, *Introduction Géométrique à la Physique*, Gauthier-Villars: Paris (1912).
E. Borel, *Introduction Géométrique à Quelques Théories Physiques*, Gauthier-Villars: Paris (1914).
See also J. M. Blatt, *Prog. Theor. Phys.* **22**, 745 (1959).
P. Morrison, *Preludes in Theoretical Physics*, (edited by A. de-Shalit, H. Feshbach and L. van Hove), p 347. North-Holland: Amsterdam (1966).
P. G. Bergmann and J. L. Lebowitz, *Phys. Rev.* **99**, 578 (1955).

[5] R. V. Wagoner, W. A. Fowler and F. Hoyle, *Astrophys. J.* **148**, 3 (1967). This paper contains references to earlier and less complete calculations.

[6] A. comprehensive review of this question is given by I. J. Danziger, *Annu. Rev. Astron. Astrophys.* **8** (1970).

[7] D. Layzer, *Astrophys. Letters*, **1**, 99 (1968).

[8] Ya. B. Zel'dovich, *J. Exp. Theor. Phys.* **16**, 1102 and 1395 (1963).

[9] D. Layzer, 'Cosmogonic Processes,' *Proceedings of the Brandeis Summer Institute in Theoretical Physics*—1968 (to be published).

THERMODYNAMICS OF HYPERON STARS

K. KOEBKE AND E. HILF

Physikalisches Institut der Universität Würzburg, W. Germany

ABSTRACT

For a single-component perfect Fermi gas we used the numerical programme for the equation of state given by Bauer. For a star of hot non-degenerate neutron gas we calculated the deviations of the internal structure with regard to a totally degenerate neutron star. For a multi-component perfect gas with an exponential-type elementary particle spectrum we present the equation of state. The highest possible temperature is $T_0 = 2 \times 10^{12}$°K, where the total mass density diverges. For the central region of hyperon stars, in contrast to other authors, we can prove that the time component of the metric tensor has no singularity, and that the velocity of sound tends to zero (instead of rising above the velocity of light).

INTRODUCTION

We are studying the internal structure of hot neutron stars, which are in fact hyperon stars. Our main interest is directed towards the peculiar singularities in the centres of these stars. For this purpose we need an equation of state which can be used up to very high total energy densities ($\rho \gg 10^{14}$ g/cm^3). The matter in such a state consists no longer of neutrons only, but also contains innumerable heavier particles and resonances. For in every elementary scattering process new particles can be produced if there is sufficient energy, and if the well-known conservation laws are not violated. Therefore detailed calculations of hot hyperon stars have to deal with the whole spectrum of elementary particles up to very high masses and their interactions. Hansen[1] has made a calculation, which is based on all particles with masses up to 1317 MeV (e, μ, π, n, p, Δ, Λ, Σ, Ξ), taking into account the conservation laws of the number of baryons, of electronic leptons, muonic leptons and of the total charge. By using a variational approach he proves that, as higher densities are approached ($\rho > 10^{17.5}$ g/cm^3) the antiparticles as well as the leptons die out.

Our aim was to continue Hansen's calculation up to even higher densities, which may occur in the central region of a heavy neutron star. We claim that it is not possible to restrict the calculations to a definite number of different elementary particles, but that the possible production of any particle of the whole particle spectrum has to be included. For example, free heavy resonances normally disintegrate quickly, yet this behaviour is no longer observed in a dense region, if the degenerate Fermi distribution of the resulting particle is already fully occupied. In the intermediate region the disintegration is greatly impeded, depending on the chemical potentials and temperature, T.

P.A.C.—22/3–4—K

Tsuruta and Cameron[2] calculated the static structure of a hyperon star with Hansen's equation of state for the degenerate case, especially for $T < 10^9\,°K$. Some of the results refer to densities up to 10^{21} g/cm^3, where Hansen's definite mass spectrum is no longer applicable since the extrapolation of the known baryon spectrum should be assumed to be exponential[3,4]. Therefore we have studied for an exponential baryon spectrum of this type the internal structure of a hyperon star up to infinite density and up to $T_0 = 2 \times 10^{12}\,°K$, which then emerges as the lightest possible temperature.

For a conventional heavy neutron star of infinite central density the possible singularities of the metric tensor as well as of the velocity of sound have often been discussed. These peculiar features do not occur in hyperon stars.

EQUATION OF STATE OF A SINGLE GAS COMPONENT

For simplicity we assume that every component of the hyperon star can be treated as a perfect gas (i.e. we neglect the interactions). First we have to compose the equation of state of a single gas component. The particle number density n, the pressure P, and the kinetic energy density ε of a perfect gas are well-known:

$$\left.\begin{aligned} n &= \frac{8\pi}{h^3}\int_0^\infty p^2\,\mathrm{d}p\,[\exp(-\gamma + E/kT) + 1]^{-1} &&= m^3 f_1\left(\gamma, \frac{T}{m}\right) \\ \varepsilon &= \frac{8\pi}{h^3}\int_0^\infty p^2\,\mathrm{d}p\,[\exp(-\gamma + E/kT) + 1]^{-1}(E - mc^2) &&= m^4 f_2\left(\gamma, \frac{T}{m}\right) \\ P &= \frac{8\pi}{h^3}\int_0^\infty p^2\,\mathrm{d}p\,[\exp(-\gamma + E/kT) + 1]^{-1}\,p\frac{\partial E}{\partial p} &&= m^4 f_3\left(\gamma, \frac{T}{m}\right) \end{aligned}\right\}, \quad (1)$$

where the total energy E of a single fermion is related to its momentum p and its rest mass m by

$$E^2 = p^2c^2 + (mc^2)^2 \quad (2)$$

and the parameter γ with the chemical potential μ by

$$\gamma := (\mu + mc^2)/kT \quad (3)$$

For our astrophysical application it is necessary to have for the integrals 1 a very fast computer programme of high accuracy and covering the whole area of the P/T plane. For some distinct regions series expansions have been given by Sommerfeld[5], Chandrasekhar[6], Guess[7], Tooper[8] and Bauer[9]. *Figure 1* shows the area where no series expansions are available. Three well-known limiting cases are of special interest. For extreme quantum degeneracy the integrals 1 can be solved analytically with the result that the equations of state $P(n, T)$ and $\varepsilon(n, T)$ respectively are independent of temperature. For the limiting cases of extreme non-relativistic degeneracy the following result is obtained:

$$\varepsilon = \tfrac{3}{2}P, \quad P \sim n^{\frac{5}{3}}, \quad P \sim \rho^{\frac{5}{3}} \quad (4)$$

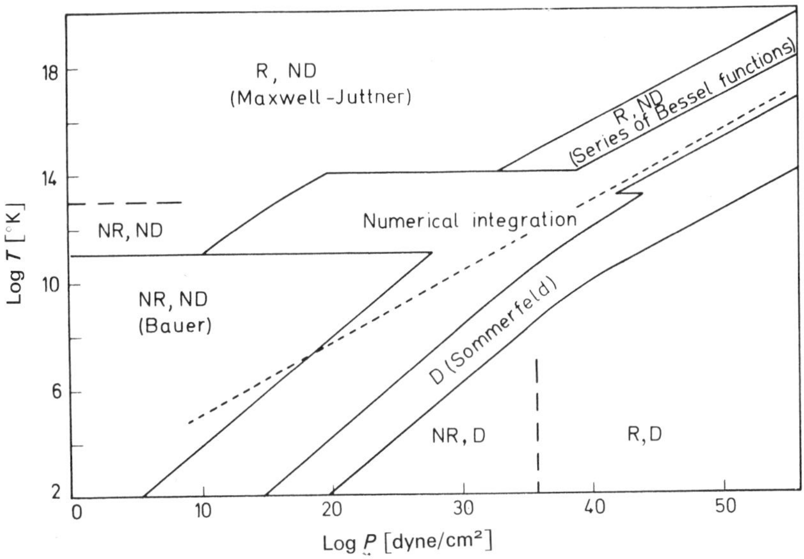

Figure 1. Areas of different ways of calculating the Fermi integrals for neutrons. Above the broken line the radiation pressure overwhelms that of a neutron gas. N non-, R -relativistic, D degenerate.

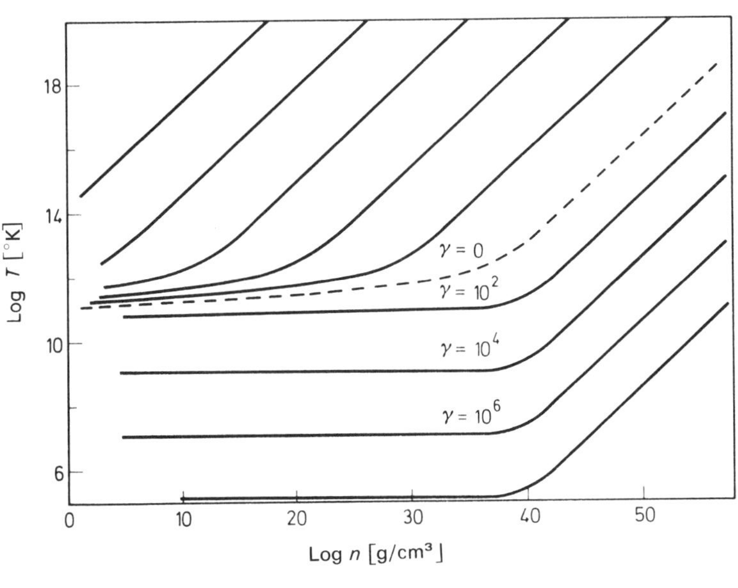

Figure 2. Lines of constant γ of a perfect neutron gas. They describe also the radial structure of the star, since the density declines monotonically and γ is constant throughout the star.

and for the extreme relativistic case:

$$\varepsilon = 3P, \quad P \sim n^{\frac{4}{3}}, \quad P \sim \tfrac{1}{3}\rho, \quad n \sim (T\gamma)^3 \tag{5}$$

and for non-relativistic non-degeneracy ($\mu/mc^2 \ll -1$, $mc^2/kT \ll 1$):

$$\varepsilon = \tfrac{3}{2}P, \quad P = nkT, \quad P = \frac{\rho}{m}kT, \quad n = \frac{2}{h^3}(2mkT)^{\frac{3}{2}} \exp\left(\gamma - \frac{mc^2}{kT}\right) \tag{6}$$

For matter in thermal equilibrium the parameter γ and $T\sqrt{g_{00}}$ (with g_{00} being the time component of the metric tensor in general relativity) are constant, as has been shown by Balazs[10], Ehlers[11] and Ebert[12]. For a neutron gas we have plotted in *Figure 2* the $T(n)$ curves as a function of γ. Since in a neutron star the density decreases with increasing distance from the centre, *Figure 2* shows that the temperature reaches a quasi-constant value in the outer parts, where $n \Rightarrow 0$.

HOT NEUTRON STARS

Oppenheimer and Volkhoff[13] discovered in 1939 that stars of a totally degenerated neutron gas are stable only if their total mass is less than a

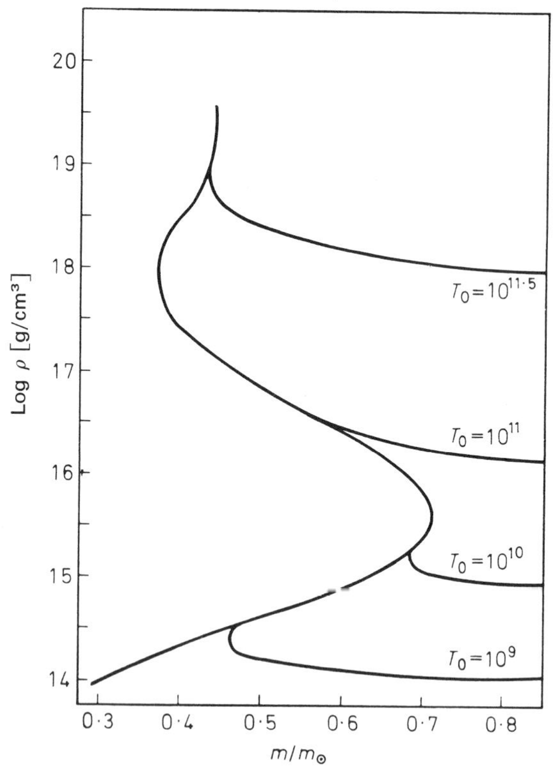

Figure 3. Derivations from the Oppenheimer–Volkoff curve (oscillating line) with increasing central temperature T_0 for a hot neutron star.

finite limiting mass m_G (see *Figure 3*). If $m > m_G$, a degenerate star will probably undergo a gravitational collapse, since in that case there is no static solution. But there are static solutions for partially degenerated neutron stars.

The structure of a radially symmetric static neutron star in thermal equilibrium is determined by the general relativistic field equations of Einstein. Using two formal parameters m^* and r they can be transformed into:

$$\frac{dm^*}{dr} = 4\pi r^2 \rho^*, \quad \frac{dT}{dr} = -T\frac{m^* + 4\pi r^3 P^*}{r(r - 2m^*)} \tag{8}$$

This system of differential equations can be solved uniquely if γ and the equations of state $\rho^* = \rho^*(\gamma, T)$ and $P^* = P^*(\gamma, T)$ are determined, and with the boundary values:

$$m^*(r = 0) = 0, \quad T(r = 0) = T_0 \tag{9}$$

In these equations the asterisk denotes that the quantity is measured in the natural units ($c = G = 1$).

m^* is chosen so that at a sufficient distance from the star the curved space becomes asymptotically flat and the motion of a sample is governed by Newton's laws. Then m^* and r can be identified by the mass and the radius respectively. Since $2m^* \ll r$ for static stars, for those stars in thermal equilibrium the temperature gradient is negative as shown in equations 8. This resembles the radial decrease of the gravitation potential which causes the temperature not to be constant in thermal equilibrium on account of the

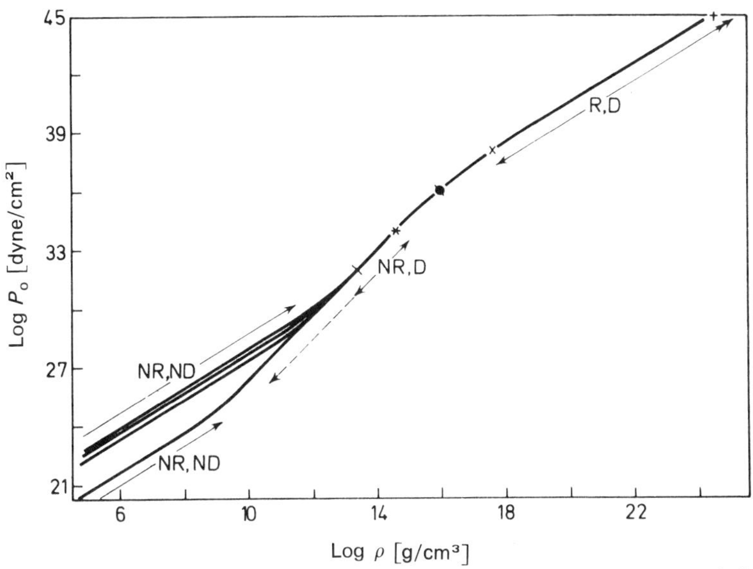

Figure 4. The equation of state of a neutron gas $P(\rho)$ for different fixed γ, or the relation of P and ρ in the radial direction of five stars with equal central temperature but different $\rho_0 = \rho(\gamma, T_0)$. Only the outer parts of the star depend on the degeneracy parameter γ.

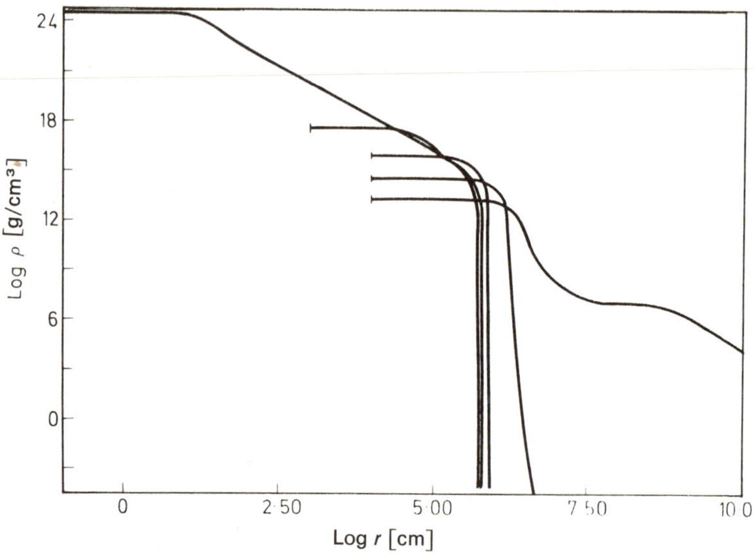

Figure 5. The density of the five selected stars as a function of r. The different behaviour in the outer parts of the star is caused by different degeneracy parameters γ. The total mass of the star becomes very great when the neutron gas in the high density region $\rho > 10^{12}$ g/cm^3 is no longer degenerate.

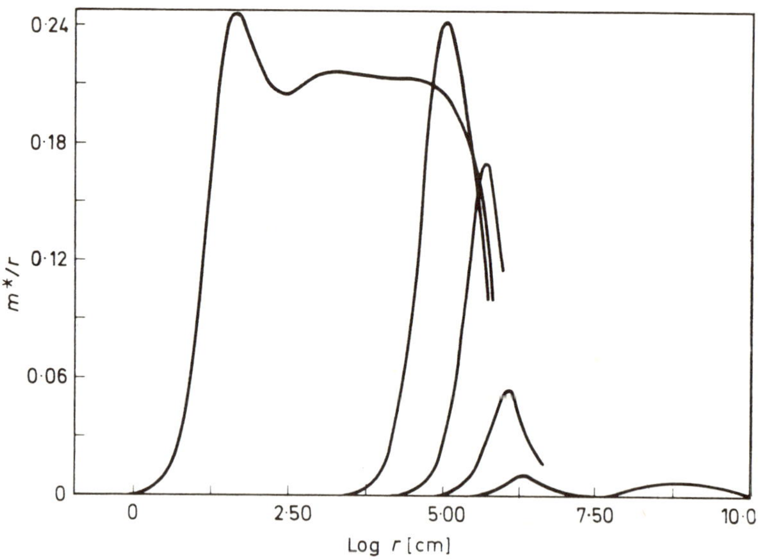

Figure 6. The parameter m^*/r as a function of the radius r for the five selected stars. In the classical theory the gravitational potential, and in Einstein's theory, the metric component g_{rr}, is related with m^*/r.

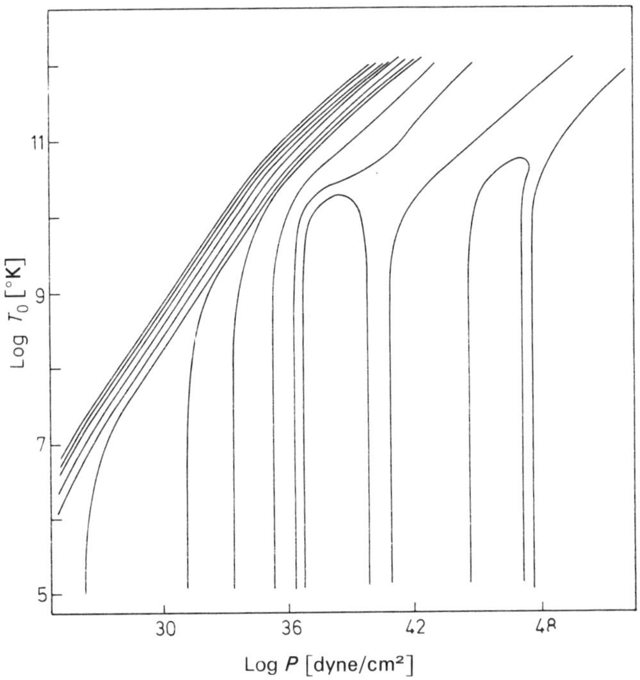

Figure 7. Contour lines of the finite radius R as a function of the central pressure P_0 and temperature T_0. It is evident that in a great area R does not depend on the temperature inside the star. This was the reason for other authors to neglect the influence of the temperature

general relativistic red shift. As a consequence of the constancy of γ throughout the star, the star structure is determined by the respective γ-line in *Figure 2*, and by the differential equation 8 the parametrization of the radius is fixed. Our results are given in *Figures 3* to *7*. In *Figure 3* the lines of equal central temperature are presented as a function of central density ρ_0 and relative total mass $m/m_\odot$ ($m_\odot$ being the mass of the sun). The great deviations of the Oppenheimer–Volkhoff behaviour are registered only at high central temperatures, when a large part of stellar matter is no longer degenerated. This is illustrated in *Figure 4*. For high central densities the equation of state remains the same. With regard to the radial structure of the density the deviations due to the increasing non-degeneracy in the outer parts of the star are shown in *Figure 5*. *Figure 6* makes the complexity of the peculiar internal structure of neutron stars quite evident. *Figure 7* shows the contour lines of the radius r in the T_0/P_0 plot. Only for low central pressure and high temperatures does the structure of the star greatly depend on T_0. This diagram is more suitable for discussing the influence of the non-degeneracy on the radius of the star than the usual vortex diagram. We would like to point out that the density decline is extremely steep in the immediate vicinity of the centre for stars with high ρ_0 and near the surface for degenerate stars.

MULTI-COMPONENT GAS

For the study of a hyperon star an equation of state for a multi-component perfect gas is needed.

The thermal equilibrium conditions T = const. and γ = const. for a multi-component gas system in a fixed volume can be derived by maximization of the entropy $S = \sum_{i=1} S_i$ of the whole system.

$$\delta S_i = \frac{\delta E_i}{KT_i} - \frac{\mu_i + m_i c^2}{KT_i} \delta N_i \tag{10}$$

If the total energy $E = \sum_i E_i$ and the total particle number $N = \sum_i N_i$ are conserved, and δE_i and δN_i can be varied independently, the above-mentioned equilibrium conditions are obtained. The equation of state of the multi-component system can be calculated directly from those of the single gas components, if γ and T are known.

In order to apply this model to the matter of hyperon stars we have to assume that the interaction between the particles can be neglected, and that the total number of particles is conserved in every elementary process. The strong interaction between some neighbouring particles can be taken into account by defining these as one new particle in terms of thermodynamics ('molecule'). Moreover, we need the abundance distribution $a\ (m)$ of the components within matter at high densities. It has been confirmed by Hansen's[1] calculation that the anti-particles and the Bosons diminish in relation to the increase of the Fermions, to which we have restricted our calculation.

The abundance distribution of the Fermions $a\ (m)$ is defined by the product

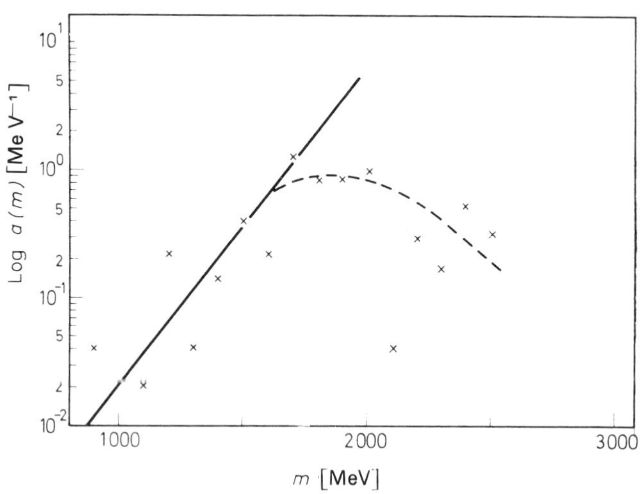

Figure 8. The abundance distribution $a(m)$, the product of the number particles per mass interval and the factor of multiplicity $(1 + 2i)\ (1 + 2j)$, is plotted against the baryon mass. The points are results of the statistics of those particles which are known, the dashed line is the interpolation line of the discrete experimental abundance distribution. The straight line is the theoretical continuous abundance distribution, suggested here.

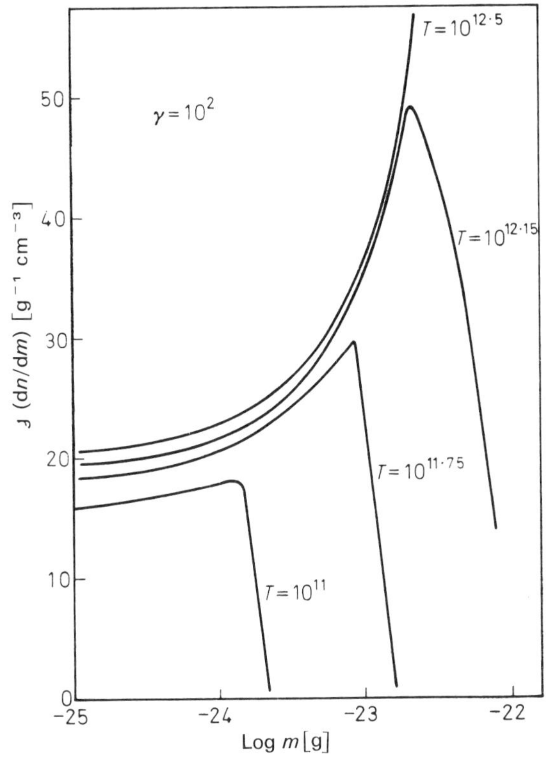

Figure 9. The particle number density per mass interval dn/dm $[g^{-1}]$ is plotted against m. While γ for all lines is fixed, the temperature varies from one line to another. Thus this series of distribution functions describes the deviations of this function if an observer moves towards the centre of the star.

of the number of baryon components per mass unit and their multiplicity $(1 + 2i) \times (1 + 2j)$ with i and j being the spin and the isospin respectively. We confine ourselves to baryons and their heavier resonances. Their abundance distribution has been measured up to about 3 GeV. In *Figure 8* interpolation of the experimental data[16] resembles very much the exponential behaviour in the area where it is likely that all baryons are detected. If it is assumed that this exponential behaviour holds for all masses—this has not been contradicted up to now by experiments, and several theoretical arguments[3,4] have been put forward in favour of it—it holds also if only strongly interacting particles are counted as new ones thermodynamically.

For the quantitative fit of $a\ (m)$ between 1 and 2 GeV we used not only particles with well-known (i, j, m) but also those where some of these quantities were lacking and were to be interpolated in a simple fashion with the result:

$$a(m) = \exp\left(\frac{mc^2}{kT_0} + 52\right), \quad T_0 = 2{\cdot}0 \times 10^{12}\,^{\circ}\mathrm{K} \tag{11}$$

This parameter T_0 is approximately the same as the 'highest possible temperature in nature' of Hagedorn[3]. After substituting the discrete abundance distribution by $a(m)$, the total number density

$$n(\gamma, T) = \int_{m_n}^{\infty} \mathrm{d}m \exp\left(\frac{mc^2}{kT_0} + 52\right) \tfrac{1}{2}m^3 f_1\left(\gamma, \frac{T}{m}\right) \tag{12}$$

together with an analogue formula for the total pressure P and for the energy density ε yields approximately the equation of state of the multi-component gas. In *Figure 9* log ($\mathrm{d}n/\mathrm{d}m$) is plotted versus log (m) for $\gamma = 10^2$ and different T. The maximum of any curve exhibits the most frequent particle component at that temperature. For $T \Rightarrow T_0$ the matter curdles, i.e. the total rest mass energy grows faster than the kinetic energy. The components of the heaviest particles for a given γ and T are not degenerated. Therefore, with the asymptotic expression 7

$$\frac{\mathrm{d}n}{\mathrm{d}m} = \frac{1}{h^3}(2\pi mkT)^{\frac{3}{2}} \exp\left\{\gamma + \frac{mc^2}{kT_0}\left(1 - \frac{T_0}{T}\right) + 52\right\} \tag{13}$$

for heavy masses, the total particle number density $n(\gamma T)$ diverges for $T \Rightarrow T_0$ (see *Figure 11*). In *Figures 10* and *11* the equation of state for fixed γ of a

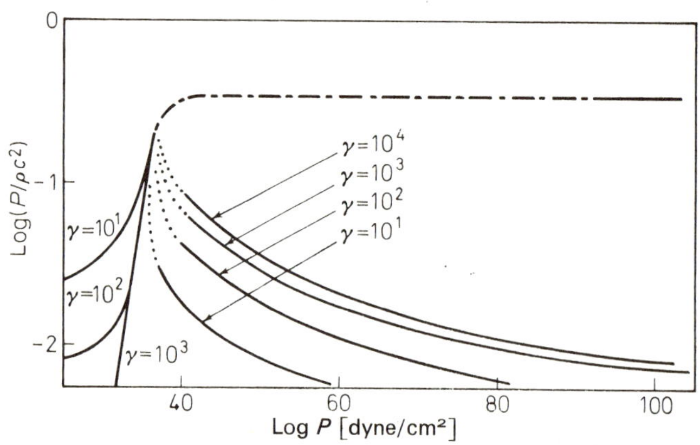

Figure 10. Differences between the perfect neutron gas (chain-dotted) and the hyperon gas (full line). For increasing pressure, P/ρ tends to the value one third for the former, in the latter case to zero. The equation of state at high pressures depends on the degeneracy parameter γ.

multi-component gas is compared with that of a neutron gas. While P/ρ for all γs in the case of the neutron gas approaches asymptotically the value one third with increasing total pressure, P/ρ declines in the case of the hyperon gas. The exponential factor z, defined by

$$P \sim \rho^z \tag{14}$$

in this case is $\ll 1$ and the equation of state depends in the high pressure region on the degeneracy parameter (the heaviest gas components are not degenerated). Although we did not calculate transition states between the multi-component gas (with the continuous abundance distribution accepted here) and the neutron gas, the qualitative behaviour as plotted in *Figures 10* and *11* (dotted lines) seems to be evident. Calculations of multi-component

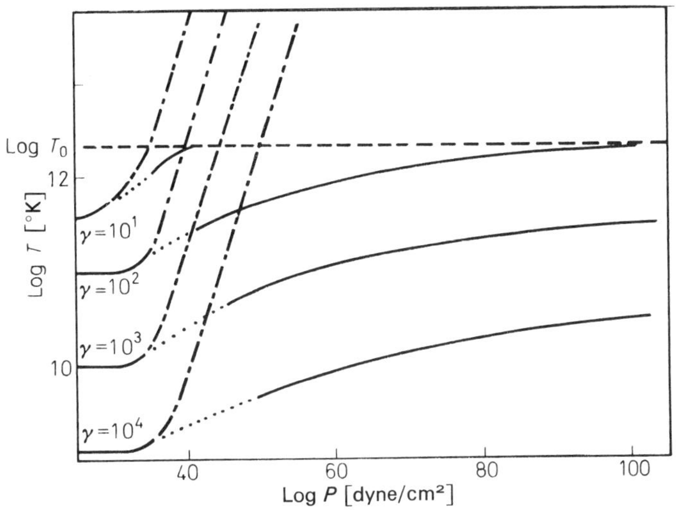

Figure 11. Pressure P dependence of the temperature of a neutron gas (chain-dotted) and a hyperon gas (full line). The derivations are important at low temperatures too.

gas systems have been made by Hagedorn[3] too, but in order to keep the calculations analytical, he restricts them to the case $\gamma = 0$. The results may therefore be applicable to the big bang[3], but not to the structure of the stars.

CENTRAL SINGULARITY OF NEUTRON AND HYPERON STARS

For neutron stars of infinite central density it is known for all equations of state applied up to now that g_{00} diverges to zero for $r = 0$ with increasing central pressure.

Using the energy–momentum tensor of a static ideal fluid $T_{00} = \rho$, $T_{ij} = -P\delta_{ij}$ and the general line element of a static radially-symmetric star

$$ds^2 = g_{00}\,dt^2 - g_{rr}\,dr^2 - r^2(d\theta^2 + \sin^2\theta\,d\phi^2) \tag{15}$$

Einstein field equations yield the equation

$$g_{00/r=0} = \left(1 - \frac{2M^*}{R}\right)^{\frac{1}{2}} \exp\left(-\int_0^{P_0} \frac{dP}{P + \rho c^2}\right) \tag{16}$$

integrated in the radial direction. The factor $(1 - 2M^*/R)^{\frac{1}{2}}$ is chosen so that g_{00} is continuous in the radial direction on the surface of the star ($P_{/r=R} = 0$). So the integral diverges in its upper limit, if the exponential factor z of equation 14 is greater than one. Then $g_{00/r=0} = 0$ for $P_0 \Rightarrow \infty$. This zero of the temporal component of the metric tensor in the centre of the star, a peculiar effect ('singularity') of general relativity, may be called 'zerolarity'. This 'zerolarity' will not occur if the exponential factor z is less one, since then the integral converges and is finite.

Ehlers[11] has proved (in a general relativistic kinetic gas theory) that the equilibrium conditions γ = const. and $T\sqrt{g_{00}}$ = const. are valid for multi-component systems too. But since in our open multi-component gas the temperature cannot rise above T_0 (even if the energy density should rise to infinity) we are sure that in thermal equilibrium g_{00} must have a positive minimum at the centre of the star and indeed the integral 14 converges for the equation of state of our system.

Other authors[15] have obtained the strange result that for very high densities the velocity of sound $v_s = (\mathrm{d}P/\mathrm{d}\rho)^{\frac{1}{2}}$ seems to surpass the velocity of light. This result has been arrived at by taking the repulsive part of the nuclear forces into account. Then the pressure rises more quickly than the total mass energy ρ. In our multi-component system such contradiction does not occur. Moreover, the sound velocity decreases with increasing density.

ACKNOWLEDGEMENTS

We wish to thank Dipl. phys. W. D. Bauer for providing the basic part of the computer programme and for his collaboration; we are very much indebted to Professor R. Ebert for initiating the neutron star project, and we would like to thank him, and Professors R. Hagedorn and J. Ehlers for stimulating discussions, and F. Schmitz and W. Langbein for helpful conversations. Mrs L. Dempsey we thank for revising the English manuscript.

REFERENCES

1 C. J. Hansen, "Composition of matter near nuclear density", in *High Energy Astrophysics*, Vol. III. Gordon and Breach: New York (1967).
2 S. Tsuruta and A. G. W. Cameron, "Some effects of nuclear forces on neutron-star models", in *High Energy Astrophysics*, Vol. III. Gordon and Breach: New York (1967).
3 R. Hagedorn, *Nuovo Cimento*, **56 A**, 1027 (1968).
4 A. Krzywicki, *Orsay-Preprint, No. 38* (1969).
5 A. Sommerfeld, *Z. Phys.* **47**, 1 (1928).
6 S. Chandrasekhar, *An Introduction to the Study of Stellar Structure*, reprinted by Dover: New York (1939).
7 A. W. Guess, "The relativistic quantum gas", in *Advances in Astronomy and Astrophysics*, ed. by Zdenek Kopal, Vol. IV. Academic Press: New York (1966).
8 R. Tooper, *Astrophys. J.* **156**, 1075 (1968).
9 W. D. Bauer, private communication.
10 N. L. Balazs and J. H. Dawson, *Physica*, **31**, 222 (1965).
11 J. Ehlers, private communication.
12 R. Ebert, to be published, and this volume, p 481.
13 J. R. Oppenheimer and G. M. Volkhoff, *Phys. Rev.* **55**, 374 (1939).
14 A. H. Rosenfeld, A. Barbaro-Galtieri, W. J. Podolsky, L. R. Price, P. Soding, C. G. Wohl, M. Roos and W. J. Willis, *Rev. Mod. Phys.* **39**, 1 (1967).
15 See references by H. Y. Chiu, *Ann. Phys. (N.Y.)*, **26**, 364 (1964).

CARNOT CYCLES FOR GENERAL RELATIVISTIC SYSTEMS

R. EBERT

Institute of Physics, University of Würzburg, Würzburg, West Germany

ABSTRACT

A generalization of ordinary Carnot cycles is given for thermodynamic systems with stationary gravitational fields. The two heat reservoirs are assumed to be located at different points in space. In addition to the standard change of thermodynamic quantities the Carnot engine is allowed to change its position during the cycle. A 'generalized Carnot cycle' is then defined by the following process: (1) Connection of the Carnot engine with the first heat reservoir (exchanging heat), (2) Change of position of the Carnot engine from the first to the second heat reservoir, (3) Connection of the Carnot engine with the second heat reservoir (exchanging heat), (4) Change of position of the Carnot engine from the second to the first heat reservoir, after which the cycle repeats. In all changes of position the presence of the gravitational field has to be considered. The special case of an ordinary Carnot cycle is obtained when there is no gravitational field or when the heat reservoirs are located at the same point. Under the assumption that gravitation can be described by general relativity the efficiency of these generalized Carnot cycles is calculated for stationary fields. Thermodynamic equilibrium exists when the efficiency of a generalized Carnot cycle operating between any two parts of the system is zero. For this case we find that $T \cdot \|\xi\|$ is a constant independent of position. As used here T is the ordinary thermodynamic temperature and $\|\xi\|$ denotes the norm field of the Killing vector field ξ, representing the stationarity of the gravitational field. The proof is independent of the field equations of general relativity. Consequently equilibrium consists of a temperature field which depends on the gravitational field. For static fields with spherical symmetry Tolman has proved this relation by using the field equation of general relativity. Our results show that this relation holds quite generally for arbitrary stationary fields.

1. INTRODUCTION

Classical thermodynamics has been developed with the assumption that either no gravitational fields are present in the system, or that the fields act on the rest-mass of the system or particles only and not on any other kind of internal energy like heat or elastic energy. Yet from special relativity we know that every kind of internal energy has inertia, and from the principle of equivalence of inertial and gravitational mass, it then follows that every kind of internal energy has (passive) gravitational mass.

In order to find the exact thermodynamic relations for systems with gravitational fields one therefore has to take into account explicitly the action of gravitation on internal energy. We assume that gravitation can be described by Einstein's theory of general relativity. The first to work on this problem were Tolman and Ehrenfest[1,2]. By a proposed generalization of the second law of classical thermodynamics to general relativistic systems Tolman[3] derived for two special cases and thermodynamic equilibrium using the field equation of general relativity the so-called Tolman relation $T \cdot \sqrt{g_{00}} =$ const. where g_{00} denotes the time component of the metric tensor of space time. Later Landau and Lifshitz[4] and Balazs[5] derived the same result by generalizing the thermodynamic relation $(\partial S/\partial E) = 1/T$, to general relativistic systems, where S and E denote entropy and energy. In the framework of general relativistic statistical mechanics Ehlers[6] and Tauber and Weinberg[7] derived the Tolman relation for an ideal gas in thermodynamic equilibrium.

The approach to general relativistic thermodynamics given here differs from those used by the above authors. It has no need of a previously defined concept of entropy for general relativistic systems but is rather an operational approach in the sense of Buchdahl[8]. The basic idea, given[9] earlier, is a straightforward generalization of the concept of Carnot cycles. With the help of these generalized Carnot cycles temperature, thermodynamic equilibrium and entropy of general relativistic systems can be defined.

For a weak gravitational field Balazs and Dawson[10] have introduced independently of us this concept of generalized Carnot cycles. Their results are the weak field approximation of the results given here.

2. DEFINITION OF GENERALIZED CARNOT CYCLES

We begin with some plausible suppositions. The considered thermodynamic system with gravitation can always be thought to be divided into arbitrary subsystems, each sufficiently small so that temperature and field quantities can be considered as constant throughout each subsystem but may be different in different subsystems. Let the Carnot engine be a machine which can convert heat into mechanical work to a certain extent, and *vice versa*, and in which the mechanical work can be stored. The engine is supposed to be smaller than each subsystem and its mass to be so small that it does not change the gravitational field when it operates between two subsystems. The subsystems can act like heat reservoirs. A *generalized Carnot cycle* is then defined by the following process:

(1) Connection of the Carnot engine with the first heat reservoir, exchange of heat and conversion of heat into mechanical work stored in the engine.
(2) Change of position of the Carnot engine from the first to the second heat reservoir and adiabatic change of its internal state.
(3) Connection of the Carnot engine with the second heat reservoir, conversion of some of the stored mechanical work into heat and exchange of heat.
(4) Change of position of the Carnot engine from the second to the first heat reservoir and adiabatic change of the internal state back to the state at the beginning of the cycle.

This generalized cycle differs from the ordinary one only by the explicit change of the position of the engine in the gravitational field. By this change heat is transported through the field and according to the equivalence of inertial and gravitational mass this needs mechanical work. The efficiency of the cycle is therefore modified by the field.

A generalized Carnot cycle is reduced to an ordinary one when there is no gravitational field or when the heat reservoirs are located at the same point.

3. EFFICIENCY OF GENERALIZED CARNOT CYCLES

We give a short and abbreviated calculation of the efficiency of a generalized cycle for stationary gravitational fields using the notation of modern differential geometry[11, 12]. For a detailed calculation see Ebert and Göbel[13].

Let **X**, **Y** be vectors of the tangent space at an arbitrary point of the Riemannian manifold space-time and let $\langle \mathbf{X}, \mathbf{Y} \rangle$ denote the inner product and $\mathrm{D}_{\mathbf{Y}}\mathbf{X}$ the covariant derivative of **X** in the direction **Y**. From the assumed stationarity of the field there follows the existence of a timelike Killing vector field ξ which satisfies the Killing equation[14].

$$\langle \mathbf{X}, \mathrm{D}_{\mathbf{Y}}\xi \rangle + \langle \mathbf{Y}, \mathrm{D}_{X}\xi \rangle = 0 \quad \text{for arbitrary } \mathbf{X}, \mathbf{Y} \tag{1}$$

A line in space-time of which all tangent vectors belong to the Killing field ξ is called a *Killing orbit*. Then by a short calculation one gets from equation 1

$$\langle \xi, \xi \rangle = \text{const. on any Killing orbit} \tag{2}$$

and, if τ denotes the unit tangent vector of a geodesic line in space-time,

$$\langle \tau, \xi \rangle = \text{const. on any geodesic line} \tag{3}$$

We now consider the Carnot engine during the generalized cycle. It will move along a world line which first coincides for a certain time interval with the world line of the first heat reservoir, then runs to the world line of the second heat reservoir, coincides with this line for a certain time interval and runs back to the world line of the first heat reservoir. The energy E_p of the Carnot engine at a point p on the world line of the engine measured by an observer for whom the gravitational field is stationary, is given by the inner product of the four-momentum of the engine at p and the unit tangent vector of the observer at the same point. Let the total mass of the engine at p be denoted by m_p and the four-velocity of the engine at p by u_p, then (sign convention: $\langle \mathbf{X}, \mathbf{X} \rangle > 0$ for time-like vectors $\mathbf{X}$; $c = 1$)

$$E_p = m_p \langle u_p, \xi_p \rangle / \|\xi_p\| \tag{4}$$

where $\|\xi_p\| := \{\langle \xi_p, \xi_p \rangle\}^{\frac{1}{2}}$ is called the norm of the Killing vector ξ at p. The mass of the engine is a scalar but it changes its value when the internal energy of the engine is changed, by heat exchange.

Without loss of generality we can accomplish the change of position of the Carnot engine (from the first to the second and from the second to the first heat reservoir) by moving it on geodesic lines (we only have to start the motion with sufficient kinetic energy). Then equation 3 applies for those parts of the world line of the engine which belong to the change of position and equation 2 applies for those parts which represent the connection of the engine with the heat reservoirs (it is assumed that the gravitational field in the whole thermodynamic system is stationary, therefore observers located at subsystems observe also a stationary field).

Let Q_α, Q_β be the absolute values of the heat exchanged with the heat reservoir α (first reservoir) and β (second reservoir) respectively and measured by an observer co-moving with the Carnot engine. Then the mass of the engine changes from the value m in the beginning of the cycle into $m + Q_\alpha$ ($c = 1$) after the first heat exchange, and into $m + Q_\alpha - Q_\beta$ after the second exchange. Taking into account all changes of kinetic and internal energy of the engine and using equations 2, 3 and 4 one finally gets for the mechanical work W gained in one cycle measured by an observer attached to the first heat reservoir α (for detailed calculations see ref. 13)

$$W = Q_\alpha - Q_\beta \frac{\|\xi\|_\beta}{\|\xi\|_\alpha} \tag{5}$$

where $\|\xi\|_\alpha$, $\|\xi\|_\beta$ denote the constant norms of the Killing vectors along the world lines of the reservoirs α and β respectively according to equation 2.

When there is no gravitational field then $\|\xi\|_\alpha = \|\xi\|_\beta = 1$ and equation 5 is reduced to the result of classical thermodynamics. If there is a field but the two heat reservoirs are located at the same point then $\alpha = \beta$ and equation 5 is again reduced to the classical result.

If we choose a coordinate system such that the time coordinate is a parameter on the Killing orbits (which is always possible), then $\langle \xi, \xi \rangle = g_{00}$. Equation 5 then becomes

$$W = Q_\alpha - Q_\beta \sqrt{[g_{00}(\beta)]}/\sqrt{[g_{00}(\alpha)]}$$

where $g_{00}(\alpha)$, $g_{00}(\beta)$ are the time-independent time components of the metric tensor at the heat reservoirs α and β respectively. For the Carnot efficiency η we get from equation 5

$$\eta := W/Q_\alpha = 1 - Q_\beta \|\xi\|_\beta / Q_\alpha \|\xi\|_\alpha$$

or in special coordinates from equation 6

$$\eta = 1 - Q_\beta \sqrt{[g_{00}(\beta)]}/Q_\alpha \sqrt{[g_{00}(\alpha)]} \tag{8}$$

4. DEFINITION OF TEMPERATURE

Before expressing the Carnot efficiency with temperatures instead of heat energies we have to define temperature in a thermodynamic system with gravitational fields. In classical thermodynamics temperature can be defined by using Carnot cycles and the principle of Kelvin[15] which may be stated: *It is impossible to convert an amount of heat completely into work by a cyclic process, without at the same time producing other changes.* By adding Carnot cycles running in opposite directions and using the above principle one gets the well-known result that the ratio of the absolute values Q_1, Q_2 of the exchanged heat energies in one cycle must be a real valued function of the temperatures T_1, T_2 of the heat reservoirs only. Taking into account a functional equation for this function, which results from the possibility of adding two cycles together to form a third one, one defines the absolute or thermodynamic temperature by

$$T_2 := T_1 Q_2/Q_1 \tag{9}$$

where T_1 has to be fixed by a physical process.

For systems with gravitational fields we define temperature in a completely analogous way. First we make the basic assumption: *Kelvin's principle also holds for systems with stationary gravitational fields.* From here we get for a generalized Carnot cycle the result that the absolute values of the exchanged heat energies Q_α, Q_β in one cycle must be a real valued function f of only the temperatures T_α, T_β of the heat reservoirs *and* of the metric tensor g_{ik} at α and β:

$$Q_\alpha/Q_\beta = \mathrm{f}[T_\alpha, g_{ik}(\alpha); T_\beta, g_{ik}(\beta)] \tag{10}$$

Because of the possibility of adding two cycles together to get a third one a certain functional equation for f has to be fulfilled[13]. Taking into account this equation we define the *thermodynamic temperature of a system with gravitation by*

$$T_\beta := T_\alpha Q_\beta/Q_\alpha \tag{11}$$

where T_α has to be fixed by some physical process.

Because Q_α, Q_β are the exchanged heat energies measured by a co-moving observer attached to the engine (or equivalently attached to the heat reservoirs α and β respectively) this temperature can be measured also by an ideal gas thermometer permanently connected with the heat reservoirs α and β respectively. This follows from the fact that a co-moving observer attached to the engine and measuring in proper units finds no difference from classical thermodynamics concerning the temperature as long as he uses the temperature definition 11. The above defined temperature is equal to the *proper temperature* introduced by Tolman[3] in a completely different way.

Using 11 we get for the Carnot efficiency 7 the relation

$$\eta = 1 - \frac{T_\beta \|\xi\|_\beta}{T_\alpha \|\xi\|_\alpha} = \frac{T_\alpha \|\xi\|_\alpha - T_\beta \|\xi\|_\beta}{T_\alpha \|\xi\|_\alpha} \tag{12}$$

or in coordinates

$$\eta = \{T_\alpha \sqrt{[g_{00}(\alpha)]} - T_\beta \sqrt{[g_{00}(\beta)]}\}/T_\alpha \sqrt{[g_{00}(\alpha)]} \tag{13}$$

Because the temperature and the norm of the Killing vector are both $\geqslant 0$ by definition and $g_{00} \geqslant 0$ (see sign convention in connection with equation 4) the relation $\eta \leqslant 1$ holds. As in classical thermodynamics η can never become greater than unity, even for the strongest gravitational fields.

5. THERMODYNAMIC EQUILIBRIUM

In classical thermodynamics two systems are in equilibrium if and only if the Carnot efficiency of a cycle operating between these two systems is zero. As can be seen this holds for systems with gravitational fields also if we use the generalized Carnot efficiency given by equation 12. Therefore equilibrium is characterized by

$$T_\alpha \|\xi\|_\alpha = T_\beta \|\xi\|_\beta \tag{14}$$

The whole system is in equilibrium if equation 14 holds for all cycles operating between any two subsystems, and therefore

$$T\|\xi\| = \text{const.} \tag{15}$$

is the temperature relation in equilibrium. In coordinates equation 15 becomes

$$T\sqrt{g_{00}} = \text{const.} \tag{16}$$

which is the Tolman relation. We see that this relation holds quite generally for arbitrary systems with stationary gravitational fields. In getting equation 15 we did not use the field equations of general relativity, we only used a Riemannian manifold for space-time, the principle of equivalence and special relativity.

REFERENCES

[1] R. C. Tolman, *Phys. Rev.* **35**, 904 (1930).
[2] R. C. Tolman and P. Ehrenfest, *Phys. Rev.* **36**, 1791 (1930).
[3] R. C. Tolman, *Relativity, Thermodynamics and Cosmology.* Oxford University Press: London (1934).
[4] L. D. Landau and E. M. Lifshitz, *Statistical Physics*, § 27. Pergamon: Oxford (1959).
[5] N. L. Balazs, *Astrophys. J.* **128**, 398 (1958).
[6] J. Ehlers, *Abh. Math.-Nat. Kl. Akad. Wiss. Mainz*, **11**, 804 (1961).
[7] G. E. Tauber and J. W. Weinberg, *Phys. Rev.* **122**, 1342 (1961).
[8] H. A. Buchdahl, *The Concepts of Classical Thermodynamics*, p 2. Cambridge University Press: London (1966).
[9] R. Ebert, 'Zum Begriff des thermodynamischen Gleichgewichtes in der allgemein-relativistischen Thermodynamik', contributed paper given at the International Conference on Statistical Mechanics and Thermodynamics, Aachen 1964 (unpublished).
[10] N. L. Balazs and J. M. Dawson, *Physica*, **31**, 222 (1965).
[11] N. J. Hicks, *Notes on Differential Geometry.* Van Nostrand: Princeton (1965).
[12] Ch. W. Misner in *Relativity, Groups and Topology*, (Les Houches 1963) ed. C. de Witt, p 892. Gordon and Breach: New York (1964).
[13] R. Ebert and R. Göbel, 'Carnot cycles in general relativity', *Phys. Rev.* in preparation.
[14] A. Schild in *Relativity Theory and Astrophysics*, Vol. I, *Relativity and Cosmology*, ed. J. Ehlers, p 48. American Mathematical Society: Providence, R.I. (1967).
[15] P. T. Landsberg, *Thermodynamics*, p 176. Interscience: New York (1961)

ENSEMBLE VERSUS TIME AVERAGE PROBABILITIES IN RELATIVISTIC STATISTICAL MECHANICS

P. T. LANDSBERG AND K. A. JOHNS

Department of Applied Mathematics and Mathematical Physics, University College, Cardiff

ABSTRACT

Three results will be reported:

(1) Reasons are advanced why discrete probabilities are not Lorentz-invariant. Such probabilities can be obtained as time average probabilities Π_i for state i in frame I. They can also be obtained from an ensemble E_1 of systems which, like the system of interest, are each on average at rest in a certain frame I_0. Such probabilities Q_i transform like the Π_i and ergodicity is then a Lorentz-invariant notion.

(2) If the ensemble is of the usual type (ensemble E_0) whose systems are all at rest in I_0, then the ensemble-based probabilities are Lorentz-invariant. If E_0 is used ergodicity is not a Lorentz-invariant notion.

(3) If entropy is regarded as invariant and entropy maximization is used, the canonical equilibrium probabilities Π_i which one finds contain an extra term which is not usually found. This term will require further discussion.

1. THE GRAND CANONICAL CONSTRAINTS

Consider within the framework of special relativity, a procedure which is familiar in statistical mechanics, namely the maximization of entropy subject to constraints. When using the grand canonical ensemble these constraints take the form:

$$\sum_i \Pi_{i0} = 1 \tag{1}$$

$$\sum_i \Pi_{i0} E_{i0} = \langle E_0 \rangle_0 \tag{2}$$

$$\sum_i \Pi_{i0} N_{i0} = \langle N_0 \rangle_0 \tag{3}$$

The system states i can change as a result of inter-particle collisions (conceived as point interactions), and collisions with the walls of the container. A state i is assumed to have probability Π_{i0}; E_{i0} and N_{i0} are the energy and particle number appropriate to state i, and $\langle E_0 \rangle_0$ and $\langle N_0 \rangle_0$ their respective mean quantities. The suffix 0 denotes that all quantities are measured in an inertial frame I_0 in which the system appears at rest, and $\langle\ \rangle_0$ means that Π_{i0} has been used in the average.

In keeping with the principle of covariance one must now seek to express these constraints in a general inertial frame I, and must also include the three components of momentum, $\mathbf{P}_i$, in the same way as the energy. Thus:

$$\sum_i \Pi_i = 1 \tag{4}$$

$$\sum_i \Pi_i E_i = \langle E \rangle \tag{5}$$

$$\sum_i \Pi_i \mathbf{P}_i = \langle \mathbf{P} \rangle \tag{6}$$

$$\sum_i \Pi_i N_i = \langle N \rangle \tag{7}$$

These new constraints, 4 to 7, must of course be satisfied in all inertial frames. Since, however, we cannot use an infinite number of them when actually maximizing the entropy, we must therefore find a finite set of constraints which ensure that equations 4 to 7 do in fact hold in all such frames.

To achieve this, it is necessary to know the Lorentz-transformation properties of the quantities involved. First, in keeping with the usual practice, probability Π_i, and particle number, N_i and $\bar{N}$, may be regarded as Lorentz-invariant. It is at once apparent that the equations 4 and 7 will be satisfied in all frames I if and only if they are satisfied in any one frame (such as I_0). The remaining quantities, energy and momentum, are not Lorentz-invariant and must be treated differently. If, as we did in a recent paper[1], one assumes the system to be *inclusive* (i.e. including the energy and momentum due to the stresses in the container) then $\langle \mathbf{P} \rangle$ and $\langle E \rangle$ are the components of a four-vector. Also, in any state i, $\mathbf{P}_i$ and E_i form a four-vector, and thus constraints 5 and 6 may be expressed by

$$\sum_i \Pi_i \{c\mathbf{P}_i, E_i\} = \{c\langle \mathbf{P} \rangle, \langle E \rangle\} \tag{8}$$

The linearity of the Lorentz transformation ensures that if an equation of this form holds in any one inertial frame then it holds in all such frames. Thus equation 8 may conveniently be expressed in the variables of I_0 as

$$\sum_i \Pi_{i0} \{c\mathbf{P}_{i0}, E_{i0}\} = \{\mathbf{0}, \langle E_0 \rangle_0\} \tag{9}$$

remembering that while the mean momentum of the system is zero in I_0, the momentum appropriate to any given state i need not be so.

Suppose, however, that it is desired to avoid taking into account the stresses in the container. One must then use the results applicable to a *confined system*[1]. In this case the mean energy is not the fourth component of a four-vector. Instead it is the enthalpy, $\langle E \rangle + pV$, which, together with the momentum, provides the four components. (Here p is the Lorentz-invariant pressure, and V is the volume, which is subject to the usual Lorentz contraction.) However, in any given state i the system has constant energy, momentum and particle number, and all its particles move freely without collisions which change these quantities. It is thus appropriate to use the transformation for a free system in this case, and to treat energy and momentum as a four-vector.

An immediate difficulty arises. On the LHS of equations 5 and 6 we have four-vectorial quantities which may be transformed to another frame of reference under the Lorentz transformation. On the right are two quantities which are not components of the same four-vector, and can only be transformed to another frame by the introduction of extra terms involving p and V. How can this difficulty be resolved? We arrive here at the notion of a probability Π_i for a discrete state i which is *not* Lorentz-invariant. This is a new suggestion since discrete probabilities are normally considered Lorentz-invariant[2].

2. IMPLICATIONS FOR STATISTICAL MECHANICS

The simple conclusion of section 1 has rather far-reaching consequences. The first of these is that it is in contradiction with any simple-minded relativistic interpretation of ensembles. If a system is on average at rest, statistical mechanics associates with it a representative ensemble of identical systems. At any one time the various available states i of the system are present in this ensemble in proportion to their probabilities Q_{i0} (say). The motion of these systems has never been discussed, as far as we know, it being assumed that they are at rest in the inertial frame I_0 in which the system of interest is at rest. If one assumes this, then one arrives at an invariant probability

$$Q_i = Q_{i0} \text{ (ensemble-based)} \tag{10}$$

For in a general frame I the number of systems in a given state i is the same when the ensemble is viewed from frame I as it is when the ensemble is viewed from frame I_0. This conclusion, based on ensemble-based probabilities, is in contradiction with the result of the preceding section.

A resolution of this paradox is, however, possible. One can consider the system of interest over a long period of time and allot probabilities Π_i to various states i according to the total time for which the system is in this state. These time-based probabilities Π_i are found to transform as one passes from I_0 to I, because of the Lorentz-transformation of the time. We shall put simply

$$\Pi_i = (1 + f_i)\Pi_{i0} \qquad \text{(time-based)} \tag{11}$$

where f_i is a function, to be discussed in section 3. It is by the use of time-based (rather than invariant ensemble-based) probabilities that one may hope to achieve agreement with section 1.

It will be appreciated that the difference between 10 and 11 implies a result about ergodicity. If one confines oneself to just one ensemble as discussed above, and considers a system which is ergodic in its rest frame I_0, then

$$\Pi_{i0} = Q_{i0}, \quad \text{i.e.} \quad \Pi_i = (1 + f_i)\, Q_i$$

It follows that with $f_i \neq 0$ the system is no longer ergodic in I. Thus one can hope to gain agreement with section 1 only by admitting *either* that ergodicity is not a Lorentz-invariant notion, *or* that the idea of an ensemble as a set of systems all strictly at rest in a certain frame of reference (an ensemble E_0) is inapplicable. We favour the latter alternative and regard it as more satisfactory to restrict the motion of the system S_0 of interest, and also the motion of the systems of the ensemble which represents it by the same condition (an ensemble E_1): the systems must be on average at rest in the same frame (I_0). Ergodicity can then become again a Lorentz-invariant notion, namely for ensembles E_1.

The second corollary of these considerations is that the invariance of the entropy cannot be inferred from the Lorentz-invariance of the probabilities Q_i, as has often been done in the past[1,2]. The reason is that it is the probabilities Π_i (not Q_i) which can agree with the considerations of section 1, and they transform as one passes from I_0 to I. The thermodynamic argument for entropy invariance is, of course, not affected by these considerations. One assumes simply that the gradual acceleration of a system from one frame to another is a reversible process which keeps the entropy unchanged.

One may ask for the constraints 4 to 7 to be amended to specify an average enthalpy. However, this does not get over the difficulty that whatever the expressions in these equations, the four-vector for the left-hand sides is $(c\mathbf{P}, E)$, while it is $(c\langle\mathbf{P}\rangle, \langle E\rangle + pV)$ for the right-hand sides.

3. THE TRANSFORMATION OF TIME-BASED PROBABILITIES

The velocity of frame I_0 in frame I is denoted by $\mathbf{w}$. The transformation of equations 5 and 6 to the frame I_0 in which the system is on average at rest will now be carried out. Using 11, one finds

$$\langle E\rangle = \gamma \sum_i \Pi_{i0}(1 + f_i)(E_{i0} + \mathbf{w}\cdot\mathbf{P}_{i0}) \tag{12}$$

$$\langle \mathbf{P}\rangle = \sum_i \Pi_{i0}(1 + f_i)\,[\gamma(\mathbf{P}_{i0\parallel} + (\mathbf{w}/c^2)E_{i0}) + \mathbf{P}_{i0\perp}] \tag{13}$$

where $\mathbf{P}_{i0\parallel}$ and $\mathbf{P}_{i0\perp}$ are respectively the components of $\mathbf{P}_{i0}$ parallel and perpendicular to $\mathbf{w}$. It will be assumed, as a restriction on the system, that in I_0

$$\langle P_0\rangle_0 \equiv \sum_i \mathbf{P}_{i0}\Pi_{i0} = 0 \tag{14}$$

Writing also

$$\langle E_0\rangle_0 \equiv \sum_i E_{i0}\Pi_{i0}$$

one knows that for a confined system[1]

$$\langle E\rangle = \gamma[\langle E_0\rangle_0 + (w^2/c^2)pV_0] \tag{15}$$

$$\langle \mathbf{P}\rangle = (\mathbf{w}/c^2)\gamma[\langle E_0\rangle_0 + pV_0] \tag{16}$$

The fact that equations 12 and 15 must be identical, and that 13 and 16 must also be identical, yields conditions on the unknown functions f_i. These are from the energy

$$\sum_i f_i(E_{i0} + \mathbf{w}\cdot\mathbf{P}_{i0})\Pi_{i0} = (w^2/c^2)pV_0 \tag{17}$$

and from the momentum

$$\sum_i f_i(E_{i0}\mathbf{w} + c^2\mathbf{P}_{i0\parallel} + (c^2/\gamma)\mathbf{P}_{i0\perp})\Pi_{i0} = pV_0\mathbf{w} \tag{18}$$

From 18 one finds

$$(w^2/c^2)\sum_i f_i E_{i0}\Pi_{i0} + \sum_i f_i \mathbf{w}\cdot\mathbf{P}_{i0}\Pi_{i0} = (w^2/c^2)pV_0 \tag{19}$$

and

$$\sum_i f_i \mathbf{P}_{i0\perp} \Pi_{i0} = 0 \tag{20}$$

Since 17, 19 and 20 hold for all $\mathbf{w}$, it follows that

$$\sum_i f_i E_{i0} \Pi_{i0} = 0 \tag{21}$$

and

$$\sum_i f_i \mathbf{P}_{i0} \Pi_{i0} = (\mathbf{w}/c^2)\, pV_0 \tag{22}$$

The equations 21 and 22 are the conditions on the functions f_i. It can be shown[3] that in a one-particle system the time-based probabilities can be transformed so as to make f_i in 11 equal to

$$f_i = \mathbf{w} \cdot \mathbf{P}_{i0}/E_{i0} \tag{23}$$

This theory also yields a pV term such that 23 satisfies 22[4]. Lastly 23 reduces condition 21 to 14 so that the solution 23 does in fact satisfy the general conditions 21 and 22.

4. DISCUSSION

The major difference between the work done here and earlier work is the rejection of the Lorentz-invariance of discrete probabilities. This apparently far-reaching alteration to basic concepts is made easier to understand by noting that the probabilities specified here represent the proportion of time which is spent in a particular state. The transformation factors for the probabilities Π_i arise because the Lorentz transformation of time depends on the velocity in each state i of the system (or particle) under consideration. We considered two possibilities based on the specific f_i expression given by equation 23. (A) If the system velocities depend on the state i, then f_i is different for the various states i. (B) If, however, the velocity is constant (i.e. the velocity is zero in frame I_0), then 23 yields $f_i = 0$, and the probabilities Π_i are Lorentz-invariant.

The difficulty of choosing between these possibilities lies in deciding on what to take as the velocity in I_0 of a system of particles in a given state. One point of view is to say that the mean velocity of all the particles in the system should be considered. Allowing for fluctuations of momentum and energy, this quantity varies between states i, and leads to the non-invariant probabilities described before, and hence to case (A). This approach corresponds to the treatment of a system as *confined*[1], in which the container is disregarded. It leads at once to a statistical description of the pressure in terms of the motion of the particles (e.g. equation 22). The complication of this method is that the time spent by a system in a state i (defined by a set of occupation numbers for particle states) is determined not by the overall system velocity but by the individual velocities of all the particles.

One arrives at case (B) for an inclusive system, defined in ref. 1, if the velocity of the system is taken to be that of the container, fixed at rest in frame I_0. The behaviour of the particles inside the container is discounted, and the momentum $\mathbf{P}_{i0}$ of the system is deduced from its zero velocity in I_0 to be itself zero. Then, by 23, f_i is zero for all states i, and the standard results with Lorentz-invariant probabilities follow at once. The pressure cannot then be calculated by this method, and must be introduced in a normalization factor.

The difference between cases (A) and (B) can most clearly be seen for a one-particle system. Here the probabilities of the system (i.e. the one particle) being in various states can undoubtedly be determined by the time intervals it spends in those states, and these are accordingly altered under a Lorentz transformation. This is case (A). It leads to the standard results for the Lorentz transformation of energy, momentum, pressure etc. of a confined system.

If entropy is regarded as invariant and if an entropy maximization technique is used, a discrepancy occurs: the canonical probability Π_{i0} is found to be

$$\Pi_{i0} = C \exp\left(-[E_{i0} + \tfrac{1}{3}\mathbf{u}_{i0} \cdot \mathbf{P}_{i0}]/kT_0\right) \tag{24}$$

instead of

$$\Pi_{i0} = C \exp(-E_{i0}/kT_0) \tag{25}$$

as given by conventional theory, and also by approach (B), treating the system as inclusive. Here C is a normalization factor, k is Boltzmann's constant, T_0 is the temperature (measured in I_0), and $\mathbf{u}_{i0}$ is the particle velocity equal to $c^2\mathbf{P}_{i0}/E_{i0}$. There is clearly a discrepancy between equations 24 and 25. This question will be discussed elsewhere.

REFERENCES

1 P. T. Landsberg and K. A. Johns, *Nuovo Cimento*, **52B**, 28 (1967).

2 R. C. Tolman, *Relativity, Thermodynamics and Cosmology*, p 158. Oxford University Press: London (1934).

A. Børs, *Proc. Phys. Soc. Lond.* **86**, 1141 (1965).

R. K. Pathria, *Proc. Phys. Soc. Lond.* **88**, 791 (1966); **91**, 1 (1967).

3 P. T. Landsberg and K. A. Johns, *J. Phys. A*, **3**, 113 (1970).

4 K. A. Johns and P. T. Landsberg, *J. Phys. A*, **3**, 121 (1970).

VARIATIONAL FORMULATION OF RELATIVISTIC FLUID THERMODYNAMICS

L. A. SCHMID

Goddard Space Flight Center, NASA

ABSTRACT

As a first step toward generalizing reversible thermodynamics from the case of a homogeneous system to that of a system whose local velocity may be a function of its position in space-time, a variational principle is derived for relativistic reversible adiabatic flow of a compressible fluid. This is done by identifying the thermodynamic internal energy function for a given sample of the fluid with its Hamiltonian function, and then invoking the canonical equations of motion. Both in order to bring the rest-mass energy into the formalism, as well as to provide a means of labelling and identifying different samples of fluid, it is necessary to introduce a new thermodynamic variable, which is just the molar initial momentum vector of the fluid sample in question. It turns out that this vector is intimately related to the vorticity of the flow, and if it had been omitted, the formalism would have been implicitly limited to a description of vorticity-free flow.

The Lagrangian density, as seen in the fixed laboratory frame, that results from identifying the Hamiltonian with the thermodynamic internal energy is just the thermodynamic pressure. This must be regarded as a function of the generalized coordinates that are canonical to the particle density, the entropy density, and the initial momentum vector (all regarded as generalized momenta). More precisely, the pressure is a function of the proper-time derivatives of these coordinates. These time derivatives are equal to the molar free enthalpy, the rest-temperature, and the initial velocity respectively. Because, in the laboratory frame, the proper-time derivative of a variable is defined as the contraction of the velocity four-vector with the four-gradient of the variable, the pressure is also a function of the fluid velocity.

This variational principle yields the correct form of the stress-energy tensor for reversible adiabatic flow of a compressible fluid (together with the necessary statements of particle and entropy conservation), and automatically gives the expression for the solution of Euler's equation of motion for the fluid in terms of the four-gradients of the generalized coordinates.

INTRODUCTION

The discussion of this article is in the spirit of well-known attempts[1] to bring continuum mechanics within the framework of thermodynamics by treating local velocity as just one more thermodynamic variable to be taken into account with all the others. The basic approach consists of identifying the appropriate thermal energy function with the Hamiltonian of the system, and the corresponding canonical equations with the mechanical and thermodynamical equations of motion of the system.

This general approach was first applied to the case of a homogeneous system by Helmholtz[2] in 1886, and adapted to relativity theory in 1907 by Planck[3]. Planck's theory was developed before four-dimensional tensor analysis and the modern covariance concept had fully evolved. Consequently, although it was form-invariant under Lorentz transformations, it fell completely outside the framework of tensor analysis, which meant that, for all but the simplest applications, it was completely unworkable. (Reviews of both the early[4] and recent[5] history of relativistic thermodynamics are available elsewhere.)

In 1939 Van Dantzig[6] constructed a manifestly covariant thermodynamics, and applied it to fluids[7], but his work failed to lift the obscurity surrounding the intimate three-way relation that binds together thermodynamics, fluid dynamics, and the canonical formalism. This relation stems from the fact that, if the right choice of variables is made, the thermodynamic energy density function plays the role of Hamiltonian density, and the thermodynamic pressure plays the role of Lagrangian density.

The identification of pressure with Lagrangian density had already been made in 1908 by Hargreaves[8] for the case of non-relativistic potential flow. Van Dantzig[7] generalized this identification to the relativistic case, but, although the point was not explicitly made, his proof was likewise limited to the case of potential flow, because he did not include the variables that are necessary for a completely general description of vorticity. (Others have since given relativistic variational principles that are free of this limitation, but these principles all involve the imposition of constraints, and do not make the identification of the Lagrangian density with the thermodynamic pressure.)

Notation

The analysis will be carried out entirely within the framework of special relativity. Boldface Latin or Greek letters will designate four-vectors, and light-face characters will designate scalars. A superior dot will designate differentiation with respect to proper-time τ, i.e. the time derivative as seen by an observer moving with the fluid. Contraction of two four-vectors will be indicated as the dot product of the corresponding boldface characters. Indices will be explicitly indicated only in the case of two-index tensors, and when indices are indicated, the summation convention will be used.

Intensive thermodynamic quantities, and extensive quantities that are referred to one mole of the fluid, will be designated by capital letters. Thus T and P are temperature and pressure respectively, and V, S, U, H and G are the molar volume, entropy, energy, enthalpy and Gibbs function (free enthalpy) respectively. The number of moles per unit volume is $n \equiv 1/V$. Extensive quantities referred to unit *volume* (not unit *mass*!) of the fixed laboratory frame will be designated by the appropriate lower-case Roman character. For example, $u \equiv nU$ is the internal energy per unit volume in the laboratory frame. Densities referred to the convected fluid frame that is based on coordinate planes embedded in the fluid and moving with it will be designated by the corresponding primed letter. Thus n' and u' are respectively the molar density and molar energy density referred to the convected frame.

ONE-DIMENSIONAL CANONICAL FORMALISM

From the point of view of an observer who remains stationary with respect to a given sample of fluid and refers all measurements to the convected frame, everything can be described as a function of a single variable—the proper-time τ of the sample of fluid under study. Because the fluid appears to remain at rest, the fluid velocity $\mathbf{v}$ does not enter into such a description. When the canonical formalism derived from such an approach is referred to the fixed laboratory frame, however, proper-time differentiation must be defined as $\mathrm{d}/\mathrm{d}\tau \equiv \mathbf{v}\cdot\partial$ where ∂ is the four-gradient operator, and this brings $\mathbf{v}$ into the formalism. Thus the development of the one-dimensional canonical formalism referred to the convected fluid frame is the first step in arriving at the desired variational principle referred to the laboratory frame.

In reversible adiabatic flow the molar entropy and the total number of particles in the fluid are conserved quantities. Our approach will consist of expressing these two conservation laws in terms of two scalar constants of motion of the fluid. The internal energy will then be written as a function of these two constants of motion and of the proper time. Identifying the internal energy with the Hamiltonian of the system and the constants of motion with generalized momentum coordinates, we are led to the canonical equations of motion.

In order to arrive at the desired statement of conservation of particles, we first note that the molar rest-volume V (not to be confused with the Lorentz-contracted molar volume $V^* = V/\Gamma$ where $\Gamma \equiv [1-(v/c)^2]^{-\frac{1}{2}}$) may be written $V = JV'$ where $V' = (V)_{\tau=0}$ is the molar volume referred to the convected frame, which is a constant of motion, and is equal to the initial value of V at $\tau = 0$, and J is the function of τ that describes the time-dependence of V that results from compression or expansion of the fluid. The Lorentz-contracted molar volume is thus $V^* = V/\Gamma = (J/\Gamma)\,V'$. Because intervals of laboratory-time $\mathrm{d}t$ and proper-time $\mathrm{d}\tau$ are related by $\mathrm{d}t = \Gamma\,\mathrm{d}\tau$ we have:

$$\mathrm{d}\mathscr{V} \equiv c\,\mathrm{d}t\,\mathrm{d}V^* = c\Gamma\,\mathrm{d}\tau(J/\Gamma)\,\mathrm{d}V' = J(c\,\mathrm{d}\tau\,\mathrm{d}V') \tag{1}$$

where $\mathrm{d}\mathscr{V}$ is the element of four-volume in the laboratory frame and $c\,\mathrm{d}\tau\,\mathrm{d}V'$ is the corresponding four-volume element in the convected frame. Thus J is just the Jacobian of the transformation between laboratory coordinates and convected coordinates.

Using $V = JV'$, the thermodynamic equation $\mathrm{d}U = T\,\mathrm{d}S - P\,\mathrm{d}V$ would become

$$\mathrm{d}U = T\,\mathrm{d}S - (PJ)\mathrm{d}V' - (PV'\dot{J})\mathrm{d}\tau \tag{2}$$

where $\dot{S} = \dot{V}' = 0$ and $U = U(S, V', \tau)$ would be the thermodynamic potential that we could identify with the Hamiltonian. However, because we are dealing with a continuum, it is more appropriate to work with densities rather than with molar quantities. For this reason, we eliminate V', U and S in favour of n', u' and s' where

$$n' \equiv 1/V' = J/V = Jn; \quad u' \equiv n'U; \quad s' \equiv n'S \tag{3}$$

Making these substitutions in 2, we find

$$\mathrm{d}u' = T\mathrm{d}s' + G\mathrm{d}n' - (P\dot{J})\mathrm{d}\tau \tag{4}$$

where $G = U + PV - TS$ is the molar Gibbs function. Thus $u' = u'(s', n', \tau)$ is a function of two constants of motion and of the proper-time.

Before identifying u' with the Hamiltonian of the system, we note that equation 4 has two deficiencies which luckily can both be removed by the addition of a single term. First, from the relativistic point of view, the rest-mass energy density $m'c^2 \equiv n'Mc^2$ (where M is the molar rest-mass) should not be isolated from all other contributions to the energy density. Hence u' should be replaced by the total energy density $\tilde{u}' \equiv n'\tilde{U}$ that includes the rest-mass energy density.

The second deficiency of equation 4 arises from the fact that, if we are to describe a fluid rather than just isolated moles of gas that in no way interact with one another, then we must in some way introduce into the formalism parameters that label and identify each mole of gas and distinguish it from all others. Because these parameters will enter into the formalism, they must have a physical significance that is essential to the description of the fluid. Both of these requirements, labelling and physical significance, are satisfied by the *initial momentum vector* $\mathbf{K} = (M\mathbf{v})_{\tau=0}$ which is the momentum possessed by the mole of gas at $\tau = 0$. In doing this we are effectively postulating that the inability to distinguish between two or more moles of gas that would result if their **K**-vectors were all equal, represents a physical degeneracy with observable consequences. (We shall, in fact, see that such a degeneracy corresponds to vorticity-free flow.) Thus **K**, like the molar entropy S, is a preserved fossil of the initial conditions of the fluid. The vector **K** is normalized to the molar mass M, i.e. $K \equiv (\mathbf{K}\cdot\mathbf{K})^{\frac{1}{2}} = Mc$, and so $M \equiv K/c$ can be used as the definition of molar mass, and $\tilde{U}$ becomes $\tilde{U} \equiv U + c(\mathbf{K}\cdot\mathbf{K})^{\frac{1}{2}}$. There exists an alternative procedure for relating $\tilde{U}$ and U that is not only more general, but also closer to the spirit of thermodynamical formalism. We may regard $\tilde{U} = U(S, n, \mathbf{K})$ as the basic thermodynamic potential and S, n (or V), and **K** as the basic variables. We then *define* M as $Mc^2 \equiv (\partial\tilde{U}/\partial\mathbf{K})\cdot\mathbf{K}$. This definition is consistent with $\tilde{U} = U(S, n) + c(\mathbf{K}\cdot\mathbf{K})^{\frac{1}{2}}$ where $c(\mathbf{K}\cdot\mathbf{K})^{\frac{1}{2}} = Mc^2$, but it is more general, and is applicable regardless of the **K**-dependence of $\tilde{U}$. Using this more general definition of M, we *define* the purely thermal energy function U as:

$$U \equiv \tilde{U} - (\partial\tilde{U}/\partial\mathbf{K})\cdot\mathbf{K} \equiv \tilde{U} - Mc^2 \tag{5}$$

Although relation 5 represents the most general way of defining U and M, in this paper we shall assume that the **K**-dependence of $\tilde{U}$ is given by $c(\mathbf{K}\cdot\mathbf{K})^{\frac{1}{2}} = Mc^2$ where M is a constant parameter. In such a case $\partial\tilde{U}/\partial\mathbf{K} = c\mathbf{K}/K = \mathbf{K}/M \equiv \boldsymbol{v}$ where $\boldsymbol{v} \equiv (\mathbf{v})_{\tau=0}$ is the initial velocity of the mole of gas in question at $\tau = 0$.

If $\boldsymbol{\kappa}' \equiv n'\mathbf{K}$ is the initial momentum density referred to the convected frame, the density relation that corresponds to relation 5 is

$$u' \equiv \tilde{u}' - \mathbf{v} \cdot \boldsymbol{\kappa}' \equiv \tilde{u}' - m'c^2 \text{ where } \mathbf{v} \equiv \partial\tilde{u}'/\partial\boldsymbol{\kappa}' = \partial\tilde{U}/\partial\mathbf{K} \tag{6}$$

Thus we see that the definition of u' in terms of $\tilde{u}'$ and $\boldsymbol{\kappa}'$ (or of U in terms of $\tilde{U}$ and $\mathbf{K}$) amounts to a Legendre transformation that replaces the variable $\boldsymbol{\kappa}'$ (or $\mathbf{K}$) with $\mathbf{v} \equiv \partial\tilde{u}'/\partial\boldsymbol{\kappa}' = \partial\tilde{U}/\partial\boldsymbol{\kappa}$.

From expressions 4 and 6 we obtain the basic thermodynamic equation of the fluid

$$\mathrm{d}\tilde{u}' = T\,\mathrm{d}s' + G\,\mathrm{d}n' + \mathbf{v} \cdot \mathrm{d}\boldsymbol{\kappa}' - (P\dot{J})\,\mathrm{d}\tau \tag{7}$$

This is to be compared with the well-known expression for the differential of the Hamiltonian $E = E(p, q, \tau)$:

$$\begin{aligned}\mathrm{d}E = \sum_p(\partial E/\partial p)_{q,\tau}\,\mathrm{d}p + \sum_q(\partial E/\partial q)_{p,\tau}\,\mathrm{d}q + (\partial E/\partial \tau)_{p,q}\,\mathrm{d}\tau \\ = \sum_p \dot{q}\,\mathrm{d}p - \sum_q \dot{p}\,\mathrm{d}q - (\partial L/\partial\tau)_{q,\dot{q}}\,\mathrm{d}\tau\end{aligned} \tag{8}$$

where the Lagrangian L is defined as follows:

$$L = L(\dot{q}, q, \tau) \equiv \sum_p p(\partial E/\partial p)_{q,\tau} - E \tag{9}$$

We now identify $\tilde{u}'$ with E. Note that although $\tilde{u}' = n'\tilde{U}$ is a density, it does in fact represent the energy of a fixed number of particles, namely the number contained in unit volume of the convected frame, and so there is no inconsistency in regarding it as the Hamiltonian of a definite dynamical system.

It turns out that, for consistency, it is necessary to identify the thermodynamic variables s', n' and $\boldsymbol{\kappa}'$ with generalized momenta, rather than with generalized coordinates. Doing this, and designating the coordinates q that are conjugate to the momenta $p = (s', n', \boldsymbol{\kappa}')$ by $q = (\mathscr{T}, \mathscr{G}, \boldsymbol{\xi})$ respectively, comparison of 7 and 8 yields the following equations:

$$T = \dot{\mathscr{T}};\ G = \dot{\mathscr{G}};\ \mathbf{v} = \dot{\boldsymbol{\xi}};\ P\dot{J} = (\partial L/\partial\tau)_{q,\dot{q}} \tag{10}$$

The fact that $\tilde{u}'$ is independent of the coordinates $\mathscr{T}$, $\mathscr{G}$ and $\boldsymbol{\xi}$ yields the desired equations of motion:

$$\dot{s}' = \dot{n}' = \dot{\boldsymbol{\kappa}}' = 0, \quad \text{which implies } \dot{\mathbf{K}} = 0 \tag{11}$$

the last equation resulting from $\boldsymbol{\kappa}' = n'K$ and $\dot{n}' = \dot{\boldsymbol{\kappa}}' = 0$.

From definition 9 we find:

$$\begin{aligned}L \equiv \sum_p p(\partial\tilde{u}'/\partial p) - \tilde{u}' = n'(G + ST + \mathbf{K} \cdot \mathbf{v} - \tilde{U}) \\ = n'(H - U) = n'PV = (n'/n)\,P = JP\end{aligned} \tag{12}$$

where use has been made of 3 and 5. Since $\dot{J} \equiv \mathrm{d}J/\mathrm{d}\tau = \partial J/\partial\tau$, equations 12, together with the last equation of 10, yield $(\partial P/\partial\tau)_{q,\dot{q}} = 0$, which means that $P = P(\dot{\mathscr{T}}, \dot{\mathscr{G}}, \dot{\boldsymbol{\xi}})$ is a function of the generalized velocities $\dot{\mathscr{T}} = T$, $\dot{\mathscr{G}} = G$, $\dot{\boldsymbol{\xi}} = \mathbf{v}$ alone, and not an explicit function of τ. For example, in the case of a perfect gas, for which $P = nRT$, where R is the gas constant and

$\gamma \equiv c_p/c_v$ = constant is the ratio of specific heats, the functional form of P is

$$P = P_0(T/T_0)^{[\gamma/(\gamma-1)]} \quad \exp \quad \{[G + Mc(\mathbf{v}\cdot\mathbf{v})^{\frac{1}{2}}]/RT\} \tag{13}$$

where P_0 and T_0 are constants.

Because $L = JP$, the Lagrangian equations of motion are:

$$\mathrm{d}[\partial(JP)/\partial\dot{q}]/\mathrm{d}\tau = \partial(JP)/\partial q \qquad \text{or} \quad \mathrm{d}[J(\partial P/\partial\dot{q})]/\mathrm{d}\tau = 0 \tag{14}$$

where use has been made of the fact that P is independent of the qs, and J is an explicit function of τ, being independent of the qs and $\dot{q}$s. To evaluate $\partial P/\partial\dot{q}$ we first note that:

$$P = n(H - U) = n(G + ST + \mathbf{K}\cdot\mathbf{v} - \tilde{U}) = nG + sT + \boldsymbol{\kappa}\cdot\mathbf{v} - \tilde{u} \tag{15}$$

where now the densities are all referred to the laboratory frame. Next we note that, from 5 and the relation $\mathrm{d}u = T\,\mathrm{d}s + G\,\mathrm{d}n$, we have

$$\mathrm{d}\tilde{u} = T\,\mathrm{d}s + G\,\mathrm{d}n + \mathbf{v}\cdot\mathrm{d}\boldsymbol{\kappa} \tag{16}$$

Taking the differential form of 15 and using 16, we find:

$$\mathrm{d}P = s\,\mathrm{d}T + n\,\mathrm{d}G + \boldsymbol{\kappa}\cdot\mathrm{d}\mathbf{v} = s\,\mathrm{d}\dot{\mathscr{T}} + n\,\mathrm{d}\dot{\mathscr{G}} + \boldsymbol{\kappa}\cdot\mathrm{d}\dot{\boldsymbol{\xi}} \tag{17}$$

Using 17 to evaluate $\partial P/\partial\dot{q}$, we arrive at the following Lagrangian equations of motion:

$$0 = \mathrm{d}(Js)/\mathrm{d}\tau = \dot{s}'; \quad 0 = \mathrm{d}(Jn)/\mathrm{d}\tau = \dot{n}'; \quad 0 = \mathrm{d}(J\boldsymbol{\kappa})/\mathrm{d}\tau = \dot{\boldsymbol{\kappa}}' \tag{18}$$

These, of course, agree with the canonical equations 11. (If we had identified some or all of the thermodynamic variables with generalized coordinates q, rather than with generalized momenta p, this agreement would not have occurred.)

In the same way that, in arriving at 5 and 6, we noted that the mass density $m' \equiv n'M$ could be *defined* in terms of the $\boldsymbol{\kappa}'$-dependence of u', we now note that the mass density $m \equiv nM$ (referred now to the laboratory frame) can be defined in terms of the $\dot{\boldsymbol{\xi}}$-dependence of P:

$$mc^2 \equiv (\partial\tilde{u}/\partial\boldsymbol{\kappa})\cdot\boldsymbol{\kappa} \equiv \mathbf{v}\cdot\boldsymbol{\kappa} = (\partial P/\partial\dot{\boldsymbol{\xi}})\cdot\dot{\boldsymbol{\xi}} \tag{19}$$

Thus, the definition of the molar mass M may be taken to be:

$$M \equiv n^{-1}(\partial P/\partial\dot{\boldsymbol{\xi}})\cdot\dot{\boldsymbol{\xi}}/c^2 = (\partial P/\partial\dot{\mathscr{G}})^{-1}(\partial P/\partial\dot{\boldsymbol{\xi}})\cdot\dot{\boldsymbol{\xi}}/c^2 \tag{20}$$

VARIATIONAL PRINCIPLE FOR FLUID

The Lagrangian equations can be obtained from the following variational principle: $0 = \delta\int L\,\mathrm{d}\tau = \delta\int JP\,\mathrm{d}\tau$. This refers to the fluid contained in unit volume of the convected frame. If the integrand were $JP\,\mathrm{d}\tau\,\mathrm{d}V'$, the principle would refer to the sample of fluid contained in the volume $\mathrm{d}V'$. Since the fluid contained in each volume element $\mathrm{d}V'$ must individually and independently satisfy the requirement $0 = \delta\int_0^\tau JP\,\mathrm{d}\tau\,\mathrm{d}V'$, then it must follow that $0 = \delta\int_{\tau,v'} JP\,\mathrm{d}\tau\,\mathrm{d}V'$ where now the integration extends over V$'$ as well as over τ. Thus, referring to relations 1, we arrive at the following variational principle for the fluid:

$$0 = \delta\int_{\tau,v'} PJc\,\mathrm{d}\tau\,\mathrm{d}v' = \delta\int_v P\,\mathrm{d}\mathscr{V} \tag{21}$$

where $\mathrm{d}\mathscr{V} \equiv c\,\mathrm{d}t\,\mathrm{d}x\,\mathrm{d}y\,\mathrm{d}z$ is the four-volume element in the laboratory frame, and $\mathscr{V}$ is the four-volume occupied by the fluid between the specified initial and final times.

Because the proper-time τ is no longer the independent variable, the operation $\mathrm{d}/\mathrm{d}\tau$ must be defined in terms of the fluid velocity $\mathbf{v}$ as $\mathrm{d}/\mathrm{d}\tau \equiv \mathbf{v}\cdot\boldsymbol{\partial}$ where $\boldsymbol{\partial}$ is the four-gradient operator. Because $\mathbf{v}$ must remain normalized ($\mathbf{v}\cdot\mathbf{v} = c^2$) during the variation process, it must be parametrized in some way so that the normalization will be guaranteed. This can be done most conveniently by introducing a vector $\boldsymbol{\rho}$ whose direction is $\mathbf{v}$

$$\mathbf{v} \equiv c\boldsymbol{\rho}/\rho \quad \text{where} \quad \rho \equiv (\boldsymbol{\rho}\cdot\boldsymbol{\rho})^{\frac{1}{2}} \tag{22}$$

It will turn out that the norm ρ does not appear in any of the Euler–Lagrange equations resulting from 21. The components of $\boldsymbol{\rho}$ are to be regarded as generalized coordinates, rather than as velocities. Using the definition 22 for $\mathbf{v}$ we have:

$$G \equiv \dot{\mathscr{G}} \equiv \mathbf{v}\cdot\boldsymbol{\partial}\mathscr{G} = \rho^{-1}\boldsymbol{\rho}\cdot\boldsymbol{\partial}\mathscr{G} \tag{23}$$

Similar expressions define $\dot{\mathscr{T}}$ and $\dot{\boldsymbol{\xi}}$.

The more detailed statement of the variational principle given in 21 is

$$0 = \sum_q \int_{\mathscr{V}} (\delta q)[(\partial P/\partial q) - \boldsymbol{\partial}\cdot\mathbf{p}]\,\mathrm{d}\mathscr{V} + \sum_q \int_{\mathscr{V}} (\delta q)\mathbf{p}\cdot\mathbf{d}\mathscr{S} \tag{24}$$

where $\mathbf{p} \equiv \partial P/\partial(\boldsymbol{\partial} q)$ and $\mathbf{d}\mathscr{S}$ is an element of the hypersurface $\mathscr{S}$ that bounds the four-volume $\mathscr{V}$ over which the integration is carried out. The variational principle is thus equivalent to the requirement that the Euler–Lagrange equations $\boldsymbol{\partial}\cdot\mathbf{p} = \partial P/\partial q$ be satisfied, and that the variables have definitely assigned values on the boundary so that $\partial q = 0$ on $\mathscr{S}$.

Referring to 17 and 23, the calculation of the generalized momenta can be illustrated by the case for $\mathscr{G}$:

$$\mathbf{p}_{\mathscr{G}} \equiv \partial P/\partial(\boldsymbol{\partial}\mathscr{G}) = (\partial P/\partial G)\,[\partial G/\partial(\boldsymbol{\partial}\mathscr{G})] = n\mathbf{v} \tag{25}$$

Similarly we find $\mathbf{p}_k = Sn\mathbf{v}$, $p^j_{\xi k} = nK_k v^j$, and $p^j_{\rho k} = 0$. Thus the Euler–Lagrange equations corresponding to variation of $\mathscr{G}$, $\mathscr{T}$ and $\boldsymbol{\xi}$ respectively are:

$$0 = \boldsymbol{\partial}\cdot(\mathbf{v}n) = \boldsymbol{\partial}\cdot(\mathbf{v}nS) = \boldsymbol{\partial}\cdot(\mathbf{v}n\mathbf{K}) \tag{26}$$

Variation of $\boldsymbol{\rho}$ yields the following equation:

$$\begin{aligned} 0 = \partial P/\partial\boldsymbol{\rho} &= \sum_q (\partial P/\partial\dot{q})\,[\partial(\mathbf{v}\cdot\boldsymbol{\partial}q)/\partial\boldsymbol{\rho}] \\ &= (c/\rho)\sum_q (\partial P/\partial\dot{q})\,[\boldsymbol{\partial}q - \mathbf{v}(\dot{q}/c^2)] \\ &= (cn/\rho)\,[\boldsymbol{\partial}\mathscr{G} + S\,\boldsymbol{\partial}\mathscr{T} + (\boldsymbol{\partial}\boldsymbol{\xi})\cdot\mathbf{K} - (M + H/c^2)\mathbf{v}] \end{aligned} \tag{27}$$

or

$$(M + H/c^2)\mathbf{v} = \boldsymbol{\partial}\mathscr{G} + S\,\boldsymbol{\partial}\mathscr{T} + (\boldsymbol{\partial}\boldsymbol{\xi})\cdot\mathbf{K} \tag{28}$$

Equations 26 are just the required conservation laws. Using the first, the second and third could also be written as $\dot{S} = \dot{\mathbf{K}} = 0$. Equation 28 is effectively the formal solution of the fluid equation of motion (Euler's equation), and thus amounts to a statement of conservation of energy-momentum.

That this is true can be verified by evaluating the stress-energy tensor w^j_k which, since our Lagrangian density is P, is given by:

$$w^j_k \equiv \sum_q p^j \partial_k q - P\delta^j_k \equiv \sum_q [\partial P/\partial(\partial_j q)]\,\partial_k q - P\delta^j_k \tag{29}$$

which, if the Euler–Lagrange equations $\boldsymbol{\partial}\cdot\mathbf{p} = \partial P/\partial q$ are satisfied, automatically satisfies the equation:

$$\partial_j w^j_k = -(\partial P/\partial x^k)_{q,\dot{q}} = 0 \tag{30}$$

where we have postulated that the pressure function possesses no *explicit* dependence on the space-time coordinates. Using the expressions for the ps that were given following 25, and making use of 28, we find that 29 becomes:

$$\begin{aligned} w^j_k &= nv^j[\partial_k\mathscr{G} + S\partial_k\mathscr{T} + (\partial_k\boldsymbol{\xi})\cdot\mathbf{K}] - P\delta^j_k \\ &= n(M + H/c^2)v^j v_k - P\delta^j_k \end{aligned} \tag{31}$$

Making use of the first equation of 26, equation 30 becomes

$$\mathrm{d}[(M + H/c^2)\mathbf{v}]/\mathrm{d}\tau = n^{-1}\boldsymbol{\partial}P \tag{32}$$

This is just Euler's equation for the fluid, and may be regarded as the determining equation for the molar energy-momentum vector $(M + H/c^2)\mathbf{v}$. But 28 gives an explicit expression for this vector in terms of the four-gradients of the canonical coordinates, so, as previously remarked, 28 constitutes the formal solution of Euler's equation.

It should be noted, incidentally, that when $\mathbf{K}$ becomes constant over any region, the term $(\boldsymbol{\partial}\xi)\cdot\mathbf{K}$ in 28 becomes the gradient of a scalar, and this corresponds to vorticity-free flow[9] in this region. As previously noted, this physically observable effect is characterized by a degeneracy resulting from the fact that the labelling vector $\mathbf{K}$ is indistinguishable for neighbouring samples of fluid. If the vector $\mathbf{K}$, and hence $\boldsymbol{\xi}$, had never been introduced into the formalism, and we had instead introduced the rest-mass energy Mc^2 simply by replacing $\mathscr{G}$ by $\tilde{\mathscr{G}}$ where $\mathrm{d}\tilde{\mathscr{G}}/\mathrm{d}\tau \equiv G + Mc^2$, we would have arrived at a variational principle implicitly restricted to the case of vorticity-free flow.

REFERENCES

1 For references, see article by C. Truesdell and R. Toupin in *Handbuch der Physik*, S. Flügge (Ed.) Springer: Berlin (1960); cf. footnote 3 on p 650.

2 H. von Helmholtz, *Jour. für Math.* (*Crelle*), **100**, 137, 213 (1886). Reprinted on pp 203–248 of *Hermann von Helmholtz, Wissenschaftliche Abhandlungen*, Vol. III. Barth: Leipzig (1895).

3 M. Planck, *S.B. Preuss. Akad. Wiss. Berlin* (1907), p 542. Reprinted in *Ann. Phys., Lpz.* **26**, 1 (1908).

4 Appendix of paper by L. A. Schmid in *Thermodynamics Symposium* (University of Pittsburgh), A. Brainard (Ed.) Mono Book Corp.: Baltimore (1970).

5 P. T. Landsberg and K. A. Johns, *Nuovo Cimento*, **52B**, 28 (1967).

6 D. Van Dantzig, *Physica*, **6**, 673 (1939); *Proc. Kon. Ned. Akad. Wet.* **42**, 601, 608 (1939).

7 D. Van Dantzig, *Proc. Kon. Ned. Akad. Wet.* **43**, 387, 609 (1940).

8 R. Hargreaves, *Phil. Mag.* **16**, 436 (1908).

9 L. A. Schmid, *Nuovo Cimento*, **52B**, 313 (1967).

F. PEDAGOGICAL PAPERS

AN ELEMENTARY INTRODUCTION TO ENTROPY VIA IRREVERSIBILITY

R. GILES

Mathematics Department, Queen's University, Kingston, Ontario, Canada

ABSTRACT

An outline is given of an approach to thermodynamics in which entropy is derived directly, from a simple postulate of direct experimental significance, without reference to temperature or thermal equilibrium. The author first shows how the irreversibility of a natural process may be quantitatively measured; entropy is then defined so that its increase in any process equals the irreversibility. Lastly equilibrium states are defined and absolute temperature is derived from entropy by differentiation.

1. INTRODUCTION

The following is an outline, necessarily very condensed, of the initial stages of a one-term course in thermodynamics which was offered for several years to the honours class at Glasgow University. The characteristic feature of the treatment is that entropy is the first and not the last of the basic thermodynamic quantities to be formally introduced; and it is not introduced *ad hoc*, but derived from a very simple and plausible postulate (4.1 below) having a direct experimental meaning. In this way its fundamental significance is made apparent and it gains for the student an aura of reality which is often not realized by conventional treatments. Another didactic advantage is that instead of having to obtain entropy from temperature and energy by a process of integration we use *differentiation*—to the average student a much simpler procedure—to define temperature in terms of entropy.

The treatment can be regarded as a very much simplified version of that developed in *Mathematical Foundations of Thermodynamics*[1] (hereafter referred to as *MFT*). At the expense of some sacrifice in rigour, mathematical sophistication has been avoided and explanations of physical concepts have been largely replaced by illustrative examples. (For reasons of space, however, the number of examples in the present account has had to be severely limited.)

2. SYSTEMS AND STATES

Space limitations preclude any proper discussion of these concepts. We denote systems by capital letters A, B, ... and states of A by A_1, A_2, ... One example must suffice for illustration. Let L denote a (particular) solid metal cube. Let L_1, L_2 and L_3 denote the states of L in which its temperature is uniform and 0°C, 50°C and 100°C respectively. Let L_4 denote the steady state attained by L when two opposite faces are maintained at 0°C and 100°C respectively. Thus in L_4 there is a uniform temperature gradient across the block. Clearly L_4 is 'different' from the other states: if L is isolated the state L_4 will change, the temperature gradually becoming more uniform until eventually L_2 will result. (This unusual behaviour of L_4 is, of course, due to the fact that it is not an 'equilibrium state'. Since we make no restriction to equilibrium states, however, we do not need to give a formal explanation of the term at this stage.)

Two systems A and B may be *thought of* as together comprising a single system which we call their *union* and denote $A + B$. To form the union is a purely conceptual process: it is not necessary that the systems interact or even be in contact. However, in practice there is little point in considering $A + B$ unless some, possibly indirect, interaction between A and B is contemplated. Occasionally we will use such an expression as $2A$. This is shorthand for $A + A$ and denotes the union of A with *a replica of itself*. Finally, $A_1 + B_1$ will denote that state of $A + B$ in which the systems A and B are separated (i.e. not in interaction) and in states A_1 and B_1.

3. PROCESSES

If the state of a system changes a *process* is said to have occurred. We name the process by giving the initial and final states: for example, (L_4, L_2) denotes the process (mentioned above) of settling down which occurs in L if it is isolated and initially in the state L_4. That this notation for processes—involving only the naming of the initial and final states—is justifiable depends on the following circumstance: classical thermodynamics is concerned primarily with those properties of processes which depend only on the initial and final states, being independent of the particular manner by which the change of state took place. Thus if two processes have these features in common they need not be distinguished and will be called *equivalent*. As an important example, any process for which the initial and final states coincide is equivalent to the trivial process in which no change whatever takes place; this trivial process we call the *zero process.*

(L_4, L_2) is an example of a process which can occur *in isolation*: i.e. while the system concerned, namely L, is isolated. It is called a *natural process* and we write $L_4 \to L_2$. On the other hand it is not *reversible* for its *reverse* (L_2, L_4) cannot occur in isolation. (Of course we can by external action *compel* the process (L_2, L_4) to occur—for instance by enclosing L, initially in the state L_2, between two heat reservoirs.) We shall call (L_2, L_4) *anti-natural* (against nature) since its reverse is natural. We have thus $L_4 \to L_2$ but $L_2 \nrightarrow L_4$.

The process (L_1, L_2) is another example of a process which cannot occur in isolation, although it can of course be compelled to occur by bringing L into contact with a suitable heat reservoir. Exactly the same applies to its reverse (L_2, L_1). (Isolation implies, in particular, perfect thermal insulation so that cooling, just as much as heating, is impossible.)

Now suppose we have two copies of the block L in the states L_1 and L_3. By bringing the two blocks together, waiting until no further change takes place, and then separating them again we obtain the final state $L_2 + L_2$. Thus $(L_1 + L_3, L_2 + L_2)$ is a natural process†. Let us denote this process by α. The process α has involved two systems, each a replica of the block L, which have experienced the processes (L_1, L_2) and (L_3, L_2) respectively. Denoting these processes by β and γ we might say that α consists in the 'simultaneous occurrence' of β and γ. We express this by writing $\alpha = \beta + \gamma$: i.e. $(L_1 + L_3, L_2 + L_2) = (L_1, L_2) + (L_3, L_2)$. Notice that neither β nor γ is natural, although their sum α is.

The *sum* of any two processes is defined in the same way: the sum of the processes (A_1, A_2) and (B_1, B_2) of the systems A and B respectively, is the process $(A_1 + B_1, A_2 + B_2)$ of the system $A + B$. Observe that the sum of a process and its reverse is (equivalent to) the zero process:

$$(A_1, A_2) + (A_2, A_1) = (A_1 + A_2, A_2 + A_1) = 0$$

since $A_1 + A_2$ and $A_2 + A_1$ are really the same state. Hence we may call the reverse of a process α the *negative* of α and denote it $-\alpha$.

We have agreed to write $A_1 \to A_2$ and to describe the process (A_1, A_2) as natural not only when the process (A_1, A_2) can occur in isolation but also when it can be caused to occur by an arbitrarily small external interference. We now make a further relaxation of these conditions. Suppose that we can envisage some apparatus K which can be used to cause A to undergo the process (A_1, A_2) and suppose, moreover, that this can be done in such a way that the final state of the apparatus coincides with its initial state, K_0 say. In this case the apparatus has in no sense been 'used up' in the process (we shall say it is not *involved* in the process)—indeed it is at once ready to be employed in the same way again. We agree to allow this sort of use of auxiliary apparatus:

3.1. Definition

We write $A_1 \to A_2$ and call the process (A_1, A_2) *natural* whenever there exists some system K and some state K_0 of K such that the process $(A_1 + K_0, A_2 + K_1)$ can occur while the system $A + K$ is isolated, the state K_1 being equal to (or at least differing arbitrarily little from) K_0.

Sometimes the system K takes the form of an *engine* which works in cycles—if a whole number of cycles has been performed the initial and final states will coincide.

Using this definition we can establish two results that we shall need later:

3.2. Theorem

Let A and B be any systems, A_1 and A_2 states of A, B_0 a state of B. Then:
(a) If $A_1 \to A_2$ then $A_1 + B_0 \to A_2 + B_0$
(b) If $A_1 + B_0 \to A_2 + B_0$ then $A_1 \to A_2$.

Proof.

(a) By hypothesis there is an 'engine' K which, starting and finishing in

† It is true that some external agency has been used to move the blocks, so the system $2L$ has not been strictly isolated. However, we still regard the process as natural since—there being no limit on the time required—the external interference can be arbitrarily slight.

some state K_0, can take A_1 into A_2. We need now merely retain B in the state B_0 during this process, and observe what has happened to $A + B$†.

(b) If K, with initial and final state K_0, can implement $(A_1 + B_0, A_2 + B_0)$ —i.e. if $(A_1 + B_0 + K_0,\ A_2 + B_0 + K_0)$ can occur in isolation—then $B + K$, with initial and final states $B_0 + K_0$, implements the process (A_1, A_2).

The proof of the following theorem is similar:

3.3 Theorem

If $A_1 \to A_2$ and $A_2 \to A_3$ then $A_1 \to A_3$.

Using these results it is now easy to prove:

3.4 Theorem

If α and β are natural processes then so is $\alpha + \beta$.

4. IRREVERSIBILITY

We now introduce the basic postulate of our formulation:

4.1. Postulate

Let A_1, A_2, A_3 be any states of any system A. If $A_1 \to A_2$ and $A_1 \to A_3$ then either $A_2 \to A_3$ or $A_3 \to A_2$ (or possibly both).

On this postulate depends the construction of an entropy function and thus the whole structure of thermodynamics. From it we deduce:

4.2. Theorem

Given any two natural processes α and β, one of them is able to 'drive the other backwards': i.e. either $\alpha - \beta$ or $\beta - \alpha$ is natural (possibly both).

The proof is simple (see *MFT*, p 34). This theorem makes it possible to measure quantitatively the irreversibility of a natural irreversible process. Indeed, we are going to assign to each possible‡ process α a scalar quantity $I(\alpha)$, the *irreversibility* of α, in such a way that:

$$\text{(i)} \begin{cases} I(\alpha) > 0 \text{ if } \alpha \text{ is natural irreversible,} \\ I(\alpha) = 0 \text{ if } \alpha \text{ is reversible,} \\ I(\alpha) < 0 \text{ if } \alpha \text{ is antinatural irreversible;} \end{cases}$$

(ii) I is *additive*: i.e. $I(\alpha + \beta) = I(\alpha) + I(\beta)$, for all possible processes α and β.

We measure the irreversibility of a natural process α by comparing it with that of a standard irreversible process γ, the irreversibility $I(\gamma)$ of γ being assigned arbitrarily§. We say α *is at least* (*most*) *r times as irreversible as* γ if

† It is necessary to assume that any state can be 'frozen', i.e. kept unchanged, when required. This may require some cunning. To freeze the state L_4, for instance, we may imagine that the block L is built out of a large number of thin square metal plates and that these are instantly separated from each other; on reassembly, the state L_4 is restored.

‡ 'Possible' means 'natural or antinatural or both'.

§ Any such process γ may be used [or, if there is none, we simply set $I(\alpha) = 0$ for all α]. However, to get the customary scales of entropy and temperature we may take for γ a natural process of the form $\gamma = (M_1 + R_1, M_2 + R_2)$ where M is a mechanical system (see § 5) with M_1 exceeding M_2 in energy by 1 erg, and R is a sealed container enclosing only a mixture of ice, water and water vapour (R is thus a 'heat reservoir' at the triple point of water); and set $I(\gamma) = (1/273{\cdot}16)$ erg/deg, *inventing* the new unit erg/deg to measure the new fundamental quantity, irreversibility.

$q\alpha - p\gamma$ is natural (antinatural) for positive integers p and q with $r = p/q$. A straightforward argument (*MFT*, p 43) based on theorems 3.4 and 4.2 shows that there is a unique real number r_0 such that, for any smaller (larger) rational number r, α is at least (most) r times as irreversible as γ. (Moreover, the principle mentioned in definition 3.1—that arbitrarily small changes in the environment are permissible— allows us to conclude that if r_0 is rational it belongs to both these classes.) We set $I(\alpha) = r_0 I(\gamma)$. It follows from the construction that the function I so defined has the property (i) above. That, for any natural processes α and β, $I(\alpha + \beta) = I(\alpha) + I(\beta)$ follows from the easily established fact (*MFT*, p 45) that if α and β are at least (most) r times and s times as irreversible as γ respectively, then $\alpha + \beta$ is at least (most) $r + s$ times as irreversible as γ.

With trivial changes the above definition of $I(\alpha)$ can be extended to every possible (i.e. natural or antinatural) process α.

5. ENTROPY

We now introduce the notion of a *mechanical system*: i.e. one of those idealized systems dealt with in elementary mechanics, from which dissipative forces (friction, viscosity, etc.) are absent. We assume as characteristic of mechanical systems that (a) the union of two mechanical systems is a mechanical system, and (b) any natural process involving only a mechanical system is reversible; thus, if α is such a process, $I(\alpha) = 0$. We define an *adiabatic process* of a system A to be a process of $A + M$ where M is any mechanical system: i.e. a process which 'involves', apart from A, only a mechanical system. Thus a *natural adiabatic process* of A means a natural process of $A + M$, and so on. Lastly, we assume (cf. Pippard[2] p 15) that any two states of a closed system can be connected, in at least one direction, by an adiabatic process. We can now prove:

5.1. Theorem

The irreversibility of a natural adiabatic process depends only on the system's initial and final states: i.e. if α and β are two natural adiabatic processes of A, both leading from A_1 to A_2, then $I(\alpha) = I(\beta)$.

Proof.

Let $\alpha = (A_1 + M_1, A_2 + M_2)$ and $\beta = (A_1 + N_1, A_2 + N_2)$, where M and N are mechanical systems. Since α and β are natural either $\alpha - \beta$ or $\beta - \alpha$ is natural, say $\alpha - \beta$. But $\alpha - \beta = (A_1 + M_1 + A_2 + N_2, A_2 + M_2 + A_1 + N_1) = (M_1 + N_2, M_2 + N_1)$ is a process involving only the mechanical system $M + N$. Being natural, it is also reversible. Thus $0 = I(\alpha - \beta) = I(\alpha) - I(\beta)$.

5.2. Theorem

If α and β are natural adiabatic processes of systems A and B, leading from A_1 to A_2 and from B_1 to B_2 respectively, then $\alpha + \beta$ is a natural adiabatic process of $A + B$ leading from $A_1 + B_1$ to $A_2 + B_2$, and $I(\alpha + \beta) = I(\alpha) + I(\beta)$.

Proof.

The first statement follows immediately from the definitions and the second is just property (ii) of §4.

An inspection of these theorems shows that 'natural' may be replaced by 'possible' without affecting the proofs.

We can now define an entropy function S. For each system A first choose arbitrarily† a reference state A_0 and assign it zero entropy: $S(A_0) = 0$. Let A_1 be any other state. By assumption there exists a natural adiabatic process connecting A_0 and A_1. If it leads from A_0 to A_1 call it α; if it leads from A_1 to A_0 call its reverse α. In either case define $S(A_1) = I(\alpha)$.

With this definition the entropy of every state of every mechanical system is automatically zero. The following theorem can now be easily proved.

5.3. Theorem

(a) For any states A_1 and B_1 of systems A and B, $S(A_1 + B_1) = S(A_1) + S(B_1)$.

(b) Let α be a natural adiabatic process of a system A leading from A_1 to A_2. Then $I(\alpha) = S(A_2) - S(A_1)$.

Now, any natural process (A_1, A_2) involving only a system A can be regarded as a special case of a natural adiabatic process of A [by writing it in the form $(A_1 + M_1, A_2 + M_1)$]. Applying this to the case when A is the union of several other systems, we have, in view of the additivity of entropy:

5.4 Corollary

In any natural process the total entropy of all the systems involved never decreases, and it remains constant only if the process is reversible.

6. EQUILIBRIUM STATES AND TEMPERATURE

The introduction of entropy in §5 involved no reference to temperature. This is not surprising since most states—e.g. L_4 or $L_1 + L_3$—do not 'have' a temperature at all. However, we now define an *equilibrium* state in such a way that every equilibrium state has a temperature. The usual meaning of 'equilibrium' is somewhat vague and involves reference to the internal structure of the state; ours is quite specific, involving only the concepts that we have already introduced. Roughly speaking, an equilibrium state is a state of 'maximum settled-down-ness':

6.1. Definition

A_1 is an *equilibrium state* of a system A if there is no state A_2 such that (A_1, A_2) is natural irreversible process.

It is easy to deduce from this definition that $A_1 + B_1$ can be an equilibrium state only if A_1 and B_1 are equilibrium states. However, this condition is not sufficient: for instance $L_1 + L_3$ is not an equilibrium state (see §3).

To introduce temperature we must first construct an internal energy function E. Our route is the usual one[2], differing only in certain details. We assume that every state of a mechanical system has a definite energy and

† Except that if A_0 and B_0 are the reference states for A and B then the reference state chosen for $A + B$ must be $A_0 + B_0$.

that for mechanical systems energy is additive and always conserved (i.e. in any natural process 'involving' only a mechanical system the initial and final energies are equal). We define the *work* W done on A in an adiabatic process to be the decrease in energy of the mechanical system involved. We can then *prove*† the first law of thermodynamics in its usual form[2, 3].

The introduction of temperature is most simply described in the case of a simple fluid, or *chemical system* in the sense of Zemansky[3]. In the present context such a system is best defined as one with the following two properties:

(a) every state of the system has a definite *volume* V,

(b) if two states A_1 and A_2 have the same energy and the same volume then either $A_1 \rightarrow A_2$ or $A_2 \rightarrow A_1$ (or both).

It follows that two equilibrium states of the same energy and volume must have the same entropy. The equilibrium states of a simple fluid thus lie on an *equilibrium surface* $S = S(E, V)$ in a 'space' with coordinates E, V, S. If we assume, as is customary in physics, that this surface is sufficiently smooth we can now define the temperature T of any equilibrium state by the equation $1/T = \partial S/\partial E$. At the same time the pressure P may be defined by the equation $P/T = -\partial S/\partial V$.

It is a simple matter to show that T and P have the qualitative and quantitative properties associated with the terms absolute temperature and pressure.

REFERENCES

1 R. Giles. *Mathematical Foundations of Thermodynamics.* Pergamon: Oxford (1964).

2 A. B. Pippard. *The Elements of Classical Thermodynamics.* Cambridge University Press: London (1957).

3 M. W. Zemansky. *Heat and Thermodynamics*, 4th ed. McGraw-Hill: New York (1957).

† That this is possible is due to the strength of the assumption stated in the previous parenthesis. The proof is practically that of theorem 5.1, with W replacing I.

A SIMPLE, UNIFIED APPROACH TO THE FIRST AND SECOND LAWS OF THERMODYNAMICS

J. KESTIN

Department of Engineering, Brown University

ABSTRACT

The article gives a new verbal formulation of the second law of thermodynamics. It is claimed that the physical content of this statement as well as the derivation of the mathematical consequences normally referred to as the first and second parts of the second law are simpler and more easily grasped by beginners than the standard formulations. The argument is so designed as to be closely modelled on one which pertains to the derivation of the mathematical formulation of the first law.

1. MOTIVATION FOR THIS ARTICLE

There is little advantage, from the point of view of advancing progress in physics, in reopening the question of the optimal formulation of the second law of thermodynamics. However, a case can be made for returning to this fundamental topic in the interests of those who are engaged in transmitting existing knowledge.

Regardless of which primary formulation of the second law is adopted, it is commonly agreed that it must lead, by an easy logical and mathematical derivation, to three statements:

(a) There exists a property called entropy, S, which is additive for subsystems and which possess the mathematical properties of a potential.

(b) There exists a variable, called thermodynamic temperature, T, which has the mathematical property of being that integrating denominator, among infinitely many, for an element of heat, dQ^0, in a *reversible process*† which turns the latter into the perfect differential of entropy

$$dS = dQ^0/T \text{ (Carnot's theorem)} \tag{1}$$

The thermodynamic temperature, T, is a unique function of any empirical temperature, t.

(c) There exists a quantity called entropy production, θ, which is positive in any irreversible process. In an adiabatic irreversible process between an initial state 1 and a final state 2, we define

$$\theta = S_2 - S_1 \tag{2a}$$

and must have

$$\theta > 0 \tag{2b}$$

† All symbols with the superscript ^{0} refer to reversible processes.

whereas in any quasistatic irreversible process we must find that

$$dS - dQ/T = d\theta \text{ with } d\theta > 0 \tag{3}$$

To the preceding three requirements one may add the pedagogical *desideratum* that the plan of derivation should be as close as possible in spirit and in the basic appeal to experiment (or intuition) to that of the first law. The present article undertakes to sketch a development of this kind for which the claim is made that it is easily grasped by beginners.

2. RECENT WORK WITH SIMILAR MOTIVATION

A similar concern is evidenced in the articles by L. A. Turner[1], P. T. Landsberg[2–4], F. W. Sears[5], and M. W. Zemansky[6], as well in the latter's recent book[7]. It may even be said to go back to M. Born[8]. In particular, M. W. Zemansky[6,7] ably proceeds to simplify the mathematical apparatus needed in the development, thus considerably reducing the amount of prior preparation required of the student. Questions of mathematical rigour which must be answered in this connection, and which are evaded here owing to present intent, have been investigated, and thoroughly answered by P. T. Landsberg[2–4,9].

3. METHODOLOGY

A review of standard textbooks reveals that there exist two fully equivalent[2–4] and yet pedagogically divergent ways of leading the student to the three conclusions. One stems from R. Clausius and Lord Kelvin, the other from C. Carathéodory and M. Born†. Broadly speaking, the first stream makes the statement that a selected irreversible process *is* irreversible, and develops the theory from a particular case by a discussion of reversible and irreversible cycles. The common objection to this development is a sense of artificiality and the impression of an unmotivated *ad hoc* reasoning given by it to a beginning student. The second method starts with an abstract, common characteristic of all irreversible processes, and derives the same three statements as a result of Carathéodory's mathematical theorem. The objection to this development turns on the fact that the theorem is not normally expounded in courses in mathematics, and that the need to grasp it diverts the student's attention from physics to mathematics.

P. T. Landsberg[3] and, later, M. W. Zemansky[6] achieved a 'reconciliation' of the two streams of thought, and the object here is to suggest a further simplification as well as a closer link with the development of the first law. Thus, in addition to statements (a) to (c) above, we must also show that

(d) There exists a universal function for all systems, called their energy, E, which has the mathematical properties of a potential.

For the sake of completeness, we must also mention the so-called postulational method which *starts* with the equivalents of statements (a) to (c). The common pedagogical objection to this mode of exposition is that it expects the student to accept statements which are alien to him without first creating an adequate intuitional and physical foundation.

† For a parallel exposition of these two streams, the reader may consult Chapters 9 and 10 in ref. 10.

4. EXPERIMENTAL BASIS

In order to provide an easy intuitive grip on the subject we propose to root the exploration in a single experiment, the famous experiment performed by J. P. Joule. The result of this experiment can be expressed in precise thermodynamic terms as follows:

A. *Given two arbitrary states of equilibrium 1 and 2 of any closed system it is either possible to reach state 2 from state 1 or state 1 from state 2, but not both, by an adiabatic process involving the performance of work only. The work so performed is independent of the details of the process.*

B. *If the process is performed at constant volume in a simple system or, generally, with constant initial and final values of the deformation coordinates, work must be performed on the system.* (Such a process cannot be carried out without work or in a manner to *produce* work.)

5. THE FIRST LAW

We consider a space of the n independent thermodynamic properties, $x_1, \ldots, x_n$, of a system. For ease of illustration, we assume in *Figure 1* that x, y are deformation coordinates and that the third coordinate is the empirical temperature, t. We now centre attention on an arbitrary state 1 and say, *by definition*, that any state 2 which can be reached from 1 or from which state 1 can be reached adiabatically without the performance of work† ($W_{12} = 0$), is called isoenergetic with it. For definiteness, we shall assume that the natural direction of all processes considered henceforth is $1 \rightarrow 2$.

We now examine all states for which the deformation variables have given values $x = x_2$ and $y = y_2$; they lie on the vertical line β. It follows immediately from statement A that there exists only one state 2 on β, denoted by e_2 in *Figure 1*, which is isoenergetic with state 1. If a second such state existed,

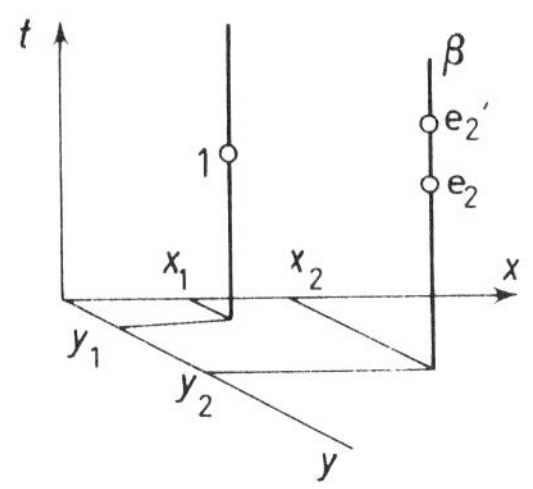

Figure 1. Uniqueness of isoenergetic point.

say at e_2', it is clear from statement B that process $e_2 \rightarrow e_2'$ or $e_2' \rightarrow e_2$ would require the performance of work. For the sake of being definite, suppose that negative work is associated with process $e_2 \rightarrow e_2'$. It follows that process $1 \rightarrow e_2 \rightarrow e_2'$ would require the performance of work. Therefore, state e_2 would not be isoenergetic with 1. By continuity‡, we now reach the conclusion that the locus of all states which are isoenergetic with an arbitrary

† We follow the convention that the work performed on the system is negative and that performed by the system is positive.

‡ The full implications of the assumption of continuity—understandably evaded in an elementary exposition—are treated rigorously in refs. 2, 3, 4 and 9. Ref. 2 examines this problem in depth.

state 1 form a surface or, more precisely, a hypersurface of $n - 1$ dimensions in the space of states of n dimensions.

Varying the temperature of state 1 along the line α for which $x = x_1$ and $y = y_1$, we can classify all such states according to the quantity of work required to reach them adiabatically from 1 or, for negative work, according to the work required to reach state 1 from them. With each such state, $1', 1'', \ldots$, there is associated an isoenergetic hypersurface of $n - 1$ dimensions, *Figure 2*.

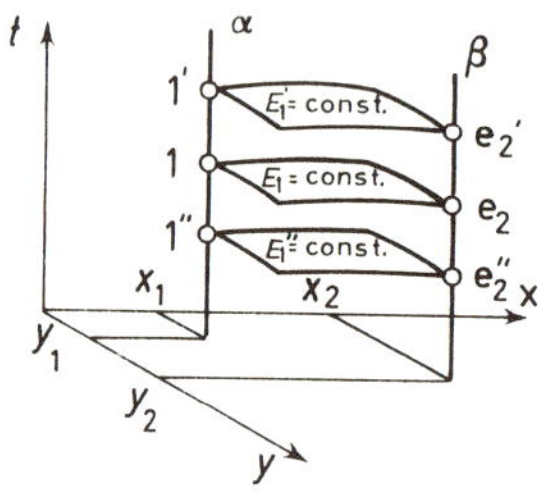

Figure 2. Surfaces of constant energy.

The preceding argument proves the existence of a potential function for any closed thermodynamic system which we can define as

$$E(t, x_1, \ldots, x_{n-1}) - E(t^*, x_1^*, \ldots, x_{n-1}^*) = -W_{\text{ad}} \tag{4}$$

This proves statement (d). Here the parameters $t^*, x_1^*, \ldots, x_{n-1}^*$, n in all, describe an arbitrary reference state, and W_{ad} is the work needed to perform the adiabatic process to (or from) the current state from (or to) the reference state. This latter is always possible, as asserted by postulate A.

The existence of a single point on a line of constant values of the deformation coordinates which is isoenergetic to a given state proves that surfaces of constant energy cannot intersect.

The generalizations

$$Q_{12} = E_2 - E_1 + W_{12} \tag{5}$$

$$\mathrm{d}Q = \mathrm{d}E + \mathrm{d}W \tag{5a}$$

to non-adiabatic general and non-adiabatic quasi-static processes, respectively, are standard and require no further comment.

6. THE SECOND LAW

Having acquired the concept of energy, we can establish an equivalent formulation of postulate B in terms of it:

B'. *It is impossible adiabatically to reduce the energy of a system when its deformation variables retain constant values.*

The equivalence follows at once from equation 4 applied to two states for which the deformation coordinates have equal values.

7. CARNOT'S THEOREM

In order to prove statement (a), we apply statement B′ to a reversible process (first part of the second law) for which

$$dQ^0 = dE + dW^0 \tag{6}$$

and *define* any state 2, denoted by s_2 in *Figure 3*, which can be reached from a given state 1 reversibly and adiabatically as *isentropic* with it. It is now easy to show, by an argument modelled on the one used earlier in conjunction with *Figure 1*, that there exists only one isentropic point on a given line β. Before we do this, however, it is necessary to point out to a beginner that an isentropic point s_2 is different from an isoenergetic point e_2. They are both reached by adiabatic processes, a reversible process now and an irreversible process before. However, the work is zero for an isoenergetic point, being different from zero, as seen from equation 6, for an isentropic point.

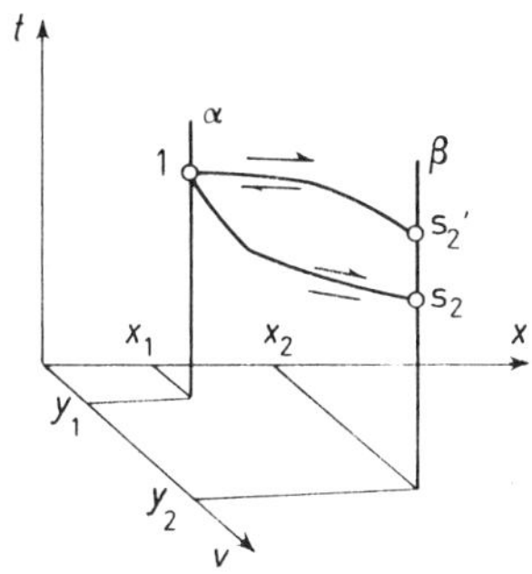

Figure 3. Uniqueness of isentropic point.

Referring to *Figure 3*, we suppose that states s_2 and s_2' are both isentropic with state 1. Further, for the sake of being definite, we suppose that

$$E(s_2') > E(s_2) \tag{7}$$

It is now clear that the system would be capable of performing some adiabatic reversible process as well as its reverse process, both symbolized by the full lines in the diagram. In the first case the energy of the system would increase at constant values of the deformation coordinates. However, in the second case, its energy would decrease with x_2 and y_2 reverting to their original values in contradiction to statement B′. Owing to the assumption of reversibility, the contradiction can be removed only by recognizing that states s_2 and s_2' must be identical. Again, by continuity[2–4, 9], it follows that with any state 1, 1′, 1″, ... along α we may associate a coherent hypersurface, *Figure 4*. The set of such hypersurfaces defines the potential. The resulting family must consist of non-intersecting hypersurfaces, because no point 1′, 1″, ... on α can be reached reversibly from point 1 without exchanging heat, as is easy to prove from equation 6 and the definition of energy. Indeed, for such points we must have $dW^0 = 0$ but $dE \neq 0$.

We can call the resulting potential the empirical entropy, σ, and assert the existence of a family of non-intersecting hypersurfaces

$$\sigma = \sigma(t, x_1, \ldots, x_{n-1}) = \text{const.} \tag{8}$$

for any system whatsoever, as shown in *Figure 4*.

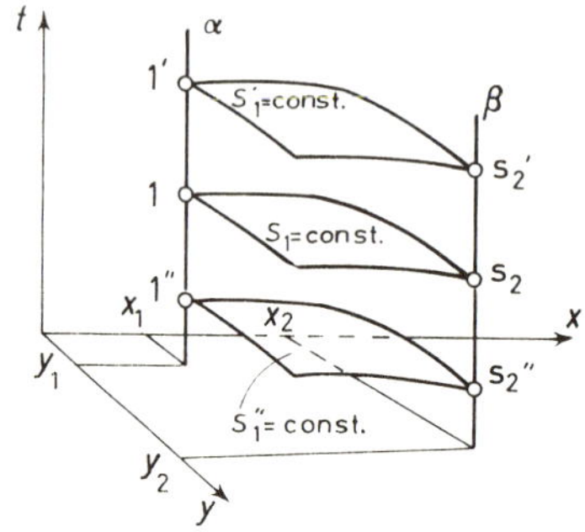

Figure 4. Surfaces of constant entropy.

This family of hypersurfaces intersects the family of hypersurfaces $E(t, x_1, \ldots, x_{n-1})$ from equation 4 along entities of $n - 2$ dimensions, proving the existence of reversible isothermal–adiabatic processes[6], as well as of intersecting isentropic *lines* whenever the number of independent variables, n, equals or exceeds three.

In the preceding derivation, unlike those in some textbook presentations, we expressly refrained from making an appeal to the statement that it is impossible to design a cycle which consists of two isentropics and one isothermal. Such cycles *are* possible if $n \geqslant 3$, as shown elsewhere[6].

To complete the argument, it is now necessary to show that the existence of non-intersecting isentropic surfaces, σ, leads to statement (b) above. The proof can be modelled on that of Carathéodory, and we refrain from giving the details, because a simple version can be found in the literature[3–5]. This reasoning leads naturally to equations 1 and 2.

8. THE SECOND PART OF THE SECOND LAW

Statement (c), or the second part of the second law, follows when we extend our inquiry to irreversible adiabatic processes. Thus, we consider an arbitrary adiabatic process which ends at point i_2 in *Figure 5*. We also consider the point s_2 which is isentropic to 1 together with the reversible process. An examination of the combined process (the reverse of which is impossible) in the light of statement B′ convinces us that

$$E(i_2) > E(s_2) \tag{9}$$

Reference to equation 1 permits us to integrate for entropy along the reversible path and reference to equation 6 with $dW^0 = 0$ shows that only positive elements of heat, $dQ^0 > 0$, must be summed. This proves that

$$S(i_2) > S(s_2) \tag{10}$$

Generally, we write

$$S_2 - S_1 > 0 \tag{10a}$$

and *define* the entropy produced as in equation 2a, so that equation 2b follows. This is the principle of entropy increase for adiabatic processes. The generalization to equation 3 is again standard. It suffices to note that $-\mathrm{d}Q/T$ is the change in the entropy of the immediate surroundings, so that $\mathrm{d}S - \mathrm{d}Q/T$ is the total change in the entropy of an adiabatic system consisting of the system proper coupled with its immediate surroundings.

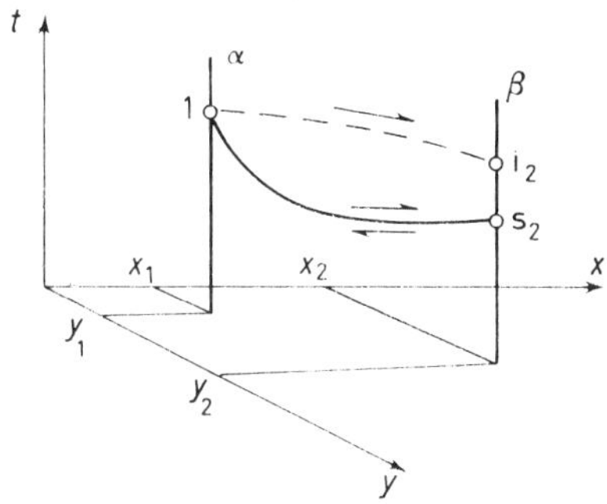

Figure 5. Characteristics of irreversible adiabatic process.

Finally, we note that equation 9 combined with equation 5 and the condition that $Q_{12} = 0$ proves that of all adiabatic processes which occur between a given initial state and prescribed values of the deformation coordinates at the final state, an isentropic process yields the maximum positive (or minimum negative) work. Indeed, along $1 \to s_2$

$$W^0_{12} = E_1 - E(s_2) \tag{11a}$$

whereas along an arbitrary adiabatic irreversible process for the same x_2, y_2, we have

$$W_{12} = E_1 - E(i_2) \tag{11b}$$

Reference to equation 9 shows that

$$W^0_{12} > W_{12} \tag{12}$$

REFERENCES

1 L. A. Turner, *Am. J. Phys.* **28**, 781 (1960); **29**, 40 (1961); **30**, 506 (1962).
2 P. T. Landsberg, *Physica Status Solidi*, **1**, 120 (1961).
3 P. T. Landsberg, *Nature, London*, **201**, 485 (1964).
4 P. T. Landsberg, *Bull. Inst. Phys. and Phys. Soc.* 150 (1964).
5 F. W. Sears, *Am. J. Phys.* **31**, 747 (1963); **34**, 665 (1966).
6 M. W. Zemansky, *Am. J. Phys.* **34**, 914 (1966).
7 M. W. Zemansky, *Heat and Thermodynamics*, Chapter 8, 5th ed. McGraw-Hill: New York (1968).
8 M. Born, *Natural Philosophy of Cause and Chance.* Clarendon Press: Oxford (1949).

[9] P. T. Landsberg, *Thermodynamics with Quantum Statistical Illustrations.* Interscience: New York (1961).
[10] J. Kestin, *A Course in Thermodynamics*, Vol. I. Blaisdell: New York (1966).
[11] J. Kestin, *Am. J. Phys.* **29**, 329 (1961).

PRINCIPLES OF CLASSICAL THERMODYNAMICS

C. GURNEY

University of Hong Kong, Department of Mechanical Engineering

'Then how should I begin,
To spit out all the butt-ends of my days and ways?
And how should I presume?' T. S. Eliot

ABSTRACT

The paper outlines the current presentation of thermodynamic principles to the combined Part I Engineering students at the University of Hong Kong. By considering all bodies taking part in a process, the first and second laws of thermodynamics are presented without the use of work or heat—terms which cannot be generally defined without anticipating the second law. The experimental data necessary to know substances thermodynamically are their internal energy functions, their isotherm functions, and their equations of chemical equilibrium—all in terms of pressure p, specific volume v, and the degree of advancement of chemical processes ξ. Thermodynamic temperature functions, affinity functions and entropy functions may be derived from these data. The paper concludes with a discussion of interactions between bodies in terms of work and heat.

1. INTRODUCTION

Neglecting metaphysical matters, we adopt the view that scientific principles are economical descriptions of events and processes. A new phenomenon is considered to be explained when it has been shown to conform to accepted principles. Thermodynamics is best regarded as an extension of rigid body Newtonian Mechanics, to deal generally with deformable bodies and chemical processes. The methodology of thermodynamics is in terms of simply measurable quantities, through relationships between functions of such quantities to the prediction of future values of the simple quantities. It is convenient to call the simple quantities coordinates. External coordinates are best taken relative to the astronomical frame of reference and include spatial coordinates x, y, z, and velocity coordinates q_x, q_y, q_z, with resultant velocity q. Confining our attention to a simple fluid, internal coordinates are conveniently taken as pressure (p) (specific force), specific volume (v) and the degree of advancement of any chemical process ξ. In addition we need the mass of each chemical species m_1, m_2, etc. Internal and external coordinates are conveniently embraced in the term thermodynamical coordinates. To make useful predictions we must include changes in all bodies which influenced each other. An isolated system of bodies is one which does not respond to changes in the environment outside the isolating wall.

2. FIRST LAW

A class of functions of the thermodynamic coordinates are called energy functions. The kinetic energy function is defined by

$$\mathrm{Kin} = \int_0^m \tfrac{1}{2}q^2\,\mathrm{d}m \tag{2.1}$$

Any other functions which when added to the kinetic energy function give a valid relationship between the coordinates are also called energy functions. In a constant long range field of strength k parallel to the z axis, a potential energy function is appropriate. This is defined as

$$\mathrm{Pot} = \int_0^m kz\,\mathrm{d}m \tag{2.2}$$

A valid description of the motion of a rigid body α in such a force field is

$$[\mathrm{Pot} + \mathrm{Kin}]^\alpha = N^\alpha = \text{constant} \tag{2.3}$$

We note that Pot and Kin are relative to external axes, and their sum may be called the external energy N. An imagined body so isolated that only changes in its external coordinates occur is conveniently called a Newtonian body.

When deformable bodies have relative motion of their parts (other than isotropic rotations), or when deformable bodies interact, the numerical value of their external energy changes. Experiment shows, however, that a valid description of the variation of the thermodynamic coordinates of an isolated system of bodies is obtained by introducing a function U of the internal coordinates. U is called the internal energy function. The behaviour of two interacting bodies α, β in an isolating envelope can be described by

$$[U + N]^\alpha + [U + N]^\beta = E^\alpha + E^\beta = \text{constant} \tag{2.4}$$

where E is the total energy function. The first law states that changes in the thermodynamic coordinates of an isolated system of bodies can be described in terms of constancy of the sum of the energy functions of all parts of the system; or more succinctly, but less informatively, 'the energy of an isolated system of bodies remains constant'. The functional form of the internal energy must be found from experiments.

2.1. To find $(\delta u/\delta p)_{v\xi}$

Let the specific internal energy be u. Let unit mass of fluid be divided into two equal parts and let these be projected horizontally at each other with equal and opposite velocities q, while contained in isolating envelopes. Let v and ξ remain constant and let the pressure rise Δp after the disturbance has died away be measured. Applying equation 2.4

$$\left(\frac{\delta u}{\delta p}\right)_{v\xi} = \underset{\Delta p \to 0}{Lt}\ \frac{\frac{1}{2}q^2}{\Delta p} \tag{2.5}$$

2.2. To find $(\delta u/\delta v)_{p\xi}$

Let a small evacuated space Δv be annexed to the fluid and let the fluid be allowed to enter it. Let the pressure change Δp be measured when the disturbance has died away. Then the energy is unchanged and with $d\xi = 0$ equation 2.4 gives

$$0 = \left(\frac{\delta u}{\delta p}\right)_{v\xi} \Delta p + \left(\frac{\delta u}{\delta v}\right)_{p\xi} \Delta v \tag{2.6}$$

Proceeding to the limit

$$\left(\frac{\delta u}{\delta v}\right)_{p\xi} = -\left(\frac{\delta u}{\delta p}\right)_{v\xi}\left(\frac{\delta p}{\delta v}\right)_{u\xi} \tag{2.7}$$

Since $(\delta u/\delta p)_{v\xi}$ has already been found from equation 2.5 and $(\delta p/\delta v)_{u\xi}$ is obtained from experiment, equation 2.7 evaluates $(\delta u/\delta v)_{p\xi}$.

2.3. To find $(\delta u/\delta \xi)_{pv}$

Let a small quantity of the fluid be separated by a partition and let the chemical process be catalysed in this quantity. Let the partititon then be broken and thorough mixing permitted. Let Δp and $\Delta\xi$ be measured. Then with u and v constant

$$0 = \left(\frac{\delta u}{\delta p}\right)_{v\xi} \Delta p + \left(\frac{\delta u}{\delta \xi}\right)_{pv} \Delta \xi \tag{2.8}$$

Thus

$$\left(\frac{\delta u}{\delta \xi}\right)_{pv} = -\left(\frac{\delta u}{\delta p}\right)_{v\xi}\left(\frac{\delta p}{\delta \xi}\right)_{uv} \tag{2.9}$$

2.4. The internal energy function

Having obtained the derivatives experimentally we have

$$du = \left(\frac{\delta u}{\delta p}\right)_{v\xi} dp + \left(\frac{\delta u}{\delta v}\right)_{p\xi} dv + \left(\frac{\delta u}{\delta \xi}\right)_{pv} d\xi \tag{2.10}$$

By repeating the experimental process a table of $u - u_0$ in terms of p, v, ξ may be enumerated or if simple, the function $u(pv\xi)$ may be found. For gases of low density experiment shows that

$$U = m[pv(a\xi + b) + u_0(1 + c\xi)] \tag{2.11}$$

where m is a measure of quantity; v is the volume per mole and a, b and c are constants for given atomic content, and over a useful range of the variables. If ξ is constant

$$U = m\frac{pv}{\gamma - 1} + u_0 = \frac{pV}{\gamma - 1} + U_0 \tag{2.12}$$

where γ is a constant.

3. THE SECOND LAW

We may imagine many impossible processes which would satisfy the energy accounting system of the first law. The admissibility or otherwise of processes may be expressed as an inequality. A burning match does not reconstitute itself, so that the process may be described as $d\xi \geqslant 0$. If otherwise in equilibrium with a constant environment, the air pressure in an inflated motor car tyre never rises so that $dp \leqslant 0$. A blackboard duster freely sliding on a horizontal table does not draw from its internal energy and increase in speed, so that $dq \leqslant 0$, or $du \geqslant 0$. The second law states that not all imagined processes actually happen and that not all imagined future states are realizable even though they would satisfy the energy accounting system of the first law.

Following Guggenheim[1], let us suppose that for any system of interacting bodies α, β etc. there is a possibility function S of the thermodynamic co-ordinates such that for an isolated system

$$\frac{dS}{dt} = \frac{dS^{\alpha}}{dt} + \frac{dS^{\beta}}{dt} \geqslant 0 \tag{3.1}$$

Here t is time, and it is in the second law that later and sooner enter scientific principles. Unfortunately Clausius gave S the confusing name of entropy. S would be much better called the Clausius function. Our proposition is illustrated in *Figure 1*.

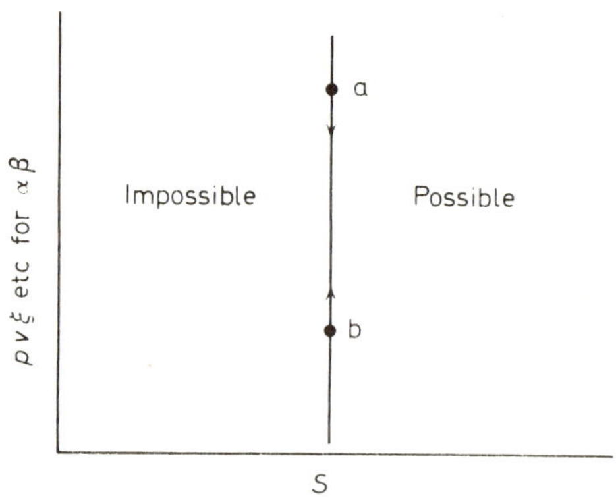

Figure 1. Possible future states of an isolated system.

Possible future states lie to the right of the line S = constant. Considering states on the line itself, any state (a) can be reached from any state (b) and vice versa. The line $dS = 0$ therefore represents reversible processes. Experience suggests that interchange in forms of external energy is unrestricted. We therefore guess that S is a function of internal coordinates only. Our statement of the second law then becomes 'changes in the internal coordinates of an isolated system of bodies are such that the sum of the entropy functions of all bodies is stationary or increases'.

As the entropy approaches a maximum, a time invariant state is reached, which is called equilibrium. We proceed to discover the entropy function, by considering a fairly general system approaching equilibrium.

Consider a fluid (α), filling a constant volume isolating cylindrical vessel, mounted on a frictionless axial vertical pivot, and in a gravitational field of strength k. Initially, let $q\ p\ v\ \xi$ all be irregular. Then changes occurring are such that the entropy of α increases to a maximum, subject to its energy, its angular momentum about the pivot, and its mass remaining constant. Taking the z axis as vertical, and using cylindrical coordinates $r\ \theta\ z$, the author[2] has shown that the equilibrium state is described by maximizing

$$I = \int_0^V J\,\mathrm{d}V = \int_0^V [s + \lambda\{u + \tfrac{1}{2}(q_\theta^2 + q_r^2 + q_z^2) + kz\} + \mu q_\theta r + \nu]\frac{\mathrm{d}V}{v} \tag{3.2}$$

where λ, μ and ν are Lagrangian multipliers, and the lower case letters are specific quantities. A stationary value entails equating to zero the partial differentials of J with respect to the independent variables conveniently chosen as $qs\xi$ and v. Differentiating with respect to the velocities yields

$$q_r = q_z = 0; \qquad \frac{q_\theta}{r} = -\frac{\mu}{\lambda} = \omega \tag{3.3}$$

where ω is angular velocity. We therefore see that the final state is a rigid body rotation. Differentiation with respect to internal coordinates yields:

$$\frac{\delta J}{\delta s} = \frac{1}{v}\left[1 + \lambda\left(\frac{\delta u}{\delta s}\right)_{v\xi}\right] = 0; \qquad \left(\frac{\delta u}{\delta s}\right)_{v\xi} = -\frac{1}{\lambda} \tag{3.4}$$

$$\frac{\delta J}{\delta \xi} = \frac{\lambda}{v}\left(\frac{\delta u}{\delta \xi}\right)_{vs} = 0; \qquad \left(\frac{\delta u}{\delta \xi}\right)_{vs} = 0 \tag{3.5}$$

$$\frac{\delta J}{\delta v} = -\frac{J}{v} + \frac{\lambda}{v}\left(\frac{\delta u}{\delta v}\right)_{s\xi} = 0 \tag{3.6}$$

These three relationships define three components of equilibrium. Equation 3.4 defines thermal equilibrium, and it is convenient to give $(\delta u/\delta s)_{v\xi}$ the symbol T, and to call T the thermodynamic temperature function. We see that at equilibrium the temperature is uniform. Equation 3.5 defines chemical equilibrium. It is convenient to give $-(\delta u/\delta \xi)_{vs}$ the symbol a and $-m(\delta u/\delta \xi)_{vs}$ the symbol A and to call A the affinity function. Chemical equilibrium is described by $A = 0$.

Equation 3.6 which includes ω^2 may be identified as describing dynamic equilibrium. Applying Newtonian mechanics to this steady state forced vortex we obtain

$$\left(\frac{\delta p}{\delta z}\right)_r = -\frac{k}{v}; \qquad v\left(\frac{\delta p}{\delta r}\right)_z = \omega^2 r \tag{3.7}$$

Comparing 3.7 with 3.6, it may be shown that identity is obtained for $(\delta u/\delta v)_{s\xi} = -p + p_0$. The constant p_0 arises as only pressure differences occur in dynamics. Putting $p_0 = 0$ defines absolute pressure.

From these identifications the generating function for S with constant quantity of matter is

$$T\,\mathrm{d}S = m[\mathrm{d}u + p\,\mathrm{d}v + a\,\mathrm{d}\xi] = \mathrm{d}U + p\,\mathrm{d}V + A\,\mathrm{d}\xi \tag{3.8}$$

and to generate the function S it remains to discover A and T as functions of the coordinates. We do this by considering cases of partial equilibrium. For many substances, chemical equilibrium may be indefinitely delayed, and experimental sets of values of $pv\xi$ for which thermal and mechanical, but not chemical, equilibrium exist, may be described by $\theta(pv\xi) = \text{constant}$. Then $T(pv\xi)$ is a function of θ and can be found from an expression due to Planck[3].

$$\ln \frac{T}{T_0} = \int_{\theta_0}^{\theta} \frac{(\delta p/\delta\theta)_{v\xi}\, d\theta}{(\delta u/\delta v)_{\theta\xi} + p} \tag{3.9}$$

For perfect gases not undergoing chemical processes, the isotherm function is found to be $\theta = pv = \text{constant}$. From 2.12 and 3.9 we easily find:

$$T = \frac{pv}{r}; \qquad S = \frac{mr}{\gamma - 1} \ln pv^{\gamma} + S_0 \tag{3.10}$$

where r is chosen so that the triple point of water–ice–vapour is allotted the temperature $T = 273{\cdot}16°$. Considering a non-ideal gas defined by 2.12 and $\theta = pv^{\beta}$, equations 3.8 and 3.9 yield:

$$T = B(pv^{\beta})^{\gamma - 1/\gamma - \beta}; \qquad S = \frac{\gamma - \beta}{B(\gamma - 1)(1 - \beta)} (pv^{\gamma})^{1 - \beta/\gamma - \beta} + S_0 \tag{3.11}$$

where B is a constant chosen as for r above.

By encapsulating different parts of the rotating fluid in light, flexible, thermally insulating walls, we may inhibit thermal equilibrium. Sets of experimental values of $pv\xi$ for which chemical and mechanical but not thermal equilibrium is attained may be described by a function $\alpha(pv\xi) = 0$. Then A must be chosen so that dS is an exact differential, and subject to $A = 0$ when $\alpha = 0$. These conditions are (J = Jacobian):

$$\frac{\delta}{\delta\alpha}\left(\frac{A}{T}\right) = \frac{1}{T^2} J\left(\frac{TU}{\alpha\xi}\right); \quad \frac{\delta}{\delta V}\left(\frac{A}{T}\right) = \frac{1}{T^2} J\left(\frac{TU}{V\xi}\right) + \frac{\delta}{\delta\xi}\left(\frac{p}{T}\right);$$

$$\frac{\delta}{\delta\alpha}\left(\frac{p}{T}\right) = \frac{1}{T^2} J\left(\frac{TU}{\alpha V}\right) \tag{3.12}$$

While generation of A from α is conceptually important, at present the entropy function with ξ as variable is not found. In current practice values of S at constant ξ are found by integration from $T = 0$. This process is made viable by the third law which states that $(\delta s/\delta\xi)_{TP} \approx 0\ (T \to 0)$. At standard pressure we may consider the entropy of all perfectly crystalline substances to approach zero as $T \to 0$. When matter is introduced from outside a body the function 3.8 does not apply, but it may rigorously be extended to show that variation in total quantity of each chemical element is accommodated by the relationship due to Gibbs

$$T\, dS = dU + p\, dV - \sum_{1}^{n} \mu\, dm \tag{3.13}$$

where μ is chemical potential and $m_1 m_2$ are quantities of each chemical substance.

So far we have only considered deductions from a statement of the stationary value of I. For a maximum in many variables see Apostol[4]. Important deductions from maximizing I are:

$$T > 0; \quad T\left(\frac{\delta S}{\delta T}\right)_{p\xi} > T\left(\frac{\delta S}{\delta T}\right)_{v\xi} > 0; \quad \left(\frac{\delta a}{\delta \xi}\right)_{vs} < 0; \quad \left(\frac{\delta p}{\delta v}\right)_{s\xi} < \left(\frac{\delta p}{\delta v}\right)_{T\xi} < 0 \tag{3.14}$$

If we have a body isolated from interaction with other than Newtonian bodies, we see from 3.1 and the first and second sets of 3.14, that only equal or higher temperatures are possible in the future, under conditions of $p\xi$ constant and $v\xi$ constant.

4. INTERACTIONS BETWEEN PARTS OF A SYSTEM

It is useful to partition interactions into work and heat, and the distinction between these quantities is the province of the second law. A heat interaction necessarily changes the entropy of a body, and thus affects the limits of its future states when subsequently isolated from other than Newtonian bodies. A work interaction may or may not generate entropy, depending on whether or not the processes occurring in the body as a result of the work interaction are reversible or not.

We may define the heat component (Q) of an interaction with an opaque body (α) of constant number of atoms and having surface temperature T^{σ} as

$$\mathrm{d}Q = [T^{\sigma}(\mathrm{d}S - \mathrm{d}S^{+})]^{\alpha}$$

where $\mathrm{d}S^{+\alpha} \geqslant 0$. $\mathrm{d}S^{+\alpha}$ is all the entropy increment additional to that due to the heat interaction at the surface. Among factors which contribute to $\mathrm{d}S^{+\alpha}$ are surface frictional effects, temperature gradients within the body, viscous effects, plastic deformation of solids, irreversible chemical processes and diffusion of chemical substances.

The work component of the interaction may then be defined as the difference between the energy change and the heat component. For mechanical work, it may be shown that the above definition can be reduced to the scalar product of a vector force and its vector displacement.

REFERENCES

1. E. A. Guggenheim, *Thermodynamics*. North Holland: Amsterdam (1949).
2. C. Gurney, *Bull. Mech. Engng Educ.* **7**, 243 (1968).
3. M. Planck, *Thermodynamics*, 3rd English Edition. Longmans Green: London (1926).
4. T. M. Apostol, *Mathematical Analysis*. Addison Wesley: Reading, U.S.A. (1963).

THERMODYNAMICS OF SOLIDS UNDER STRESS

T. H. K. BARRON

School of Chemistry, University of Bristol

AND

R. W. MUNN

Division of Chemistry,
National Research Council of Canada, Ottawa

ABSTRACT

This expository paper discusses how the equilibrium thermodynamics of an ideal elastic solid differs from that of a fluid. Because at least some species are immobile, there is no unique Gibbs energy. There are six independent finite strain parameters, η_λ, referred to a chosen reference state. Thermodynamic manipulations are straightforward if stresses t_λ are defined conjugate to the η_λ, but these can be identified with the Cauchy stresses, σ_λ, only in the reference state. Typical experimental conditions are described directly by the σ_λ, which have to be related to the t_λ before experimental quantities like C_p can be related to formally derived quantities like C_t.

---

1. INTRODUCTION

This paper discusses the thermodynamics of ideal solids which can sustain a permanent shear stress. The theory is now well established, and is of increasing importance in solid state physics. However, rigorous discussions are available only in some specialized texts[1,2], which are not always easy to read. In addition to results given in such texts, this account presents some new results relating experimental quantities to ones of theoretical application.

A rigid solid must contain a framework of atoms in which neighbours remain neighbours throughout any deformation (although mobile species can also be present). Processes in which there is transport of the immobile atoms, including exchange with another phase, are forbidden. Consequently the solid need not be in a state of thermochemical equilibrium; in general a chemical potential can be defined only for mobile species. There is therefore no unique Gibbs energy G, although it may be convenient to define analogous functions (see §2). True thermochemical equilibrium is reached only under isotropic stress (hydrostatic pressure)†. Consequently, normal fluid thermodynamics can be applied to solids under isotropic stress (e.g. a solid immersed

† McLellan[3] has derived a chemical potential by assuming thermochemical equilibrium. His result thus appears to be valid only under hydrostatic pressure.

in a fluid). However, care is needed, especially for non-cubic solids. For example, the relation $C_V = C_p - \beta^2 VT/\kappa_T$‡ gives not the heat capacity of a solid whose dimensions are kept constant, but that of a solid whose shape changes to keep the stress isotropic at constant volume[4].

When transport is negligible during the time of an experiment, a solid under shear stress is in a metastable state with a well defined entropy S and Helmholtz energy A, which are functions of the strain and temperature. The thermodynamics of this state is our main topic. We deal only with thermo-elastic properties, omitting electric and magnetic effects.

2. THERMODYNAMIC THEORY FOR SOLIDS

Description of strain

We treat the solid as a continuum, and consider only strains which are effectively uniform over distances of several atomic spacings. The strain can be specified by the displacement of each point in the solid from its position $(\mathring{x}_1, \mathring{x}_2, \mathring{x}_3)$ in some *reference configuration.* A superposed circle will denote properties of this configuration. For a uniform strain the new positions are given in tensor notation[5] by

$$x_i = (\delta_{ij} + u_{ij})\, \mathring{x}_j \tag{1}$$

where summation from 1 to 3 is implied over a repeated suffix. If the displacements u_{ij} are small, their symmetric and antisymmetric parts give, to first order, pure strains and pure rotations

$$e_{ij} \equiv \tfrac{1}{2}(u_{ij} + u_{ji}), \qquad \omega_{ij} \equiv \tfrac{1}{2}(u_{ij} - u_{ji}) \tag{2}$$

Thus infinitesimal strains are specified by the six e_{ij}.

However, finite strains are not specified by the e_{ij}. An arbitrary vector $\mathring{\mathbf{r}}$ in the reference configuration which becomes the vector $\mathbf{r}$ in the strained configuration changes in length by[6]

$$r^2 - \mathring{r}^2 = (u_{ij} + u_{ji} + u_{ki}u_{kj})\, \mathring{r}_i \mathring{r}_j \tag{3}$$

where the second-order terms depend on the ω_{ij} as well as the e_{ij}. We can thus specify arbitrary strain by the symmetric *Lagrange finite strain tensor*

$$\eta_{ij} \equiv \tfrac{1}{2}(u_{ij} + u_{ji} + u_{ki}u_{kj}) \tag{4}$$

which can be reduced to e_{ij} for infinitesimal strains. We shall use the *Voigt abbreviated notation*[7]

$$\eta_1 = \eta_{11}, \eta_2 = \eta_{22}, \eta_3 = \eta_{33}; \tag{5}$$

$$\eta_4 = 2\eta_{23}, \eta_5 = 2\eta_{31}, \eta_6 = 2\eta_{12} \tag{6}$$

the factors of two in equation 6 are introduced for later convenience. Voigt subscripts will be denoted by Greek letters λ, etc. ($\lambda = 1, .. 6$). A similar scheme defines infinitesimal strains e_λ.

‡ β is used for the volumetric expansion to avoid confusion with the Grüneisen function γ.

Description of stress

The stress is most directly described by the well-known *Cauchy stress tensor*[8], which in the absence of couple stresses is symmetric[9]. However, the Cauchy stress has the disadvantage for thermodynamic purposes that it does not determine the strain unless the orientation is specified (or unless the stress is isotropic). We therefore define other stress parameters, t_λ, which are thermodynamically conjugate to the strain parameters η_λ, and so depend on this prior choice of strain parameter.

The energy U and Helmholtz energy A are functions of the strain and one other variable. The stresses t_λ are defined by

$$\mathring{V} t_\lambda \equiv (\partial U/\partial \eta_\lambda)_{\eta',S} = (\partial A/\partial \eta_\lambda)_{\eta',T} \tag{7}$$

where the subscript η' denotes that all the η_μ except η_λ are kept constant during differentiation. The t_λ have the dimensions of (negative) pressure, and are sometimes called the *thermodynamic tensions*[1,2]. One may also retain tensor notation to define stresses t_{ij} by equations like 7, with this convention[10] for differentiation with respect to the components of a symmetric tensor. Write the function to be differentiated symmetrically in η_{ij} and η_{ji} and then differentiate treating all nine η_{ij} as independent. The resulting tensor is symmetric, and is related to the t_λ by a scheme like equations 5 and 6 without the factors of two.

The relation of t_{ij} to the Cauchy stress σ_{ij} is discussed in §3.

Energy functions and Maxwell relations

We define quantities analogous to the enthalpy and Gibbs energy

$$H' \equiv U - \mathring{V} t_\lambda \eta_\lambda, \qquad G' \equiv A - \mathring{V} t_\lambda \eta_\lambda \tag{8}$$

where the primes remind us that these cannot be identified with the functions H and G defined under hydrostatic pressure. The repeated subscript λ denotes summation from 1 to 6; by virtue of the factors of two in the abbreviated notation for strains but not for tensions, $t_\lambda \eta_\lambda$ is equal to $t_{ij}\eta_{ij}$.

The differentials of the functions U, A, H' and G' are given to first order by

$$\mathrm{d}U - T\,\mathrm{d}S = \mathring{V} t_\lambda\,\mathrm{d}\eta_\lambda = \mathrm{d}A + S\,\mathrm{d}T \tag{9}$$

$$\mathrm{d}H' - T\,\mathrm{d}S = -\mathring{V}\eta_\lambda\,\mathrm{d}t_\lambda = \mathrm{d}G' + S\,\mathrm{d}T \tag{10}$$

Maxwell relations follow as for fluids, e.g.

$$(\partial S/\partial \eta_\lambda)_{\eta',T} = -(\partial^2 A/\partial \eta_\lambda\,\partial T)_{\eta'} = -\mathring{V}(\partial t_\lambda/\partial T)_\eta \tag{11}$$

$$(\partial T/\partial t_\lambda)_{t',S} = (\partial^2 H'/\partial t_\lambda\,\partial S)_{t'} = -\mathring{V}(\partial \eta_\lambda/\partial S)_t \tag{12}$$

and two similar expressions derived from $\mathrm{d}U$ and $\mathrm{d}G'$. Relations of this type also establish the symmetry of the isothermal elastic stiffnesses, analogous to the bulk modulus B for fluids:

$$C^T_{\lambda\mu} \equiv \left(\frac{\partial t_\lambda}{\partial \eta_\mu}\right)_{\eta',T} = \frac{1}{\mathring{V}}\left(\frac{\partial^2 A}{\partial \eta_\mu\,\partial \eta_\lambda}\right)_{\eta',T} = C^T_{\mu\lambda} \tag{13}$$

and similarly for the adiabatic stiffnesses $C^S_{\lambda\mu}$ and the compliances $S_{\lambda\mu} = (\partial\eta_\lambda/\partial t_\mu)_{t'}$. The same full symmetry is possessed by simple higher-order elastic constants, e.g.

$$C^T_{\lambda\mu\nu} \equiv (\partial C^T_{\lambda\mu}/\partial\eta_\nu)_{\eta',T} = (\partial^3 A/\partial\eta_\lambda\,\partial\eta_\mu\,\partial\eta_\nu)/\mathring{V} \tag{14}$$

but mixed constants like $(\partial C^S_{\lambda\mu}/\partial\eta_\nu)_{\eta' T}$ have lower symmetry[11].

From these generalized energy functions a selfconsistent thermodynamic theory can be developed[1, 2] in much the same way as for fluids. The development again depends strongly on the elementary theory of partial differentiation, suitably extended to seven independent variables instead of two. For instance, just as $(\partial p/\partial T)_V$ is equal to $-(\partial p/\partial V)_T(\partial V/\partial T)_p$, so

$$(\partial t_\lambda/\partial T)_\eta = -(\partial t_\lambda/\partial\eta_\mu)_{\eta',T}(\partial\eta_\mu/\partial T)_t \tag{15}$$

where it should be recalled that by the summation convention the RHS is a sum of six terms. In the present brief account this example must suffice; it is used in obtaining equation 32.

3. USE OF THE CAUCHY STRESS TENSOR

The problem

We come now to a major source of difficulty and confusion. Although the t_λ are by their definition convenient for thermodynamic analysis, it is the Cauchy stresses σ_λ which are most simply related to experimental conditions. We have therefore to relate properties defined in terms of the σ_λ to the thermodynamic results obtained in terms of the t_λ, which can require rather complex expressions.

Elementary treatments[8] attempt to avoid this difficulty by always choosing the instantaneous (often unstressed) configuration for the reference configuration and considering only infinitesimal strains. Then

$$dU(e, \omega, S) = T\,dS + \mathring{V}\mathring{\sigma}_\lambda\,de_\lambda \tag{16}$$

Since to first order $de_\lambda = d\eta_\lambda$ we see from equation 7 that $\mathring{\sigma}_\lambda = \mathring{t}_\lambda$. We may then define coefficients of thermal expansion

$$\alpha_\lambda \equiv (\partial e_\lambda/\partial T)_{\sigma,\omega} \tag{17}$$

stiffnesses

$$c^T_{\lambda\mu} \equiv (\partial\sigma_\lambda/\partial e_\mu)_{e',\omega,T} \tag{18}$$

and other properties directly related to physical measurements. The difficulty is that since these relations hold only at the reference configuration, we cannot differentiate a second time with respect to strain or stress. So, for example, there is no Maxwell relation like equation 13.

Relation of σ_{ij} to t_{ij}

By treating the configuration reached by displacements u_{ij} as a new reference configuration and then applying the result $\mathring{\sigma}_{ij} = \mathring{t}_{ij}$, we can show[2, 10] that

$$\sigma_{ij} = (V/\mathring{V})(\delta_{ip} + u_{ip})(\delta_{jq} + u_{jq})\,t_{pq} \tag{19}$$

To first order in the displacements this gives

$$\sigma_{ij} = t_{ij} + P_{ijkl}e_{kl} + Q_{ijkl}\omega_{kl} \tag{20}$$

where

$$P_{ijkl} = \tfrac{1}{2}(\mathring{\sigma}_{jl}\delta_{ik} + \mathring{\sigma}_{il}\delta_{jk} + \mathring{\sigma}_{ik}\delta_{jl} + \mathring{\sigma}_{jk}\delta_{il}) - \mathring{\sigma}_{ij}\delta_{kl} \tag{21}$$

$$Q_{ijkl} = \tfrac{1}{2}(\mathring{\sigma}_{jl}\delta_{ik} + \mathring{\sigma}_{il}\delta_{jk} - \mathring{\sigma}_{ik}\delta_{lj} - \mathring{\sigma}_{jk}\delta_{il}) \tag{22}$$

This result, together with the observations that to first order

$$de_{\lambda} = d\eta_{\lambda}, \quad (dt)_{\eta} = (d\sigma)_{e.\,\omega} \tag{23}$$

can be used to derive thermodynamic relations for measured quantities.

A particularly simple example following directly from equation 20 by differentiation with respect to strain relates $c^{T}_{\lambda\mu}$ [equation 18] to $C^{T}_{\lambda\mu}$ [equation 13] by

$$c^{T}_{\lambda\mu} = C^{T}_{\lambda\mu} + P_{\lambda\mu} \tag{24}$$

where $P_{14} = P_{1123}$, etc. Hence $c^{T}_{\lambda\mu} = C^{T}_{\lambda\mu}$ only for a solid under zero stress[6]. Examination of equation 21 shows that $P_{\lambda\mu}$ is symmetric only when $\mathring{\sigma}_{ij}$ represents a hydrostatic pressure, so that for an anisotropic stress $c^{T}_{\lambda\mu} \neq c^{T}_{\mu\lambda}$. The compliance $s^{T}_{\lambda\mu} = (\partial e_{\lambda}/\partial\sigma_{\mu})_{\sigma',\,\omega,\,T}$ is inverse to $c^{T}_{\lambda\mu}$ (i.e. $s^{T}_{\lambda\mu}c^{T}_{\mu\nu} = \delta_{\lambda\nu}$), so that $s_{\lambda\mu}$ too is symmetric only under isotropic stress.

Maxwell relations

Because of equation 23, the analogues of equation 11 and the similar relation deduced from dU remain valid:

$$(\partial S/\partial e_{\lambda})_{e',\,\omega,\,T} = -\mathring{V}(\partial\sigma_{\lambda}/\partial T)_{e.\,\omega} \tag{25}$$

$$(\partial T/\partial e_{\lambda})_{e',\,\omega,S} = \mathring{V}(\partial\sigma_{\lambda}/\partial S)_{e,\,\omega} \tag{26}$$

Let us now try to derive the analogue of equation 12. We have

$$(\partial T/\partial\sigma_{\lambda})_{\sigma',\,\omega,S} = (\partial T/\partial e_{\mu})_{e',\,\omega,S}(\partial e_{\mu}/\partial\sigma_{\lambda})_{\sigma',\,\omega,S} \tag{27}$$

$$= \mathring{V}(\partial e_{\mu}/\partial\sigma_{\lambda})_{\sigma',\,\omega,S}(\partial\sigma_{\mu}/\partial S)_{e,\,\omega} \tag{28}$$

by equation 26. The RHS of equation 28 will equal $-\mathring{V}(\partial e_{\lambda}/\partial S)_{e,\omega}$ if the compliance $s^{S}_{\lambda\mu}$ is symmetric, that is, if the stress $\mathring{\sigma}_{ij}$ is isotropic. So the analogue of equation 12 is valid only under hydrostatic pressure. It follows that expressions derived assuming the validity of this Maxwell relation under anisotropic stress involve errors of the order of the fractional difference between $s_{\lambda\mu}$ and $s_{\mu\lambda}$. Similarly, the analogue of the relation like equation 12 derived from dG' is valid only under isotropic stress.

Some further results

The heat capacity most readily measurable is that at constant stress, $C_{\sigma} \equiv T(\partial S/\partial T)_{\sigma,\,\omega}$. It differs from its analogue at constant tension, C_t, by a power series in the stress components. To first order in an isotropic pressure $\sigma_{ij} = -p\delta_{ij}$, when $C_{\sigma} \equiv C_p$, the result becomes

$$C_t - C_{\sigma} = pVT(\beta^2 - 2\alpha_{\lambda}\alpha_{\lambda}) \tag{29}$$

for cubic solids $C_t \geqslant C_\sigma$, but for non-cubic solids C_t may be less than C_σ.

Although in general analogous quantities defined in terms of the physical variables e and σ differ from those in terms of the thermodynamic variables η and t, some quantities are the same in both systems, and so form an important link between them. One such quantity is the heat capacity at constant strain, $C_\eta \equiv T(\partial S/\partial T)_\eta = T(\partial S/\partial T)_e$, as follows from equation 23. C_η is related to C_σ and C_t by[4]

$$C_\eta = C_\sigma - TVc^T_{\lambda\mu}\alpha_\lambda\alpha_\mu = C_t - T\mathring{V}C^T_{\lambda\mu}a_\lambda a_\mu \tag{30}$$

where the a_λ are thermal expansion coefficients $(\partial\eta_\lambda/\partial T)_t$. Another unique quantity is the Grüneisen function defined by[12]

$$\gamma_\lambda \equiv (\partial S/\partial\eta_\lambda)_{\eta', T}/C_\eta = (\partial S/\partial e_\lambda)_{e', \omega, T}/C_\eta \tag{31}$$

Through the Maxwell relations 11 and 25 and equation 15 these equations can be transformed to

$$\gamma_\lambda = \mathring{V}C^T_{\lambda\mu}a_\mu/C_\eta = Vc^T_{\lambda\mu}\alpha_\mu/C_\eta \tag{32}$$

which can be shown to be equal to

$$\gamma_\lambda = \mathring{V}C^S_{\lambda\mu}a_\mu/C_t = Vc^S_{\lambda\mu}\alpha_\mu/C_\sigma \tag{33}$$

4. CONCLUSION

In general, an extended thermodynamic theory is required for solids. It is most naturally developed in terms of the conjugate variables η_λ and t_λ, but experiments are often more readily described by e_λ and σ_λ. Analogous quantities in the two systems differ by amounts depending on the magnitude and anisotropy of the stress. Consequently, care is needed in comparing theory and experiment at high pressures, particularly for highly compressible solids like helium.

REFERENCES

1 C. Truesdell and R. A. Toupin, *Handb. der Phys.* **III/1**, 226 (1960).
2 R. N. Thurston, *Physical Acoustics*, **1A**, 1 (1964).
3 A. G. McLellan, *Proc. Roy. Soc. A*, **307**, 1 (1968).
4 T. H. K. Barron and R. W. Munn, *J. Phys. Chem.* **1**, 1 (1968).
5 H. Jeffreys, *Cartesian Tensors*, p 71. Cambridge University Press: London (1963).
6 T. H. K. Barron and M. L. Klein, *Proc. Phys. Soc.* **85**, 523 (1965).
7 K. Brugger, *Phys. Rev.* **133**, A1611 (1964).
8 H. B. Callen, *Thermodynamics*, p 213. Wiley: New York (1960).
9 M. Lax, *Lattice Dynamics*, p 583. Pergamon Press: London (1965).
10 F. D. Murnagham, *Finite Deformation of an Elastic Solid*, p 43. Dover Publications: New York (1967).
11 M. J. Skove and B. E. Powell, *J. Appl. Phys.* **38**, 404 (1967).
12 T. H. K. Barron and R. W. Munn, *Phil. Mag. VIII*, **15**, 85 (1967).

G. DISCUSSION REPORTS

DISCUSSION ON THE LAWS OF THERMODYNAMICS

Chairman: M. L. McGLASHAN
Reporter: J. S. ROWLINSON, F.R.S.

The discussion was in two parts: in the first, four provocative statements which had been submitted in advance were defended by their proposers, whilst the second was a discussion of the axiomatic foundations of thermodynamics. These foundations were a recurring theme at the conference and their place in the teaching of the subject was taken up again in the third discussion session.

In the first of the four provocations it was contended by T. Bidard (Paris) that Carnot's cycle is inadequate for the discussion of processes at high temperatures in which chemical reactions replace an external source as the origin of the heat supplied to the system[1]. E. J. Le Fevre (London) agreed that this was so, but was no cause for surprise. He distinguished three types of device of interest to engineers. In the first, we put in heat, extract shaft work, and take out heat; in the second, we put in material, extract shaft work, and take out material; in the third, we put in reactive material, extract shaft work and, maybe, heat, and take out the products of the reactions. The Carnot cycle is relevant to the discussion of efficiency of the first, the concept of isentropic efficiency to the second, and that of availability to the third. He was supported by J. Kestin (Providence, USA) who said that the first of the three devices is only an artificially separated part of the third, which is all that engineers are ultimately interested in.

The second statement was made by A. J. Brainard (Pennsylvania, USA) who observed that the principle of maximum entropy does not apply to compound systems subject to internal constraints[2]. He postulated a closed adiabatic cylinder containing two samples of gas at different pressures and temperatures separated by an adiabatic piston[3]. Initially the piston is held in position by a peg, and when this is removed the piston oscillates. Dissipative processes in the gases bring it to rest in a position in which the two pressures are equal. The final state of the system cannot be calculated by the methods of classical thermodynamics, although the pressure alone can be so calculated if the gases are perfect and have heat capacities independent of temperature. He showed, however, that ΔS is not a maximum (which would require a diathermal piston), nor is it as large as could be conceived with an adiabatic piston since such a value of ΔS requires that there is zero change of entropy on one side of the piston.

The results were not disputed. R. L. Scott (California, USA) said that Brainard's weaker conclusion was certainly no occasion for surprise, namely that the ΔS was not as large as that found with a different constraint, namely a diathermal piston. The final state is one of maximum entropy in the sense that, once it has been achieved, a small arbitrary displacement of the piston causes S to increase. Kestin said that since a complicated flow pattern of gas develops during the approach to the final state the system is not describable within the implied assumption of uniform states, and so it is not surprising that the process cannot be described by the laws of classical thermodynamics alone.

As a postscript to this discussion Le Fevre said that Planck's views on the principle of maximum entropy were often misunderstood because of a fault in Ogg's translation. A correct version of Planck's text[4] reads

> The second law of thermodynamics thus says that in nature there exists, for every material system, a property such that for all changes in which the system alone participates, this property either remains constant (reversible processes) or increases (irreversible processes).

Ogg's version[5] omits the words 'in which the system alone participates'.

In the third statement, C. Mascré (Paris) said that Lindemann's treatment of melting is not reconcilable with the zeroth law. He discussed a melting solid by means of the extensive function $S(U, V)$ and suggested that a common tangent to the solid and liquid branches in a plane of fixed V might have a slope $(\partial S/\partial U)_V$ different from the reciprocal of the absolute temperature of one of the phases if this phase has no metastable states. This contention was not accepted by L. Tisza (Massachusetts, USA), the reporter, and several others who maintained that the proper construction of Gibbs's primitive surface $S(U, V)$ is not a tangent in a plane of fixed V but a rolling tangent-plane. The discussion was brought to a close by the chairman's observation that if Lindemann's treatment of melting was irreconcilable with the zeroth law, then it was so much the worse for Lindemann's theory.

The final provocation was that of J. K. Tyldesley (Glasgow) who suggested that if the particles discussed in statistical thermodynamics were not molecules of fixed mass but were entities of variable mass then the techniques of that subject could, perhaps, be extended to discuss turbulent flow. He was supported by E. Ascher (Geneva), whilst Le Fevre thought that Burgers had already explored this extension forty years ago.

The second part of the discussion, on the proper role of axioms, was opened by M. W. Zemansky (New York), whose complaint was that those who put thermodynamics entirely on an axiomatic basis never made it clear where the physics came in. An experimentalist, when trying to explain his results in terms of a theory, is expected to make clear his mathematical assumptions, and will be rightly criticized if, say, his argument depends on a function being continuously differentiable and if he fails to make this clear. There is a reciprocal obligation on the part of the axiomatizers to say clearly where 'nature' enters into their system. Thus Landsberg's paper made no mention of work, which seemed to be anathema to him. The place of work and heat in the subject should be made clear, not concealed in this way. The chairman added that P. T. Landsberg (Cardiff) had used the word adiabatic repeatedly, and so had assumed implicitly the existence of what he called the (-1)th law, namely, that there are systems such that they can be changed only by doing work on them.

Landsberg protested that he was not himself an axiomatizer, he had attempted only to review this approach to the subject. He admitted that axiomatics rarely yielded new science [a point later questioned by L. L. Whyte (London)], but said that nevertheless the search for axiomatic foundations was defensible. He cited Euclidean geometry as a case in point, for here the mathematics and the physics had been so mixed that the very possibility of non-Euclidean geometries had not been suspected until the nineteenth century. Similarly in thermodynamics the mathematics and physics are usually mixed in happy confusion, and it is very proper that some people should try to find the abstract mathematics that lies behind the subject as we know it.

He was supported by W. J. Hornix (Nijmegen, Netherlands) who pointed out that many have tried to reduce thermodynamics to mechanics, but have failed because the concept of heat defies such reduction. An axiomatic approach has made clear the reason for this by showing that an adiabatic process is necessarily a truly primitive term in thermodynamics; it cannot be derived from mechanics.

Tisza argued that all theories are axiomatic in some degree. We put in our axioms, we work on them, and we take out our theory. The real test of the worth of what we have done is the value of what we have added by this operation. [Had the reporter not been so busy scribbling, he would have said here that it has been generally accepted since the days of Kant that the conclusions of any formally valid argument are contained already in its premisses[6]. It follows that from a set of axioms we cannot extract a theory of greater content; we can only reveal what is already there, although this revelation can, of course, have greater value to us than the raw axioms.]

A. Katz (Rehovoth, Israel) regretted that the word adiabatic was being confined in this discussion to its thermodynamic use. He said that it had a closely related use in quantum mechanics. In a system in which the Hamiltonian is changing slowly the process is mechanically adiabatic if the system remains at all times in an eigenstate of the changing Hamiltonian. Such a process is also thermodynamically adiabatic. T. H. K. Barron (Bristol) objected to this conflation of the two ideas, and cited the case of a system of phonons whose dimensions could be changed. In a thermodynamically adiabatic process the occupation numbers change, whilst in a mechanically adiabatic process they do not.

A final diversion was introduced by Kestın who questioned whether the discussion of the thermodynamics of 'materials with memory' was a useful innovation. He said that the 'memory' of materials was not a property, but a state resulting from past actions on them. Every material, natural or artificial, is properly described by the thermodynamics of irreversible processes in terms of the concept of a local state specified by the appropriate internal variables. I. Müller (Templergraben) disagreed. He found the concept of memory to be useful in just the same way as the Navier–Stokes equations are useful as constitutive equations. They are not obeyed exactly but they are a useful idealization of the behaviour of real fluids. B. D. Coleman (Pennsylvania) closed the discussion by saying that it was his experience that materials such as molten polymers did not fit into Kestin's scheme; they were not describable by a finite (or even by a discrete) set of internal variables.

REFERENCES

[1] R. Bidard, *Entropie*, No. 15, 13 (1967).
[2] H. B. Callen, *Thermodynamics*, pp 23 and 321. Wiley: New York (1960).
[3] A. J. Brainard, *Nuovo Cimento*, **62B**, 88 (1969); *Chemical Engineering Education*, in press.
[4] M. Planck, *Vorlesungen über Thermodynamik*, 10th ed., p 87. de Gruyter: Berlin (1954).
[5] M. Planck, *Treatise on Thermodynamics*, translated by A. Ogg, p 88. Dover Publications: New York (n.d.).
[6] F. P. Ramsey, *The Foundations of Mathematics*, p 185–188. Kegan Paul: London (1931).

DISCUSSION ON TEMPORAL ASYMMETRY IN THERMODYNAMICS AND COSMOLOGY

Chairman: R. O. DAVIES
Reporter: O. COSTA DE BEAUREGARD

The reporter has felt that, in a delicate subject where many arguments have already a long history and, moreover, can often have different shades of meaning, the best procedure was to produce a 'reader's digest' of the actual Cardiff discussion. So, after carefully listening to the tape recording and slightly rearranging the order, he has attempted to extract the essence of what each speaker had to say, and to preserve authenticity and flavour by using, whenever feasible, the acutal words spoken. He thus hopes that the discussion will unfold like a drama. He also apologizes if some contributors feel that what has, perforce, been left out was precisely what should have been included.

Chairman—A rough and ready test of the importance of any subject is the amount of nonsense that has been written about it (laughter). When applied to temporal asymmetry this test would place it as somewhat less important than religion and more important than information theory (laughter).

If we lay aside what elementary particle physicists are now telling us, all the elementary laws of physics are time reversible. The question then arises, why is it that for processes that can actually be seen there is in fact a greater variety of behaviour with respect to the time reversed transition? It seems that the central strands of the thing we are concerned with here are precisely how, and how firmly, the thermodynamic arrow may be associated with some other source of directiveness (perhaps a unique, perhaps not a unique, association). In an attempt to subdivide what is a very wide and perhaps indivisible field, I have entered on the board (Appendix A) a few categories which attempt to classify, among other things, the fascinating quotations prepared by Landsberg (See Appendix B).

Now I invite first those speakers who have, as it were, stuck their neck out by making statements they are willing to defend. Dr Collins, you have written that '*With a proper definition of a clock, the second law might be seen as a tautology*'.

R. Collins (Salford)—What prompted my statement was a remark by Zwanzig that, given long enough, any clock in a closed system will eventually wind itself up. It seems that at no point in the papers we have heard is there an analysis of what you require a clock to be. Statistical mechanicians would say it has to be larger than the system you are looking at, while cosmologists would say smaller (laughter). Do you have to suit the clock to the problem? And if not, why?

R. Zwanzig (Maryland, USA)—By international convention a clock is an atomic oscillator operating under the time reversible laws of quantum mechanics, so the time arrow is not built into it.

J. Lewis (Oxford)—Is it not possible that the time arrow is built into the clock through the process that counts the ticks?

P. T. Landsberg (Cardiff)—How would you know which tick is earlier and which later?

Lewis—By counting them.

Landsberg—Ha! *Then* you are using the biological arrow of time.

K. G. Denbigh (London)—No, sir, you are doing more than that. Something *occurs* and, in the very definition of the word *occurs*, a time arrow is assumed. It was the same this morning with Narlikar's oscillating universe: in order to speak of a reversal occurring (note the word *occurring*) you have to assume some reference according to which that occurring occurs. In other words you would have to postulate a supertime.

A. Katz (Rehovoth, Israel)—I would refute that point. Time plays two distinct roles. The interval between two events can be measured by a reversible apparatus, while to know which is earlier or later is provided by the human sense of time.

Chairman—Let us pass on to a subject where attention is drawn to the essential aspect of measuring, implying perhaps those biological or psychological aspects just mentioned. Costa de Beauregard's statement was: '*To state the Einstein–Podolsky–Rosen "paradox" is to state that telegraphing into the past occurs on an elementary quantum level. And this happens in any quantum measurement.*'

O. Costa de Beauregard (Paris)—In Pittsburgh* I argued that the root of

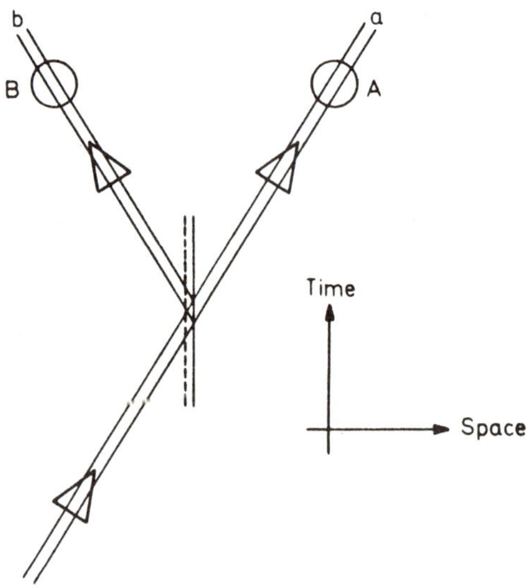

* A symposium on 'A critical review of the foundations of relativistic and classical thermodynamics' was held in April 1969. The proceedings are in course of publication.

stochastic irreversibility lies in the nature of a boundary condition which states that *blind retrodiction is forbidden* and that, provided one uses a theory implying both statistics and waves (namely quantum mechanics), this boundary condition can be connected with the one stating that *advanced waves are forbidden*. My demonstration consisted in a mere rewording of von Neumann's irreversibility proof for the quantum process of measurement. To put it briefly, in quantum mechanics retarded and advanced waves respectively are used in prediction and retrodiction—whence my Pittsburgh statement.

It thus seems that Einstein's prohibition to telegraph into the past might well be of a macroscopic rather than of a microscopic character, so that, on the elementary quantum level, there would remain only a prohibition to telegraph outside the light cone. This I believe is shown by the so-called E–P–R 'paradox'. Suppose we have a wave which is split by a semi-transparent mirror and which we assume for simplicity to carry just one particle. If an observer A operating on beam *a* either finds the particle is present or absent in his beam, then he knows it is respectively absent or present in the other beam *b*, and an observer B operating on *b* is bound to find it so. The point is that the AB vector is spacelike and, moreover, that it can be quite large. Now, the calculation shows quite clearly that the logical inference from A to B (or from B to A) (or, if you prefer, the *telediction* along AB, because it is neither prediction nor retrodiction) is *not* telegraphed *directly* along AB, but along two timelike vectors, AS and SB, with S in the space-time domain where the separation occurs.

And I insist that this is a very general procedure occurring each time a quantum measurement is performed. Then *b* corresponds to the outgoing quantum object and *a* to the measuring device which observer A reads.

L. Tisza (Massachusetts, USA)—We are not really sending a message into the past. We get a message from the past, and what we project into the past is our information. As I said this morning, we make an inference from our present knowledge into the past. So, is it a good thing to call this 'telegraphing into the past'?

Costa de Beauregard—I had to make a provocative statement, you see (laughter).

D. Layzer (Massachusetts, USA)—This seems to me an attempt to discuss issues of information theory. When A gets the message he has all the information there is in this particular issue, so there is no transmission of information at all. My difficulty is that I do not see that the inference is drawn anywhere else than at the site of the measurement.

Costa de Beauregard—But it could be drawn at B just as well.

Katz—This type of telegraphing has nothing to do with causality. Causality (a rather shaky concept in general) would require that A or B could transfer (to B or A) a signal at will, and that he decides at some moment what to transfer. No such possibility exists.

Costa de Beauregard—I am glad you raise this question, which has been left pending since the Bohr–Einstein controversy. According to the accepted version of quantum theory, performing a measurement contributes producing the result of it. Thus it is definitely *not* at the surface of the mirror that the decision is made, but later.

P.A.C.—22/3–4—N

Katz—Even so, the measurement does not produce an *arbitrary* result.

Costa de Beauregard—That is true, but either A or B *does* have control on the type of measurement he chooses to perform (spin or anything else). In *this* sense there is some kind of telegraphing between A and B.

R. J. Heaston (Germany)—In Pavlovian terms, a response is the result of a stimulation. So it is a matter of temporal ordering to know which event is stimulation and which response.

Costa de Beauregard—This simply won't work here, due to the spacelike character of the AB vector.

Chairman—Perhaps, we ought to tackle this from another point of view. Would Landsberg like to add something to his nine-years old quotation '*This illustrates clearly how the entropy of a system or text depends not only on the system or text, but also on our knowledge of it, and the questions we ask about it*'.

Landsberg—In my opinion entropy is not an absolute quantity, but it depends on the available information. The Gibbs paradox (as I said this morning) is a good example of this, and the simplest.

J. A. Wilson (Cardiff)—I do not see why the entropy of a system should bear any relation to what you think it is. A system may well have an entropy defined by its own characteristics quite distinct from the one you assign to it by your theory, your calculations and (possibly) your measurements.

J. S. Rowlinson (London)—A short answer is to contemplate what happened before isotopes were known. All through the nineteenth century engineers and chemists were making perfectly accurate calculations with entropy, not knowing that there were isotopes. When these became known, then, as a matter of convention, all entropies could be redefined, and we now have them all larger than they were. And I can see no limit in such a process.

H. S. Robertson (Florida, USA)—My point of view also is that entropy really is our measure of the uncertainty regarding a system. When we describe a system thermodynamically we choose (or are forced) to give up our dynamical knowledge. That we say entropy increases as a system evolves to equilibrium, I regard as a statement of our knowledge. Also, my theory is time symmetric: we are just as unable to predict a (detailed) future as to retrodict a (detailed) past. Therefore the time arrow is not within the problems of thermodynamics or statistical mechanics.

Chairman—So, in your lecture, jiggling of the walls was not really the cause of irreversibility?

Robertson—I did use the outside world to bring in the uncertainty, but I can do it just as well by other means.

Chairman—It seems we have reached the end of this question. So I come back to another statement by Layzer: '*The phenomenon of irreversibility in isolated physical systems has its origin in the absence of microscopic information about initial states. The assumption that initial states have this property singles out a direction of time.*' Do you assign the time directiveness to the very form of the assumption pertaining to the initial state, or are you simply pointing to an initial state subject to previous remarks (in this discussion)?

Layzer—That's it. The time directiveness is away from that state [taken in itself].

Chairman—So 'initial' has to be understood by reference to something outside the system you are talking about.

Anon.*—Your statement specifies 'isolated systems'. How can you draw any information from a system without putting yourself in some kind of interaction with it? What can you say about time development in an isolated system of which you are not part?

Layzer—That is one idealization among many that one makes when analysing experiments. 'No interaction with the rest of the universe' is another one. I am free to leave out these interactions and see whether I am able to secure agreement with experiment.

Chairman—Perhaps the time has come to pass on to the cosmic question, with reference to another of Landsberg's statements: '*If entropy increase determines the direction of coarse grained time, then observers in an oscillating universe have their sense of time reversed during the contraction, and a new principle of impotence results: a contracting universe is unobservable.*' Would you like to defend that?

Landsberg—No. I have given the argument.

J. V. Narlikar (Cambridge)—In a model I discussed this morning, retarded and advanced potentials are respectively consistent with expansion and contraction of the universe. Thus, in an oscillating model, observers will always have their time arrow pointing towards expansion.

Costa de Beauregard—Boltzmann made an analogous statement in his well-known book. It may be, he says, that in the universe there are regions A where the entropy is going up and others, B, where it is going 'down' (with respect to some common time coordinate the direction of which is irrelevant, but which must be thought of as 'time extended'). He then feels that living beings are bound to experience increasing entropies in both the A and B regions.

D. Park (Massachusetts, USA)—It seems that we get our sense of time direction very much more from the radiation of the sun and the energy processes we take part in, than from anything the universe is doing. Why on earth should non-radiative living processes be bound up with the ultimate fate of radiation? This is not clear in Landsberg's statement, but Narlikar has his own answer. According to it, if suddenly the universe started to contract, then it would seem to us that, as a result of distant events, the sun would start re-absorbing radiation.

Landsberg—It would seem so to God, not to living things. God would say, ah, the universe is contracting and everybody is getting younger while I, God, am getting older.

Costa de Beauregard—No! Eternity is time-extended!

Landsberg—Mon Dieu (laughter)! I didn't really mean God.

Robertson—May I suggest that this being Landsberg requires for observing the oscillations of his Universe be hereafter called 'Landsberg's demon' (laughter)?

Katz—Statistics alone, as Zwanzig and others have stressed, will not produce a time arrow. Some other assumption is needed, which could be one of the many in Davies's list, or it could be Narlikar's, namely, retarded potentials.

* It was not possible to identify this contributor.

Retarded potentials, like the other things in the list, would have no effect on the immediate approach to equilibrium, but would have a great effect in the time range which obtains for the recurrences.

Zwanzig—It seems to me that retarded potentials are irrelevant here. Consider the decay of an excited hydrogen atom inside a closed box with perfectly reflecting walls. This is a closed quantum mechanical system with well-defined eigenvalues. Everything can be done in complete detail without any reference to retarded actions: it is straightforward quantum electrodynamics. Assuming that at some time the atom is in an excited state, the calculation shows that, provided the box is big enough, the probability goes down with the decay time appropriate for spontaneous emission of a photon. Eventually, when a photon bounces from the wall enough times, this curve may come up again. Nevertheless, as I have explained, for a long time everything looks like a standard decay process, which gives us the basis for our human direction of time.

[Katz and Zwanzig are reviving here the old Ritz–Einstein controversy where, the reporter believes, both were saying the same thing in *reciprocal* forms. Why they could not see it clearly was that, if photons were then known, matter waves were not. Today it is clear that particle scattering (in the sense of statistical mechanics) and wave scattering go hand in hand, so that the two *macroscopic* principles of 'blind retrodiction forbidden' and of 'advanced waves forbidden' are just two different wordings for one and the same statement. This being granted, it remains to understand why living beings are bound to follow the time arrow of increasing probabilities and retarded waves. Could it be, in the context of the generalized entropy principle of information theory, that they must gain information?]

Tisza—May I put a question to the cosmologists. Is it not conceivable that we notice a contracting universe by the violet shift as otherwise our biological feeling of time would remain unchanged?

Layzer—Not only is it conceivable, but it is what happens in the framework of accepted cosmological and physical theories. There is no reason why there should be any connection whatever between the expansion, and the direction of processes in the laboratory or in biological organisms. On these same grounds I would question Landsberg's assumption.

Narlikar—Of course I disagree with both Layzer and Zwanzig. And that is logical, because our basic assumptions are different. They are using a local field theory, while I am using a direct interaction theory which is non-local, and *does* bring in cosmology.

Tisza—I would say that the question of origin of irreversibility is biased by philosophical prejudice. I believe irreversibility is an inherent feature of Nature which need not be reduced to something else (laughter). I don't quite say there is no problem, because the very fact that it has been thought to be a problem is in itself a problem, and a problem that should be exorcised in some way.

As I understand it, in some future stage of the true quantum dynamics which we do not have yet, but which is already shaping up, the problem would appear as the rich interplay of dynamics and stochastic elements, both of which are inherent, but appear on very different grounds.

[Dr Tisza's wish looks extremely like a modernised form of what has been Boltzmann's and Gibbs's in their own days. What has become of it, Zwanzig, Robertson, Davies and others have told us today—not to mention Loschmidt, Zermelo and the Ehrenfests. So, exorcising the demon in irreversibility theory might be not an easy task.]

B. A. Pethica (Cheshire)—Thermodynamics is a first class science. Mechanics is only a second class science and we should stop pretending it comes prior to thermodynamics. Any attempt to provide an excuse for deriving from mechanics the arrow of time is faith. It is faith because the equations of mechanics are time symmetrical while mechanical events are irreversible. Thus thermodynamics is stronger than mechanics, and if mechanics will agree with thermodynamics, so much the better for it.

Rowlinson—I regard the fact that time has an asymmetry as a fact of Nature which does not worry me any more than does the fact that there are two kinds of electricity and not three (laughter). Where I think there is a problem, one that should be discussed and has at least been partially resolved, is of course between the time symmetric equations we use in certain parts of physics and the time asymmetric ones we use in others. This is a difficulty worthy of conferences of this kind. But the early problem I regard as a metaproblem.

Katz—I would express the view that the problem of the direction of time is outside the framework of either thermodynamics or statistical mechanics (as has been explained by Zwanzig and Robertson). But I would also submit that problems that are outside a certain science at a certain time should be studied nevertheless in a larger framework.

[Thus we have the 'agnostic minded', for whom temporal asymmetry is a natural fact needing no more explanation than Nature itself. 'Exorcism', 'faith', 'metaproblem' are the words they would use to qualify the 'religious-minded' who keep on asking 'why'? Why is it that we can at will enclose an excited atom inside a perfectly reflecting box, but we cannot *at will* open the box and pick out the atom in its excited state?]

APPENDIX A

Some of R. O. Davies's statements on the black board

Which reversed processes happen?—A Classification

		Examples
Required		
	Must happen	Fluctuations in isolated systems
	Does happen	Rolling balls; simple particle processes
	Does not happen	Emission of waves; cosmic evolution
	Must not happen	Thermodynamically irreversible processes
Forbidden		

APPENDIX B

This appendix reproduces the part of a paper, circulated to all participants, to which the Chairman referred in his opening remarks. It is based on Appendix A of the paper by P. T. Landsberg, 'Time in statistical physics and special relativity'. *Stadium Generale* (1970), to be published.

Quotations on irreversibility and entropy
(selected by P. T. Landsberg)

(i) *Irreversibility not yet understood*

It is not very difficult to show that the combination of the reversible laws of mechanics with Gibbsian statistics does not lead to irreversibility, but that the notion of irreversibility must be added as an extra ingredient . . . . the explanation of irreversibility in nature is to my mind still open.

P. G. BERGMANN, 1967 (ref. 1, p 11)

. . . causality plus statistics means irreversiblity. I think that is nonsense.

P. G. BERGMANN, 1967 (ref. 2, p 190)

(ii) *Entropy increase due to non-isolation of systems*

The failure of S to increase with time is due to the fact that we have overidealized an 'isolated' system . . . The momentum and energy transferred between outside molecules and the system proper then acts as a source of true randomness influencing the dynamical behaviour of the system inside the walls. We maintain that this is the origin of randomness and increasing entropy in statistical mechanics.

J. M. BLATT, 1959 (ref. 3, p 751)

(iii) *Time direction due to measurement*

In any observation process there must be a signal coming from the observed system to the recording apparatus, and since the propagation of any signal requires a finite time interval, this gives the possibility of defining the arrival of the signal to be 'later' than the time of emission. This specification of the sense of time is perfectly general.

L. ROSENFELD, 1967 (ref. 2, p 193; see also ref. 4, p 3)

(iv) *Irreversibility due to large systems*

. . . irreversible evolution towards equilibrium is an asymptotic property of large systems, for long times, derivable from mechanics alone.

R. BALESCU, 1967 (ref. 5, p 434)

. . . a necessary condition for a rigorous transition from statistical mechanics to thermodynamics consists in taking the so-called thermodynamic limit $N \to \infty$, $V + \infty$, N/V finite, where N represents the number of particles and V the volume of the system.

E. J. VERDOVEN, 1967 (ref. 6, p 49)

(v) *The need for coarse-graining and macro-observables*

Thus we have arrived at the crucial question of how to choose the set of macroscopic variables $A^{(v)}$. This seems to me the main problem in statistical mechanics of irreversible processes.

N. G. VAN KAMPEN, 1961 (ref. 7, p 183)

Any really satisfactory demonstration of the second law must therefore be based on a different approach than coarse graining.

E. T. JAYNES, 1965 (ref. 8, p 392)

(vi) *The importance of measurement and knowledge*

The increase of entropy comes where a *known* distribution goes over into an *unknown* distribution.

R. M. LEWIS, 1930 (ref. 9, p 573)

This illustrates clearly how the entropy of a system or text depends not only on the system or text, *but also on our knowledge of it, and on the questions we ask about it.*

P. T. LANDSBERG, 1961 (ref. 10, p 237)

For it (entropy) is a property, not of the physical system, but of the particular experiments you or I choose to perform on it.

E. T. JAYNES, 1965 (ref. 8, p 392)

..... the irreversibility exhibited by this system consists in the information becoming less relevant to the experiments which can be performed on the system.

A. HOBSON, 1966 (ref. 11, p 411)

(vii) *Irreversibility due to causality conditions*

... one may say that *irreversibility appears as a special aspect of the physical causality requirement,* which states that the distribution function at a given point is influenced only by the distribution function at points which correspond to earlier times on the trajectory.

I. PRIGOGINE, 1962 (ref. 12, p 296)

(viii) *Irreversibility due to ignorance concerning initial conditions*

The phenomenon of irreversibility in isolated physical systems has its origin in the absence of microscopic information about initial states. The assumption that *initial* states have this property singles out a direction of time.

D. LAYZER, 1967 (ref. 13, p 258)

... I presume that most of us would agree ... that the initial conditions generate thermodynamics ... The striking asymmetry of the dynamics originates from this asymmetry in the boundary conditions.

J. A. WHEELER, 1967 (ref. 2, p 233–234)

(ix) *Irreversibility due to smoothing*

Die Irreversibilität ist eine Folge der Reduktion der exakten mechanischen Gleichung (3) durch Mittelung auf die statistische Gleichung (8)... Diese Mittelung... stellt eine absichtliche 'Fälschung' der Mechanik dar, und angesichts dieses Umstandes ist es klar, dass kein Widerspruch zwischen Mechanik und Thermodynamik besteht; sie beruhen auf verschiedenen Grundannahmen.

M. BORN, 1948 (ref. 14, p 109)

The total probability density function W, even for a thermodynamically isolated system, does not obey the Liouville equation, $\partial W/\partial t = LW$, since small fluctuations due to its contact with the rest of the universe necessarily 'smoothe' *W*, by smoothing the direct many-body correlations in its logarithm. This smoothing is the cause of the entropy increase...

J. E. MAYER, 1961 (ref. 15, p 1207)

REFERENCES

[1] P. G. Bergmann in *Delaware Seminar in the Foundations of Physics*, pp 1–14 (Ed. M. Bunge), Springer: Berlin (1967).

[2] T. Gold (Ed.), *The Nature of Time*. Cornell University Press: Ithaca (1967).

[3] J. M. Blatt, *Progr. Theor. Phys.* **22**, 745–756 (1959).

[4] P. Caldirola (Ed.) *Ergodic Theories*. Academic Press: New York (1961).

[5] R. Balescu, 'Velocity inversion in statistical mechanics'. *Physica*, **36**, 433–456 (1967).

[6] E. J. Verboven, 'Quantum thermodynamics of an infinite system of harmonic oscillators', in *Statistical Mechanics* (Ed. T. A. Bak), pp 49–54. Benjamin: New York (1967).

[7] N. G. van Kampen in *Fundamental Problems in Statistical Mechanics*, pp 173–202. North Holland: Amsterdam (1962).

[8] E. T. Jaynes, *Am. J. Phys.* **33**, 391–398 (1965).

[9] R. M. Lewis, *Science*, **71**, 569–577 (1930).

[10] P. T. Landsberg, *Thermodynamics with Quantum Statistical Illustrations*. Interscience: New York (1961).

[11] A. Hobson, *Am. J. Phys.* **34**, 411–416 (1966).

[12] I. Prigogine, *Non-equilibrium Statistical Mechanics*. Interscience: New York (1962).

[13] D. Layzer, in *Lectures in Applied Mathematics*, Vol. VIII (Ed. J. Ehlers), pp 237–262. American Mathematical Society (1967).

[14] M. Born, *Ann. Phys. (Lpz)*, **3**, 107–114 (1948).

[15] J. E. Mayer, *J. Chem. Phys.* **34**, 1207–1223 (1961).

DISCUSSION ON THE TEACHING OF THERMODYNAMICS

Chairman: K. G. DENBIGH, F.R.S.
Reporter: M. W. ZEMANSKY

Several of the papers presented at this conference were concerned with axiomatic treatments of thermodynamics. The following statement by H. Buchdahl (Australia), namely, '*Even if axiomatic thermodynamics is physics, it should not be taught*', evoked considerable comment. The discussion was started by Veazey (Luton) who complained that the statement is incomplete in that it does not say at what level, or for what students, the subject is to be taught. To understand axiomatics, the students must have some previous knowledge. One can see no reason why advanced level students cannot profit from such a study. It would be a good topic for an advanced degree. There is, however, no place for axiomatics for introductory students. It was then argued by Le Fevre (London) that axiomatics is just as much legitimate physics as statements about engines you can't build. Even if one does not like to *start* the subject with an axiomatic treatment, there is no need to say that such a treatment should not be taught.

It was then suggested by Silver (Glasgow) that Buchdahl's statement should be altered to read, 'Even if axiomatic thermodynamics is *not* physics, it *should* be taught'. Giles (Canada) who only half an hour previously had described an axiomatic treatment, admitted that he would not teach such a subject in a first course to elementary students. It was, however, pointed out by Landsberg (Cardiff) that all young people must learn the rules of inference, which *are* axiomatic. No matter what method is used in teaching, the theoretical principles cannot be exact physics. Even Euclidean geometry fails to take into account the curvature of the earth; the second law in its Carathéodory formulation breaks down when a system is very small, etc. He therefore suggested another rearrangement of the words of Buchdahl's statement, namely, 'Even if axiomatic thermodynamics is taught, it could not be physics'.

The statement made by B. Cimbleris of the Nuclear Energy Commission in Brazil, namely, that '*the quantity* $\Delta(U + P_0V - T_0S + \Sigma\,\mu_i n_i)$, *which represents the maximum amount of work during a reversible process in which the system exchanges heat and mass with its environment, should be given a prominent role in the teaching of thermodynamics to engineers*', evoked the comment from Le Fevre that the last two words, 'to engineers' should be eliminated, inasmuch as thermodynamics is a single subject. Zemansky (New York) felt that the statement, although a useful one to engineers, referred to a situation that was not fundamental, but he was immediately opposed by Horne (Michigan, USA) on the ground that perhaps we need a new definition of thermodynamics. In his opinion the expression in question dealt with a real process of a real system in which the temperature inside and that outside differ. This is also true of a living system, to which thermodynamics ought to apply as well.

Tisza (M.I.T., USA) made the point that the origins of thermodynamics are very diversified, and that open systems are just as simple and important to treat as closed systems. He objected to the statement that the expression in question was of value to engineers only, on the grounds that Landau and Lifshitz made use of it extensively and that engineers often contribute (sometimes in a sloppy manner) excellent notions that prove of value to mathematicians and mathematical physicists. He concluded by pointing out that the word 'fundamental' is often used to mean 'what we are used to'.

Landsberg then proceeded to defend the statement, which appears in his textbook published in 1961, to the effect that '*In so far as the time co-ordinate is absent, nothing happens in thermodynamics*'. He maintained that real processes are not always discussed in thermodynamics, sometimes one deals only with ideal abstractions. A quasistatic process, for example, being a succession of equilibrium states, may go forwards or backwards and is therefore reversible. In the real world, one may approach this condition as close as one pleases, but if one postulates that one reaches it, then nothing could happen. He would prefer to think of a quasistatic process as a curve in phase space.

It was pointed out by Kestin (Providence, USA) that in real or irreversible processes the initial state might be an equilibrium state and also the final state, but between the two terminal states, there may be no possibility of drawing a curve in phase space. For irreversible processes the concept of a field is essential. Zemansky tried to give an experimental interpretation of a quasistatic process as one in which instruments behave in a manner that enables the experimenter to take meaningful readings. Such a process is slow and a good enough approximation to a quasistatic process to enable one to use the appropriate equations.

Gurney (Hong-Kong) objected to Landsberg's equating the words 'quasistatic' and 'reversible'. He pointed out that the motion of a blackboard eraser across the table at constant velocity is quasistatic but, because of the large amount of friction, is hardly reversible. A testing machine may stretch a sample of material beyond the elastic limit in a process that is quasistatic but not reversible. [There are some writers who regard a reversible process as one which satisfies two conditions: quasistatic and non-dissipative.]

Landsberg reiterated his belief that his statement was true because an *approximation* to a quasistatic process is not to be confused with an *ideal* quasistatic process. Tisza maintained that time *does* appear in thermodynamics but in hidden fashion. Thermodynamics is always in close contact with experiment. When an event occurs in an experiment, such as the opening of a valve, this involves the removal of a constraint within a given duration of time. Time is really there and plays a role in the consideration of the relation between the thermodynamic equations and the experimental realization. A similar situation occurs in quantum mechanics, where there is a closer connection between a measurement and the quantum-mechanical description of the system undergoing the measurement.

Zemansky then proceeded to defend his 'provocative statement' reading as follows: '*The expression for the work of a thermodynamic system should be chosen so that the definition of internal energy should* not *include external potential energy.*'

He pointed out that the expression for the work in increasing the magnetization of a stationary paramagnetic bar is $-H\,dM$, whereas the work in moving a paramagnetic bar from one point in a field H to another point where the field is $H + dH$ is $+M\,dH$. Since a change of internal energy is defined to be adiabatic work, the first point of view provides an energy U and the second the energy $U - HM$. Since $-HM$ is the external potential energy of a system of magnetic moment M by virtue of its presence in a field H, the second expression is seen to contain as part of the internal energy the external potential energy. It has been the custom to accept these two expressions for work as equally legitimate because all thermodynamic equations based on the two points of view are identical.

Zemansky said that two expressions for work exist also in the case of gas. The first is the well-known one, $+P\,dV$. The second is the work needed to move a small gas balloon from one point in a pressure field (provided, for example, by a tall cylinder containing a dense liquid) to another point of higher pressure, namely, $-V\,dP$ where the minus sign signifies that work must be done *on* the gas. The first expression provides an internal energy U and the second an internal energy $U + PV$. Again, both expressions yield identical equations, but no one would consider seriously the adoption of the second point of view.

Kestin pointed out that the expression $-V\,dP$ is known in engineering as 'technical work' and is widely used. He emphasized that, if one tells him the system and what the system does, he will accept any expression that fits the conditions so specified. R. O. Davies (London) supported the expression $-H\,dM$ for the magnetic work on the ground that it allowed a more acceptable correlation between the statistical mechanics of paramagnetic systems and thermodynamics. Zemansky agreed whole-heartedly.

There then ensued a discussion of the statement by Hornix, namely, '*It is desirable to replace the Kelvin and Clausius formulations of the second law through a set of statements which expresses the "accessibility structure" of phase space in a simplified physical way*'. Hornix (Nijmegen) objected to the classical statements on the ground that they are the result of engineering experience, whereas thermodynamics requires a more sophisticated mode of presentation.

Barron (Bristol) wondered whether the Hornix statement would be of much value to the thousands of students that we are soon to confront in the classroom. He felt that the very words 'accessibility structure of phase space' would be enough to finish at least three quarters of the first-year students. The laughter that ensued indicated considerable sympathy toward Barron's point of view. Silver clinched this point by saying that he was sick and tired of the patronizing attitude of some physicists, expressed by the contention that 'the engineering background of thermodynamics is of historical interest only'. He went on to say that anyone who believes the preceding contention fails to understand some important parts of thermodynamics. Le Fevre hoped that the Hornix method gave clearer statements to students concerning the states to which systems tend to go, and Hornix replied that this is what he accomplished with students. The sophisticated language that Barron decried was meant for teachers, not for students.

The final argument arose over the statement of Silver, '*In the teaching of engineering thermodynamics*: (a) *irreversibility should be introduced at the outset, and* (b) $dQ = du + p\,dv - dW_f$ (*where* dW_f *is the work done against friction*) *should be deduced from mechanics and conservation of energy subsequently identifying* dQ *as the energy transferred by virtue of a temperature difference*'. Just how this is done was shown in a careful and explicit manner by Silver in the presentation of his thesis. Discussion was started by Frank (Bristol) who agreed with Giles that in the presentation of thermodynamics entropy should come early and temperature later. One of the troubles indicated by students, he said, is that they believe they know what temperature is and they believe that they will *never* know what entropy is. Entropy is what Carnot called heat, and entropy is what is conserved in reversible processes. Frank avoids the concept of quantity of heat because even Carnot was not sure what he meant, although he did state that 'chaleur' and 'calorique' meant the same thing.

Le Fevre agreed with most of the ideas contained within the formulation of Silver but disagreed with regard to the moment when the increasing property of entropy should be introduced. He felt that one should arrive as quickly as possible at the entropy statement and then use it to infer the existence of frictional forces and other causes of irreversibility, instead of bringing in friction first and then entropy changes.

After congratulating Silver on his presentation and expressing complete agreement, Zemansky asked permission to object to some of the remarks made by Frank. He said he had no patience with the point of view that entropy should be brought in early. If it could, it would be most desirable, but there are so many concepts such as temperature, work, energy, heat, engine cycles, the second law, etc., that *must* be understood first. To deal with the foundations of thermodynamics *as though* you don't know what temperature, work and heat are is nonsense. The entropy change should be introduced in an *operational* manner, so that the student will know how to measure it and how to calculate it. If a reservoir at T parts with heat Q, the entropy change is calculated on the basis of a knowledge of T and Q, not on statistical considerations. McGlashan (Exeter) indicated his agreement with Zemansky.

Landsberg agreed that the concept of temperature should be taught first and the difficult concepts of entropy and chemical potential last. He suggested that there was a big difference between the order of events when teaching large numbers of ordinary students, and the reformulation of logical structures of thermodynamics such as those suggested by Giles and others, which may be suited to research workers, or possibly to advanced students.

Silver referred to a remark by Landsberg to the effect that one could look at thermodynamics in a variety of ways, analogous to the ways in which a man might look at a woman. Silver insisted that he looks at thermodynamics as an engineer who wants to produce children, so he looks at her in a very definite and pragmatic way (Laughter). Hornix emphasized that, in the introduction of the concept of empirical temperature and its measurement everyone concedes that it is necessary to associate with each isotherm a definite number. In a similar manner, it is necessary to associate numbers with adiabatics. These numbers are analogously empirical entropies. When this is done in such a manner that entropy is additive, one gets a system similar to that of Giles. What we have to learn from the axiomatic point of view is that things are in some respects more simple than we suspected at first, because of the history of the subject. We have to try to become more independent of the historical approach.

Frank struck back at Zemansky by objecting to the latter's insistence that 'simple' ideas like temperature and heat be treated first, and the difficult concept of entropy be reserved for later. Frank insisted that the only thing simple about heat is the fact that it is treated early in Zemansky's book, whereas what makes entropy difficult is that it appears late in this book (Laughter). In the first really valuable publication on this subject, namely in Carnot's book, entropy was called heat, and if it was called heat, it would seem to be the simple concept that I believe it could be made. [Only a few people believe that Carnot anticipated Clausius by having an idea of the meaning of entropy. They maintain that when Carnot used the word 'chaleur' or 'feu', he meant ordinary heat; whereas with the word 'calorique' he meant 'entropy'. This belief is more hero worship of Carnot than practical sense because in *Le Pouvoir Motrice du Feu*, Carnot states definitely once and for all that 'chaleur' and 'calorique' mean the same thing.]

AUTHOR INDEX